2009 Annual IEEE Compound Semiconductor Integrated Circuits Symposium

(CSICS 2009)

Greensboro, North Carolina, USA
11 – 14 October 2009

IEEE Catalog Number:	CFP09GAA-PRT
ISBN:	978-1-4244-5190-6

Copyright © 2009 by the Institute of Electrical and Electronic Engineers, Inc
All Rights Reserved

Copyright and Reprint Permissions: Abstracting is permitted with credit to the source. Libraries are permitted to photocopy beyond the limit of U.S. copyright law for private use of patrons those articles in this volume that carry a code at the bottom of the first page, provided the per-copy fee indicated in the code is paid through Copyright Clearance Center, 222 Rosewood Drive, Danvers, MA 01923.

For other copying, reprint or republication permission, write to IEEE Copyrights Manager, IEEE Service Center, 445 Hoes Lane, Piscataway, NJ 08854. All rights reserved.

This publication is a representation of what appears in the IEEE Digital Libraries. Some format issues inherent in the e-media version may also appear in this print version.

IEEE Catalog Number: CFP09GAA-PRT
ISBN 13:   978-1-4244-5190-6
ISSN: 1550-8781

Additional Copies of This Publication Are Available From:

Curran Associates, Inc
57 Morehouse Lane
Red Hook, NY 12571 USA
Phone: (845) 758-0400
Fax: (845) 758-2633
E-mail: curran@proceedings.com
Web: www.proceedings.com

TABLE OF CONTENTS

IMT-Advanced - Objective and Challenges 1
K. Jay Miyahara

RF Waveform Measurement and Engineering 5
Paul J. Tasker

The Evolution and Importance of Composition in RF Compound Semiconductors 9
C. A. Barratt

GaN Technology for RF Electronics - Development Status in Europe 12
H. Blanck, J. Splettstößer, D. Floriot

A 140 GHz Heterodyne Receiver Chipset for Passive Millimeter Wave Imaging Applications 16
S. Koch, M. Guthoerl, I. Kallfass, A. Leuther, S. Saito

Ultra Low Power 60 GHz ASK SiGe Receiver with 3-6 GBPS Capabilities 20
Woorim Shin, Mehmet Uzunkol, Gabriel M. Rebeiz

A Fully Integrated, Compound Transceiver MIMIC Utilizing Six Antenna Ports for 60 GHz Wireless Applications 24
S. Koch, I. Kallfass, R. Weber, A. Leuther, M. Schlechtweg, S. Saito

A 12-Gb/s, Direct QPSK Modulation SiGe BiCMOS Transceiver for Last Mile Links in the 70-80 GHz Band 28
I. Sarkas, S. T. Nicolson, A. Tomkins, E. Laskin, P. Chevalier, B. Sautreuil, S. P. Voinigescu

Normally-Off Operation GaN Based MOSFETs for Power Electronics 32
Yuki Niiyama, Shinya Ootomo, Hiroshi Kambayashi, Nariaki Ikeda, Takehiko Nomura, Sadahiro Kato

On-Wafer Seamless Integration of GaN and Si (100) Electronics 36
Jin Wook Chung, Bin Lu, Tomás Palacios

High Linearity AlGaAs/InGaAs Pseudomorphic HEMT Driver Amplifier Using Tunable Field-Plate Voltage Technology 40
Chia-Shih Cheng, Shao-Wei Lin, Jeffrey S. Fu, Hsien-Chin Chiu

High Aspect Ratio CPW Fabricated Using a Micromachining Process Combining DRIE, Thermal Oxidation, Electroplating, and Planarization 44
Shane T. Todd, Xiaojun T. Huang, John E. Bowers, Noel C. MacDonald

Design Method for UHF Class-E Power Amplifiers 48
Néstor D. López, John Hoversten, Zoya Popovic

0.5-2.5 GHz, 10W MMIC Power Amplifier in GaN HEMT Technology 52
K. Krishnamurthy, D. Green, R. Vetury, M. Poulton, J. Martin

High Efficiency Digital GaN MMIC Power Amplifiers for Future Switch-Mode Based Mobile Communication Systems 56
S. Maroldt, C. Haupt, R. Kiefer, W. Bronner, S. Mueller, W. Benz, R. Quay, O. Ambacher

A Pulsed Load Modulation (PLM) Power Amplifier with 0.35μm pHEMT Technology 60
Shuhsien Liao, Yuanxun Ethan Wang

10-Gbit/s Wireless Transmission Systems Using 120-GHz-Band Photodiode and MMIC Technologies 64
Naoya Kukutsu, Akihiko Hirata, Toshihiko Kosugi, Hiroyuki Takahashi, Tadao Nagatsuma, Yuichi Kado, Hiroshi Nishikawa, Akihiko Irino, Toshihiro Nakayama, Naohiro Sudo

Low-Cost CMOS-Based Receive Modules for 60 GHz Wireless Communication 68
Piet Wambacq, Kuba Raczkowski, Valéry Ramon, Alexander Vasylchenko, Amin Enayati, Michael Libois, Jonathan Borremans, Karen Scheir, Stephane Bronckers, André Bourdoux, Bertrand Parvais, Bob Verbruggen, Steven Brebels, Wim Van Thillo, Christophe Pavageau

A Passive W-Band Imager in 65nm Bulk CMOS 72
A. Tomkins, P. Garcia, S. P. Voinigescu

A Compact Cascode Power Amplifier in 45-nm CMOS for 60-GHz Wireless Systems 76
Torgil Kjellberg, Morteza Abbasi, Mattias Ferndahl, Anton de Graauw, Edwin v. d. Heijden, Herbert Zirath

A 2.5-V Low-Reference-Voltage, 2.8-V Low-Collector-Voltage Operation, HBT Power Amplifier for 0.8-0.9-GHz Broadband CDMA Applications 80
Kazuya Yamamoto, Atsushi Okamura, Takayuki Matsuzuka, Yutaka Yoshii, Nobuyuki Ogawa, Masatoshi Nakayama, Teruyuki Shimura, Naohito Yoshida

Dual Transformer Injection Locked Frequency Divider Using GaAs E/D-Mode PHEMTs Process 84
Po-Yu Ke, Hsien-Chin Chiu, Jeffrey S. Fu

Modeling of an InGaP/GaAs BiFET VVR Device 88
William Clausen, Brian Moser

2x2 and 4x4 CMOS Switching Matrices for 0.01-12 GHz Applications 92
Donghyup Shin, Gabriel M. Rebeiz

High-Resistivity SOI CMOS Cellular Antenna Switches 97
M. Carroll, D. Kerr, C. Iversen, A. Tombak, J.-B. Pierres, P. Mason, J. Costa

Advanced InP HBT Technology at Northrop Grumman Aerospace Systems 101
Augusto Gutierrez-Aitken, Cedric Monier, Pablo Chang, Eric Kaneshiro, Dennis Scott, Beckie Chan, Matt D'Amore, Steven Lin, Bert Oyama, Ken Sato, Abdullah Cavus, Aaron Oki

V-Band Amplifier MMICs Using Multi-Finger InP/GaAsSb DHBT Technology 105
Jean Godin, Virginie Nodjiadjim, Muriel Riet, Philippe Berdaguer, Stéphane Piotrowicz, Olivier Jardel, Jean-Christophe Nallatamby, André Scavennec, Christophe Gaquière, Matthieu Werquin

Sub-mW Operation of InP HEMT X-Band Low-Noise Amplifiers for Low Power Applications 109
C. H. Lin, X. B. Mei, Y. C. Chou, L. S. Lee, J. M. Yang, M. Y. Nishimoto, P. H. Liu, R. To, A. Cavus, R. Tsai, M. Wojtowicz, R. Lai

Accurate HEMT Switch Large-Signal Device Model Derived from Pulsed-Bias Capacitance and Current Characteristics 113
Shinichiro Takatani, Cheng-Duan Chen

An Image Reject Mixer for High-Speed E-Band (71-76, 81-86 GHz) Wireless Communication 117
Marcus Gavell, Mattias Ferndahl, Sten E. Gunnarsson, Morteza Abbasi, Herbert Zirath

Robust AlGaN/GaN Low Noise Amplifier MMICs for C-, Ku- and Ka-Band Space Applications 121
E. M. Suijker, M. Rodenburg, J. A. Hoogland, M. van Heijningen, M. Seelmann-Eggebert, R. Quay, P. Brückner, F. E. van Vliet

Compact and Broadband Millimeter-Wave Mixer Based on the New Phase Relationship 125
Yu-Ann Lai, Shih-Han Hung, Yeong-Her Wang

X/Ku-Band SiGe BiCMOS Phased Array Chips with Simultaneous 2- and 4-Beam Capabilities 129
Dong-Woo Kang, Gabriel M. Rebeiz, Kwang-Jin Koh

A 2x22.3Gb/s SFI5.2 SerDes in 65nm CMOS 133
N. Nedovic, A. Kristensson, S. Parikh, S. Reddy, W. Walker, S. McLeod, N. Tzartzanis, H. Tamura, K. Kanda, T. Yamamoto, S. Matsubara, M. Kibune, Y. Doi, S. Ide, Y. Tsunoda, T. Yamabana, T. Shibasaki, Y. Tomita, T. Hamada, M. Sugawara, J. Ogawa, T. Ikeuchi

A 0.25μm InP DHBT 200GHz+ Static Frequency Divider 137
M. D'Amore, C. Monier, S. Lin, B. Oyama, D. Scott, E. Kaneshiro, A. Gutierrez-Aitken, A. Oki

A 32-GS/s 6-Bit Double-Sampling DAC in InP HBT Technology 141
Munehiko Nagatani, Hideyuki Nosaka, Shogo Yamanaka, Kimikazu Sano, Koichi Murata

Ultra-Low-Power 500-MSPS 12-bit A/D Converter Using Interleaving and CMOS Charge-Domain Technology 145
Michael P. Anthony, G. Sollner

Ultra Low-Loss 50-70 GHz SPDT Switch in 90 nm CMOS 149
Mehmet Uzunkol, Gabriel M. Rebeiz

A 4-Bit Passive Phase Shifter for Automotive Radar Applications in 0.13 μm CMOS 153
Sang Young Kim, Gabriel M. Rebeiz

Broadband, Thin-Film, Liquid Crystal Polymer Air-Cavity Quad Flat No-Lead (QFN) Package 157
Morgan J. Chen, Seyed A. Tabatabaei

MMIC LNAs for Radioastronomy Applications Using Advanced Industrial 70 nm Metamorphic Technology 161
W. Ciccognani, E. Limiti, P. E. Longhi, M. Renvoisè

A Low Power Ka-Band Receiver Front-End in 0.13μm SiGe BiCMOS for Space Transponders 165
Firooz Aflatouni, Hossein Hashemi

Author Index

Chair's Message

On behalf of the organizing committee and the IEEE Electron Devices Society, the Microwave Theory and Techniques Society, and the Solid-State Circuits Society, welcome to the 2009 IEEE Compound Semiconductor IC Symposium (CSICS). This year's symposium is being held October 11th – October 14th at the Sheraton Greensboro Hotel at Four Seasons in Greensboro, NC.

CSICS has been going strong for 31 years. This year we chose to locate in what has become a hub of Compound Semiconductor research and manufacturing. I don't think that anyone at the 1979 meeting of the GaAs IC Symposium (the first) would have imagined that the largest manufacturer of GaAs in the year 2000 would be in the cellular (or in late 70's parlance... "portable") telephone business. Yet by 1990 it was clear that this was one of the most promising applications of this fledgling technology. Through the years, the GaAs IC Symposium grew in size and breadth as GaAs integrated circuits spread into defense and commercial products. Corporate and academic programs in GaAs research led to exciting advances in materials growth, device physics, higher integration levels and commercial applications. As GaAs technology matured, other III-V materials and other compound semiconductors came into the mix. In 2004 in Monterey, CA, the Symposium changed its name to IEEE Compound Semiconductor IC Symposium (CSICS) to reflect the evolution of the III-V industry and the interests of its participants. Now compound semiconductors play increasingly important roles in many aspects of our wireless world.

The CSIC Symposium is the preeminent international forum on developments in integrated circuits using compound semiconductors such as GaAs, InP, GaN, SiGe and other materials. Throughout our history we have featured the initial demonstrations of circuits in emerging technologies ("the first", "the fastest") and we embrace all aspects of the technology, from materials issues and device fabrication, through IC design and testing, high volume manufacturing, and system applications.

Several social events are planned that allow our attendees to interact in a relaxed setting. Events include the Sunday Evening Opening Reception, the Monday evening Technology Exhibition Opening Reception, the Tuesday Technology Exhibition Luncheon, and the Tuesday Theme Party. This year's Theme Party, "Hospitality-Southern Style", has a distinct Southern flavor that I'm sure you will enjoy. We also offer daily breakfast and AM/PM coffee breaks Monday through Wednesday.

We are glad that you could join us again this year!

Marko Sokolich, Chair
2009 IEEE CSICS

2009 IEEE CSIC Symposium Organizers

EXECUTIVE COMMITTEE

Marko Sokolich
Symposium Chair
HRL Laboratories, LLC
Malibu, CA

Dave Halchin
Technical Program Chair
Peregrine Semiconductor
Greensboro, NC

Dan Scherrer
Technical Program Vice Chair
Northrop Grumman
Redondo Beach, CA

Sorin Voinigescu
Local Arrangements Chair
University of Toronto
Toronto, Canada

Francois Colomb
Symposium Treasurer
Raytheon
Andover, MA

Douglas McPherson
Symposium Publicity Chair
Zarlink Semiconductor
Ottawa, ON, Canada

Charles Campbell
Symposium Publications Chair
TriQuint Semiconductor
Richardson, TX

Mohammad Madihian
Electronic Program Management
NEC Corporation of America
Princeton, NJ

William Peatman
Chair, Emeritus
ANADIGICS
Warren, NJ

OVERSEAS ADVISORS

Toshihide Kikkawa
Fujitsu Laboratories Ltd.
Japan

Jan-Erik Mueller
Infineon Technologies
Munich, GERMANY

Chul Soon Park
School of Engineering, ICU
KOREA

Huei Wang
National Taiwan University
Taiwan

Marc Rocchi
OMMIC
Limeil Brevannes, FRANCE

Freek van Straten
NXP Semiconductors
The Netherlands

IEEE ADVISORS

Mohammad Madihian
MTT-S Liaison
NEC Corporation of America,
Princeton, NJ

Lisa Boyd, Supervisor
IEEE
Meeting & Conference Management
Piscataway, NJ

Herbert S. Bennett
EDS Liaison
NIST
Gaithersburg, MD

CORPORATE BENEFACTORS

This year, we are pleased to continue with the IEEE Compound Semiconductor IC Symposium Corporate Benefactors Program. This program allows companies interested in compound semiconductors to show their support of the Symposium by making contributions towards the cost of some of our social events.

These additional resources enable the Symposium to increase the quality of our event, as well as allowing companies an opportunity for some tasteful promotional activities. To discuss any of the benefactor opportunities in more depth, please contact:

>Marko Sokolich
>Tel: +1-310-317-5148
>E-mail: msokolich@hrl.com

As of this printing, the Corporate Benefactors for the 2009 Compound Semiconductor IC Symposium are as follows:

Gold Level Benefactor
RF MICRO DEVICES, INC.

SilverLevel Benefactors
TriQuint Semiconductor
OMMIC
HRLLaboratories, LLC
AWR Corporation

The Symposium Web Site www.csics.org has become a critical tool for the dissemination of information to prospective attendees, committee members and sponsors of the Symposium. Every year, the web site must be updated and maintained to effectively serve this purpose. We would like to acknowledge the following benefactor for providing the Symposium web site support for the 2009 CSIC Symposium:

Comments regarding the web site or any publicity materials should be directed to the Publicity Chair, Douglas McPherson (douglas.mcpherson@zarlink.com). Links to our corporate benefactors appear on our symposium website.

IMT-Advanced - Objective and Challenges

K. Jay Miyahara
Mobile Network Operations Unit
NEC Corporation
Tokyo, Japan
Kj-miyahara@cw.jp.nec.com

Abstract—**This paper describes drivers and objectives for IMT-Advanced mobile broadband system defined by ITU-R as well as technical challenges.**

Keywords-component; IMT-Advanced; LTE; IEEE802.16; WiMAX

I. INTRODUCTION

The life with "internet" have evolved from simple eMail and file transfer, web surfing to enriched web-based communities and hosted services, such as social-networking sites, video sharing sites, wikis, blogs, and folksonomies. It can be said that most of these services are "user centric" and "user driven" and gradually changing the way we communicate and the business model.

LTE and WiMAX started with solving clear technical objectives: improve radio spectrum efficiency and provide economical Mobile Broadband Wireless network. We should also consider how this evolution will enable new service to be created and how it may create new business models similar to what has been happening with Web 2.0

Obvious advantage LTE and WiMAX will offer is high speed transport. Higher bandwidth availability means less time to download (or transfer) given amount of data. This will significantly increase variety of information download/push services because minimizing the time to wait for information is one critical issue when assessing the usability of any service.

In addition to the higher radio transport capability, introduction of all IP network architecture and various Open Interfaces will allow more flexibility in developing and deploying new services. Like evolution commonly know as Web 2.0 in the fixed internet world, user driven, user centric development of new services will likely to take place and will bring about new business models.

To meet the ever increasing demand for wireless broadband communication (e.g. increased number of users, higher data rates, video or gaming services which require increased quality of service, etc.), IMT-Advanced is currently under discussion in ITU-R, 3GPP, and IEEE802.16.

II. TWO APPROACHES TOWARD IMT-ADVANCED; WIMAX AND LTE

Although current WiMAX and LTE may seem similar, it is different in many ways. The difference in the current specification is mainly due to the intended deployment scenario and the timing of specification definition. IEEE 802 LAN/MAN Standards Committee has developed IEEE802.16 which defines WirelessMAN™ air interface specification for wireless metropolitan area networks (MANs). Since the initial issue in October 2001, it has evolved from fixed broadband access system to mobile broadband access system. LTE standards have been developed by 3GPP as evolution of Third Generation Mobile System to support growing needs for high speed data traffic. Although LTE can support VoIP, it is assumed to coexist with existing mobile system (2G and/or 3G) for some time.

Both WiMAX and LTE advanced are currently candidate technology of IMT-Advanced. In order for these technologies to be chosen as technology for IMT-Advanced, it must satisfy common requirements and support services specified by ITU-R [1][2].

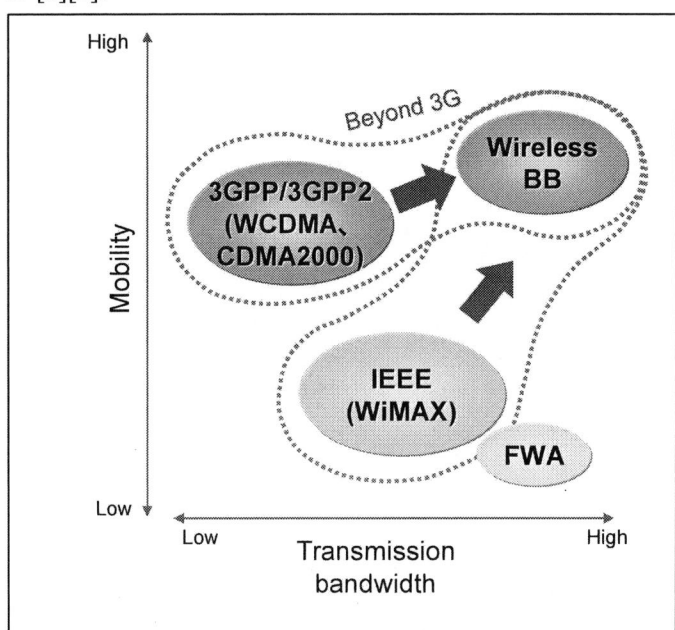

Figure 1. Evolution of Wireless Broadband Access technologies

A. Spectrum efficiency

Spectrum efficiency (Cell spectral efficiency, Peak spectral efficiency, Cell edge user spectral efficiency) and scalable bandwidth up to and including 40 MHz are required to support various broadband services. These requirements are essential to meet the growing need for higher speed by efficiently utilizing the limited frequency resources available.

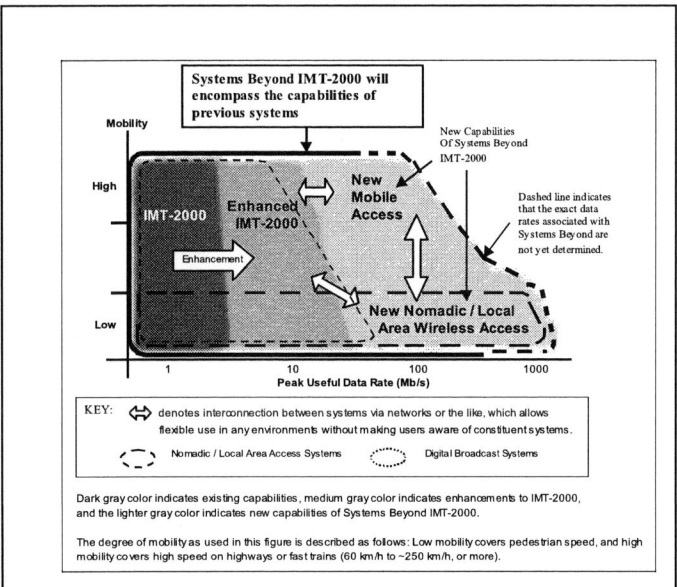

Figure 2. **Illustration of capabilities of IMT-2000 and IMT-Advanced**

B. Latency

Latency is another important criteria where requirements are specified. User Plane latency (packet transfer delay) will significantly impact the perceived end-to-end transmission throughput especially using TCP/IP. IMT-Advanced systems shall be able to achieve a user plane latency of less than 10 ms in unloaded conditions (i.e., a single user with a single data stream) for small IP packets (e.g., 0 byte payload + IP header) for both downlink and uplink. Control Plane latency is also specified as less than 100 ms to reduce the time required to establish/tear down a packet session.

C. Mobility

For Mobility and Handover, speed of movement up to 350 km/h and maximum handover interruption time is defined to provide minimum service quality of IMT-Advanced. (The handover interruption time is defined as the time duration during which a user terminal cannot exchange user plane packets with any base station.)

III. KEY TECHNOLOGIES

A. Spectrum issues

It is well know that both OFDMA and MIMO are two key technologies providing the throughput and spectrum efficiency for WiMAX, LTE, and IMT-Advanced system.

Due to the practical limitations such as portability requirements of end devices (i.e., size and power consumption limitations) and the mobility requirements, we are reaching the practical limit of achieving spectrum efficiency using these technologies.

The following frequency bands have been identified by ITU-R [3] for IMT and/or IMT-2000 by WARC-92, WRC-2000 and WRC-07:

- 450-470 MHz
- 698-960 MHz
- 1 710-2 025 MHz
- 2 110-2 200 MHz
- 2 300-2 400 MHz
- 2 500-2 690 MHz
- 3 400-3 600 MHz

ITU-R has developed frequency arrangements for the bands identified by WARC-92 and WRC-2000, which are described in Recommendation ITU-R M.1036-3. For the frequency bands that were identified at WRC-07, work on the frequency arrangements is ongoing in ITU-R.

Although lower spectrum is preferable considering the indoor penetration loss, the difficulty is anticipated to obtain bandwidth up to and including 40 MHz. Aggregation of spectrum in different frequency band may be required to support minimum service requirements and traffic coverage.

Femto cell deployment is another approach to effectively improve spectrum efficiency. By reducing the size of radio cell and increase the reusability of the same frequency bands, it is an attractive solution for indoor application. (e.g., home, small to medium enterprises, enclosed public areas, etc.)

B. Migration issues

IMT-Advanced system will most likely be deployed along side of or as an over-lay to the existing wireless mobile network. For some combination, seamless handover to/from other technology will be required. In certain deployment scenario, concurrent operation in different network technology may be required.

When deploying IMT-Advanced network in a region where voice telephony traffic is predominant and small data user is expected, it may be more efficient, from network resource perspective, to use earlier generation mobile system (e.g., GSM) for voice and use IMT-Advanced network for data services. In this scenario, end devices most likely must support concurrent operation in both network environments. If IMT Advanced network is deployed only in selected areas initially and if VoIP based telephony service is offered, it will be mandatory requirement to support seamless handover from IMT Advanced to earlier generation mobile network which has much better radio coverage. The handover will be required for both voice and data traffic.

For the Basestations, the migration requirement is slightly different. In short, smooth evolution scenario is often a requirement to reduce capital investment by up-grade or re-use the equipment as much as possible. Due to the similarity in key technology adopted by LTE, WiMAX, IMT-Advanced up-grade for some deployment can be done by simple software upgrade.

978-1-4244-5190-6/09 $25.00 © 2009 IEEE

ITU-R System Name	IMT			
	IMT-2000	*Enhanced IMT-2000*	*Enhanced IMT-2000*	*IMT-Advanced*
Generation	3G	3.5G	3.9G/Pre 4G	4G
Concept	High capacity/quality voice Mobile Internet (Wireless Multimedia)	High speed Mobile Internet Mobile Broadcast/Multicast	Toward Mobile Broadband	Mobile Broadband NGN/NGMN
RAN — Radio Technology	CDMA	CDMA TDM scheduling	OFDMA TDM·FDM scheduling	OFDMA (?)
RAN — 3GPP	WCDMA (FDD) TD-CDMA (TDD) TD-SCDMA (TDD)	HSPA(FDD, TDD) HSPA+	LTE (FDD, TDD)	LTE+ (?)
RAN — 3GPP2	cdma2000 1x (FDD) (Rel.0, Rev.A, Rev.B)	EV-DO(FDD) (Rel.0/Rev.A/Rev.B)	UMB (FDD, TDD)	UMB+ (?)
RAN — IEEE	OFDM TDD WMAN (802.16e)			802.16m
NW — Network Architecture	CS Domain PS Domain	PS evolution for IMS/MMD, Network Interoperability	Flat Architecture All-IP, Network Interoperability	NW Convergence
NW — 3GPP	UTRAN GSM-evolved CN	UTRAN CN	E-UTRAN EPC(SAE)	E-UTRAN (?) EPC(SAE)(?)
NW — 3GPP2	cdma2000 RAN IS-41 evolved CN and PCN	EV-DO RAN EV-DO CN	CAN Enhanced CN	CAN (?) Enhanced CN (?)

TABLE I.   EVOLUTION OF IMT SYSTEM

IV.   Technical Challenges

A.   Power Efficiency

Reduction of Basestation power consumption is a requirement both from reducing OPEX as well as reducing the size of equipment for easy installation and less conspicuous installation. For a high power output amplifier, power efficiency is especially important to reduce the size of the equipment. From the manufacturer's perspective, low cost of production is also essential.

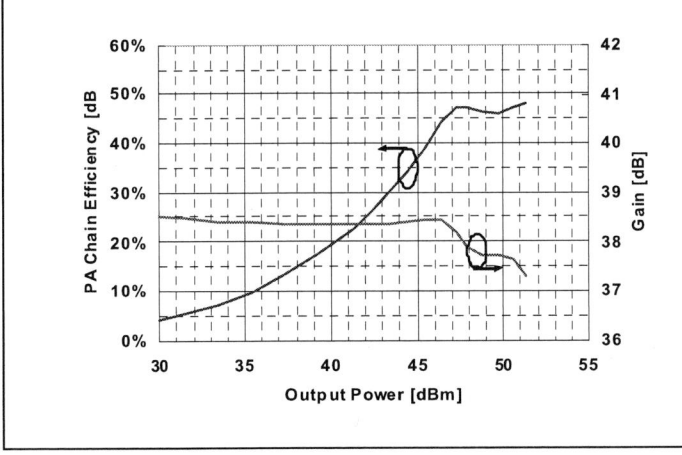

Figure 3.   Output Power vs Efficiency

For WiMAX (802.16e based system) and LTE, LDMOS (Laterally Diffused Metal Oxide Semiconductor) or Gallium nitride (GaN) amplifier with Digital Pre-distortion technique can achieve better than 30% power efficiency and meet all RF requirements specified by IEEE802.16 or 3GPP specifications. For future application, HEMT (High Electron Mobility Transistor) devices are promising but commercially available devices are currently limited to an application for certain frequency spectrum.

NEC has developed transmitter amplifiers for mobile base stations, a 2.1GHz model that produces 45W of output power per 100W of power consumption by adopting high performance and highly reliable RF transistor technologies, in addition to independently optimizing Doherty RF circuitry with harmonic tuning.

B.   RF Filter

Any equipment which radiate radio signal must conform to spurious emission requirement specified. Although some attempts are made to define uniform specifications in certain frequency bands, it is often different in each country/regions. Depending on the system using the adjacent spectrum, spurious emission requirement may even be different within the same spectrum band. If the entire spectrum intended for the system is available initially, band pass filter providing the sufficient elimination of spurious emission can be provided.

The traditional filter approach is inconvenient for a scenario where desired spectrum bandwidth is not initially available. The Basestation must initially conform to the spurious emission requirement of the allocated narrow channel. When adjacent spectrum become available, it is preferred to be used as one contiguous wider channel but it will have different spurious emission requirement and filter cut off frequency will be different from the initial configuration. In some cases, it may be more economical to replace the amplifier and filter section of the Basestation. The Remote RF Head configuration where RF Unit is physically separated from the base band processing portion of the Basestation will allow such replacement with ease. For situations where physical replacement of RF Unit is not an economical solution, tunable filter or easily replaceable filter will be necessary.

C. Smart Antenna Technology

As mentioned previously, Mobile device has disadvantage due to power consumption and physical size limitation. The up-link performance characteristics often become the limiting factor of coverage. In MIMO configuration, Maximum ratio combining (MRC) is used at the receiving end to improve performance.

MRC is a simple processing technique that uses pilot or control channel signals to estimate channel characteristics for multiple antennas and then apply weights to each antenna to maximize signal to noise ratio for the summed signal. MRC captures diversity and combining gains but does not involve active interference mitigation or spatial multiplexing in any way. MMSE, or minimum mean squared error, is another variation on this receive-processing approach.

Adaptive antenna systems (AAS) take the basic concept of MRC a few steps further — by building a richer model of the channel using training data embedded in the traffic channel that enables focusing more closely on users of interest and de-focusing on interferers in both transmit and receive. AAS captures diversity, combining, and interference rejection gains.

V. Evaluation of radio interface technologies for IMT-Advanced

ITU-R has issued a guidelines [4] for both the procedure and the criteria (technical, spectrum and service) to be used in evaluating the proposed IMT-Advanced radio interface technologies (RITs) or Sets of RITs (SRITs) for a number of test environments and deployment scenarios for evaluation..

The evaluation criteria specified are: Cell spectral efficiency, Peak spectral efficiency, Bandwidth, Cell edge user spectral efficiency, Control plane latency, User plane latency, Mobility, Intra- and inter-frequency handover interruption time, Inter-system handover, VoIP capacity, Deployment possible in at least one of the identified IMT bands, Channel bandwidth scalability, Support for a wide range of services

VI. Conclusion

This paper described drivers and objectives for IMT-Advanced mobile broadband system defined by ITU-R. With considerations of some realistic deployment scenarios, technical challenges and some solutions are proposed to realize economical solutions to realize IMT-Advanced system to provide Mobile Broadband environment.

REFERENCES

[1] "Background on IMT-Advanced", Document IMT-ADV/1-E

[2] "Requirements, evaluation criteria and submission templates for the development of IMT-Advanced", REPORT ITU-R M.2133

[3] "Requirements related to technical performance for IMT-Advanced radio interface(s)", REPORT ITU-R M.2134

[4] "Guidelines for evaluation of radio interface technologies for IMT-Advanced", REPORT ITU-R M.2135

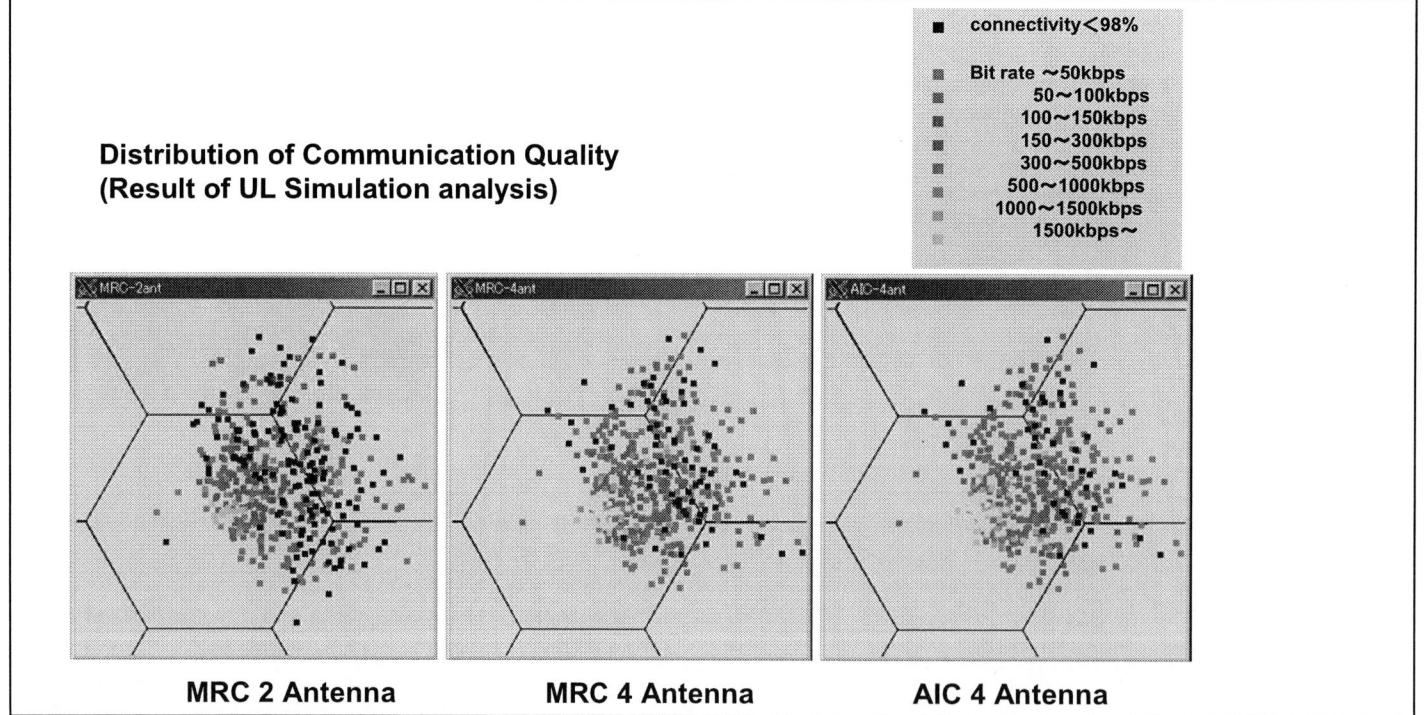

Figure 4. Uplink simulation analysis of MRC and Adaptive Interefence Cancelation

RF Waveform Measurement and Engineering

Invited

Paul J. Tasker

Centre for High Frequency Engineering, School of Engineering
Cardiff University, Cardiff CF24 3AA
tasker@cf.ac.uk

Abstract—**RF I-V Waveform measurement and engineering Systems are finally enabling practical waveform engineering to be directly undertaken within the RF power amplifier (RFPA) design cycle. RFPAs are a critical component in many systems; e.g. mobile communications, satellite communications and radar systems**.

RF Waveform measurement, RF Waveform Engineering, Non-Linear Measurements, Load-pull, Power Amplifiers.

I.   INTRODUCTION

RF Power Amplifier (RFPA) performance, output power, conversion efficiency and linearity, etc., is influenced by the transistor terminal voltage and current time varying waveforms. Thus *waveform engineering* should be driving the RFPA design process at all levels; transistor optimization, circuit design and system requirements. In practical RFPA design, waveform engineering is only a guiding principle, its direct application previously hindered by the lack of appropriate RF waveform measurements tools. The past 15 years has seen the maturing of RF voltage and current waveforms measurements systems. Coupling with impedance control hardware also enables experimental control (engineering) of these terminal RF waveforms during measurements; *thus providing a practical RF Waveform Measurement & Engineering solution.* Application involves either the direct utilization of the measurement system in the design investigation/evaluation loop, or its indirect use by providing CAD accessible datasets or behavioral model parameters.

II.   RF WAVEFORM MEASUREMENTS

The most utilized microwave measurement tools supporting CAD design measure s-parameters, small signal parameters, are only of limited use in non-linear CAD design. Non-linear CAD design must include mixing (intermodulation distortion), compression (harmonic generation), etc.   While this information, ideal for system evaluation, is provided by (Vector) Spectrum Analyzer and/or Power Meter; in this form it is again of limited use in non-linear CAD design. This explains the prevalence of "build and test" in the RFPA design cycle.

RF waveform measurement capability was first demonstrated using a sampling scopes in the late 1980's [1]. The Microwave Transition Analyzer (MTA) triggered the rapid developments of RF waveform measurement systems in the 1990's [2]. Figure 1 shows the standard, full two-port, architecture of RF waveform measurement system, often referred to as a Non-Linear Vector Network analyzer (NLVNA) now typically employed [3]. Presently receivers employed include four channel digital sampling scopes (i.e. Tektronix DSA + Mesuro M20), sampling down converters or 5 channel network analyzers (i.e. Agilent PNA-X).   All systems are fully vector calibrated and provide for error corrected measurement of the time varying voltage and current waveforms present at the device under test (DUT) terminals, as shown in figure 2.

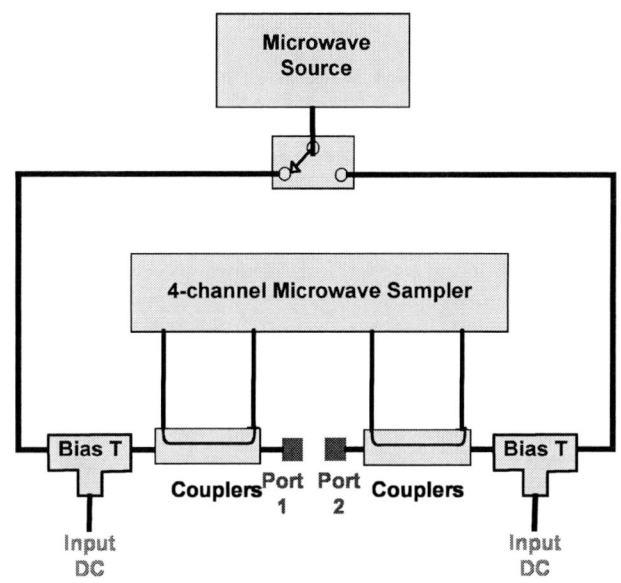

Figure 1.   Basic schematic of an RF waveform measurement system [3].

Calibration involves a two-step process [3]. First a VNA calibration process is employed to determine all but one of the required calibration coefficients. VNAs only measures the ratio of quantities hence one error term is not required and is set to unity.   The second calibration step involves the determination of this residual calibration term. Magnitude calibration is done with a calibrated Power Meter. Phase calibration is done by either assuming that the sampling receiver has an ideal phase response [3] or by employing a reference harmonic phase generator [4].

III. RF WAVEFORM ENGINEERING

S-parameter measurements are generally performed into a known, fixed 50 ohm reference impedance. This does not impose any constraints in linear CAD design since the performance into arbitrary terminal impedances is computed via linear algebraic transformations. This is not the case for non-linear CAD design, hence the limited usability of s-parameters. This constrain also applied to RF waveform measurements. While the RF waveforms contain non-linear information they cannot be used to simply compute non-linear performance into arbitrary terminal impedances.

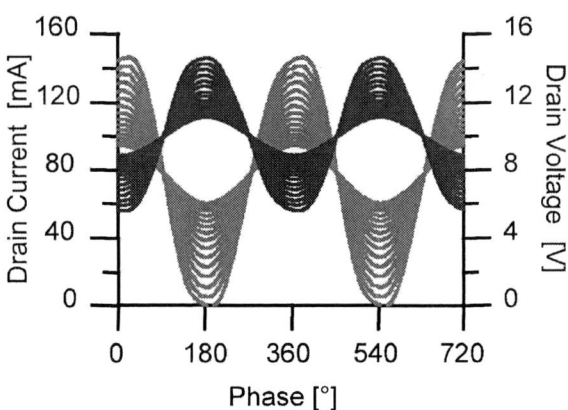

Figure 2. Typical Measured, as a function of input power, RF Voltage and Current Waveform for a HFET.

This explains the initial use of RF waveform measurement systems simply as a tool to help develop or optimize non-linear transistor models [5,6]. Such models can then be used in non-linear CAD simulators, mainly utilizing harmonic balance concepts, to compute non-linear performance into arbitrary terminal impedances.

Unfortunately the RFPA design community has often found the model accuracy, hence usefulness of these CAD simulations, inadequate. They have placed more reliance on load-pull; the direct measurement of the key non-linear performance parameters, output power, gain, efficiency, linearity, etc. as a function of load impedances. Typically passive mechanical systems based on multiple stub tuners are used to achieve this variation of terminal impedances. Coverage of the impedance plane is limited by loses in these passive mechanical systems. To overcome these loses a number of active load-pull systems have been investigated [7]. However, they can often be prone to oscillations, hence limiting their uptake.

RF waveform measurement systems can, however, be integrated with such impedance control hardware [3,8]. The practical consequence of integrating experimental control (Engineering) of the terminal impedances is that RF incident traveling voltage waveforms can be modified (Engineered) during measurements; hence the term RF Waveform Engineering rather than source- or load-pull. It is important to note, however, if the stimulus signal is a single-tone CW signal then the RF waveform generated by the non-linear DUT will also contain harmonics components. RF waveform engineering requires not only fundamental load-pull but also harmonic source- and load-pull. This highlights the fact that when investigating the behavior of a non-linear DUT, impedance variation investigation should not be limited to fundamental frequency alone.

Reviewing the fact that the role of this impedance variation is to engineer the RF incident traveling voltage waveforms led to the development of a relatively simple implementation; the open loop utilization of multiple RF sources [8]. Multiple RF sources are required since the impedance must be engineered at all of the RF waveforms spectral components. If the stimulus signal is a single-tone CW signal, confining the problem to the third harmonic allows for band-limited RF waveform engineering with a practical number of RF sources, i.e. six. Multi-tone stimulus will result in a significant increase in spectral components; RF waveform engineering of these signals cannot be addressed by simply increasing the number of RF sources. The recent introduction of a Microwave Arbitrary Waveform Generator (Mesuro M20 + Tektronix AWG) addresses this limitation.

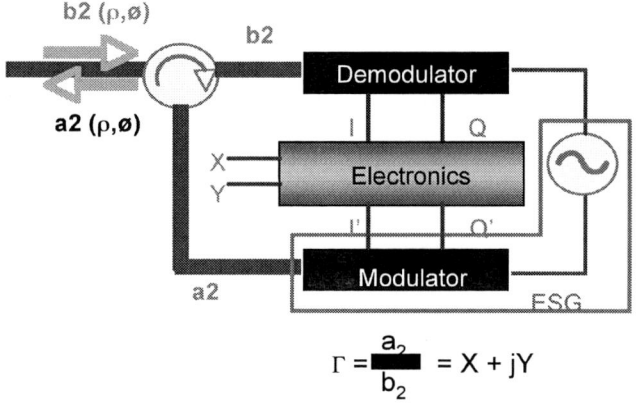

Figure 3. Typical basic architecture of a Envelop Load-Pull System. [9].

The open loop active load-pull solution is not prone to oscillations [8], however, to emulate a given load termination requires an iterative adjustment of the additional RF sources power levels and relative phases; this is time consuming. This can be overcome by providing, as shown in figure 3, some base-band feedback (envelop signal) to allow for automatic adjustment of the additional RF sources magnitude and phase; this approach is referred to as envelop load-pull [9]. It can also address the multi-tone signal issue provided that the feedback loop has sufficient envelop bandwidth; this is typically 10 times the modulated bandwidth of the stimulus multi-tone signal [10].

IV. TRANSISTOR CHARACTERIZATION

RF waveform measurement and engineering systems can be used to fully investigate the transistors dynamic non-linear response at microwave frequencies. It can thus support technology optimization, selection, modeling and utilization as well as the investigation of trapping problems and reliability.

Consider for example the optimization of the emerging GaN HFETS. Fabricated RF Power GaN transistors often have disappointing RF power performance; generally associated with both surface and buffer trapping. Conventional microwave measurements, whether linear or large signal, only measure their consequences; transconductance and output conductance dispersion, RF power loss and/or decreased efficiency. This is not the case with an RF waveform measurement and engineering systems, as shown in figure 4 [11]. Here CW measurements are performed while sweeping the RF fundamental load impedance from a low (short) to a high (open) impedance. The RF knee walkout is clearly observed associated with surface traps. Additionally, if this measurement is repeated for increased drain voltage the increased knee walkout is clearly identified.

The results in figure 4 also shown that this particular technology has an additional buffer problem, hence poor pinch-off that also leads to an identical degradation of RF power performance. Via the optimization of the iron doped buffer design this pinch-off problem was eliminate [11].

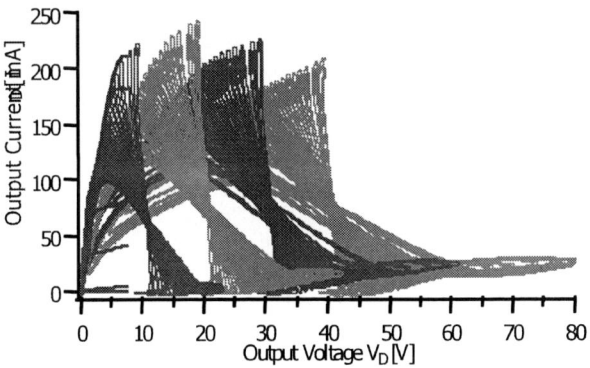

Figure 4. RF fan diagrams, showing measured knee walkout [11].

V. AMPLIFIER DESIGN

High efficiency power amplifier design is actually quantified in terms of waveform engineering. Amplifier efficiency can be improved if circuit termination impedances can be found that allow the transistor to generate or support half rectified sinusoidal waveforms and/or square waveforms. For example, the inverse class F solutions requires simultaneous a square waveform output current and a half rectified waveform output voltage. Figure 5 shows the measured (engineered) RF waveforms achieved using a 10W GaN HFET, clearly indicating that the transistor can support this mode of operation, hence deliver the RF output power required (>10watts) with a very high power added efficiency (>80%) [12].

In addition to confirming the correct (desired) waveforms it also provided the designer with the desired input and output

matching circuit impedances necessary at the fundamental, second harmonic and third harmonic. The designer now has all the information necessary to design an appropriate microwave matching circuit and assemble the amplifier, once assembled the amplifier was simply characterized giving a measured performance identical to that predicted; i.e. a first pass design success [12].

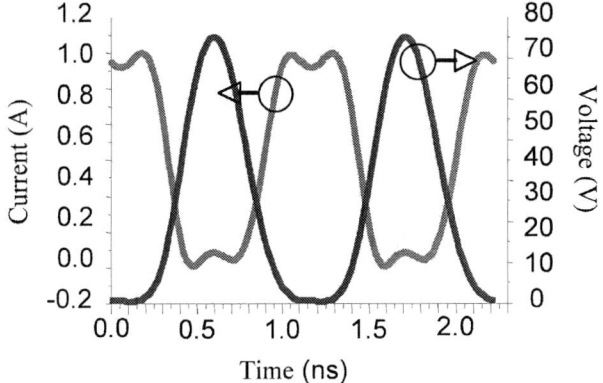

Figure 5. Engineered RF waveforms showing GaN HFET operating in inverse Class F [12].

VI. TRANSISTOR MODELLING

The initial use of RF waveform measurement systems was to help develop or optimize non-linear transistor models. Its ability to support model development is advanced considerably by the addition of RF waveform engineering. For example, this enables the full device I-V plane to be investigated via load-impedance variation rather than DC bias point when validating conventional analytical models or extracting the transistors non-linear state functions [13].

RF waveforms, measured as a function of the load impedance, can also provide datasets that be used directly, via a waveform data lookup model within the CAD design tool [14]. Alternatively experimental waveform datasets are transformed into behavioral model datasets [15-17]. All these models can then be used in non-linear CAD simulators, mainly utilizing harmonic balance concepts, to compute non-linear performance into arbitrary terminal impedances.

VII. CONCLUSIONS

RF I-V waveform measurement and engineering systems are now finally enabling practical waveform engineering to be directly undertaken. This measurement capability extends the characterization opportunities for both high frequency/speed transistor technology developers and circuit/system designers; *terminal waveforms are the unifying theoretical link between transistor technology, circuit design and system performance.*

ACKNOWLEDGMENT

This author would like to acknowledge all his present and previous co-workers from the FhG IAF, Freiburg, Germany, University of Vigo, Spain and the Centre for High Frequency Engineering, Cardiff University, UK.

REFERENCES

[1] Sipilä, M.; Lethinen, K.; Porra, V., "High-Frequency Periodic Time-Domain Waveform Measurement System". *IEEE Trans. MTT*, Vol. 36, No 10, Oct. 1988, pp.1397-1405.

[2] M. Demmler et al, "A vector corrected high power on-wafer measurement System with a frequency range for the high harmonics up to 40 GHz.", Proc. 24th EuMC, Cannes, France, Sept. 1994, Page(s): 1367-1372.

[3] J. Benedikt et al, "High-power time-domain measurement system with active harmonic load-pull for high-efficiency base-station amplifier design.", IEEE Trans. MTT, Vol. 48, No. 12, Dec. 2000, Page(s):2617 – 2624.

[4] J. A. Jargon, et al. "Repeatability study of commercial harmonic phase standards measured by a nonlinear vector network analyzer.", ARFTG Conf. Digest, 2003. Fall 2003. 62nd, 4-5 Dec. 2003 Page(s):243 – 258.

[5] M.C. Curras-Francos, et al.; "Direct extraction of nonlinear FET Q-V functions from time domain large signal measurements.", IEEE MGWL, [see also IEEE MWCL] Vol. 10, No. 12, Dec. 2000 Page(s):531 – 533

[6] D. Schreurs, et al, "Straightforward and accurate nonlinear device model parameter-estimation method based on vectorial large-signal measurements.", IEEE Trans. MTT, Vol. 50, No. 10, Oct. 2002 Page(s):2315 - 2319

[7] B. Noori, et al, "Comparison of Passive and Active Load-Pull Systems In High Power Amplifier Measurement Applications", 72nd ARFTG Conf. Digest Fall, 2008.

[8] D.J. Williams and P.J. Tasker, "An automated active source and load pull measurement system.", High Frequency Postgraduate Student Colloquium, 2001. 6th IEEE, 9-10 Sept. 2001 Page(s):7 – 12.

[9] T. Williams, et al, "Experimental evaluation of an active envelope load pull architecture for high speed device characterization.", Microwave Symposium Digest, 2005 IEEE MTT-S International, 12-17 June 2005

[10] S.J. Hashim, et al, "Active Envelope Load-Pull for Wideband Multi-tone Stimulus Incorporating Delay Compensation.", EuMC 2008. Oct. 2008 Page(s):317 – 320

[11] C. Roff, et al "Analysis of DC–RF Dispersion in AlGaN/GaN HFETs Using RF Waveform Engineering.", IEEE Trans. ED, Vol. 56, No. 1, 2009 Page(s):13 – 19

[12] P. Wright, P et al, "Highly efficient operation modes in GaN power transistors delivering upwards of 81% efficiency and 12W output power.", Microwave Symposium Digest, 2008 IEEE MTT-S International, 15-20 June 2008 Page(s):1147 – 1150

[13] R. Gaddi, et al, "LDMOS electro-thermal model validation from large-signal time-domain measurements.", Microwave Symposium Digest, 2001 IEEE MTT-S International, Volume 1, 20-25 May 2001 Page(s):399 - 402 vol.1

[14] H. Qi, et al From Direct Data Lookup to Behavioral Modeling.", to be published, IEEE Trans. MTT.

[15] S. Woodington, et al, "A novel measurement based method enabling rapid extraction of a RF Waveform Look-Up table based behavioral model.", Microwave Symposium Digest, 2008 IEEE MTT-S International, 15-20 June 2008 Page(s):1453 – 1456

[16] G. Simpson, et al, "Load-pull + NVNA = enhanced X-parameters for PA designs with high mismatch and technology-independent large-signal device models.", ARFTG Conf. Fall. 2008 Page(s):88 - 91

[17] D.E. Root, et al, "Broad-band poly-harmonic distortion (PHD) behavioral models from fast automated simulations and large-signal vectorial network measurements.", IEEE Trans. MTT, Vol. 53, No. 11, 2005 Page(s):3656 - 3664

The Evolution and Importance of Composition in RF Compound Semiconductors

C.A. Barratt
RFMD, Greensboro, NC 27410, USA

History

Compound semiconductors found their origins in the discovery of the transferred electron or "Gunn" effect by J.B. Gunn in 1962 while working at the Royal Signals and Radar Establishment in the UK[1]. This phenomenon provided a net reduction in electron drift velocity with increased electric field at specific carrier concentrations. By utilizing this negative impedance, a transit time device useful as an oscillator could be fabricated. These two terminal devices while called diodes, do not contain a semiconductor junction but when properly designed and placed in a resonant cavity, can produce low efficiency RF energy directly from DC.

The Gunn Effect

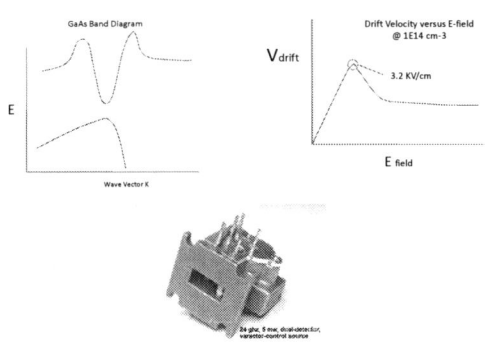

Figure 1: GaAs band diagram and drift velocity plot showing the Gunn Effect

Figure 1 shows the band diagram of GaAs and the drift velocity versus electric field for low carrier concentrations along with a picture of a wave guide Gunn Diode source. As in the past, these devices still find homes in military and commercial applications. This property of GaAs and InP was not available from silicon based microwave diodes leading to investment in compound semiconductor radio frequency diode technology. These investments led to complimentary device applications and the higher electron mobility of compound semiconductors provided enhanced performance over similar devices produced in silicon. Mixer diodes, tuning varactors, impatts and even PIN diodes[2] were all developed as the development of Gunn Diodes spread to other device applications The similarities were that these devices were all made using chlorine transport vapor phase epitaxy (VPE), and mesa diodes fabricated on N+ substrates. The substrates were grown using Horizontal Bridgman bulk crystal growth and yielded a "D" shaped wafer that were very irregular ingot to ingot. This was a severe limitation in the ability to produce these devices at low cost in an automated factory setting.

The MIMIC Program

In the period starting in the late 1970s and continuing onward into the early 1990s a very large amount of investment and interest went into compound semiconductors due to the development of planar processing and round, SEMI standard, undoped semi-insulating substrates. For RF applications the primary investments were US Government and the interest was in the potential use of compound semiconductors for producing monolithic microwave integrated circuits (MMICs). The MIMIC Program was a seven year, $570 million program begun in 1986[3]. This program provided teams of contractors with funding to pursue low cost microwave integrated circuits for electronic warfare, communications and smart weapons for the US DoD. This effort was the hub of a very significant hype cycle that drove investment and expectations to a high degree. Figure 2 shows the Gartner Hype Cycle concept along with approximate times and graphical depictions of associated events driving the hype.

The GaAs Hype Cycle

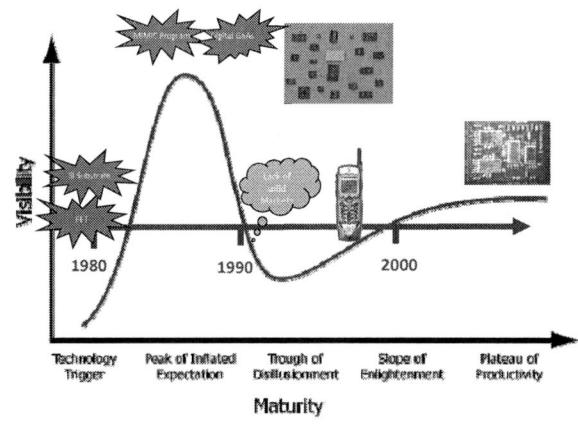

Figure 2: The Gartner Hype Cycle with superimposed time and certain relevant developments.

While beginning the movement to utilize different compound semiconductor compositions in the same device, the MIMIC program began with the pursuit of a more silicon model as a predominant technical direction.

Liquid Encapsulated Czochralsky (LEC) growth of SEMI standard round, undoped, semi-insulating wafers was available. This provided for the ability to do direct ion implantation of impurities for doping and therefore spatial control over where that doping would be placed on the wafer surface. Automated processing of multi-function microwave integrated circuits along a silicon model could be envisioned and was pursued. Compound semiconductors possessed of substantially higher electron mobility than silicon should have an inherent performance advantage and while not driven by the MIMIC program even logic applications were targeted. The early MIMIC program's technical direction gave rise to what can be described as a technology node, the 0.5 micron gate length, ion implanted MESFET complete with on-chip passives, air bridges and through substrate via hole grounds[4]. Figure 3 shows an example of an MMIC typical of the 0.5 micron ion implanted MESFET technology of the time.

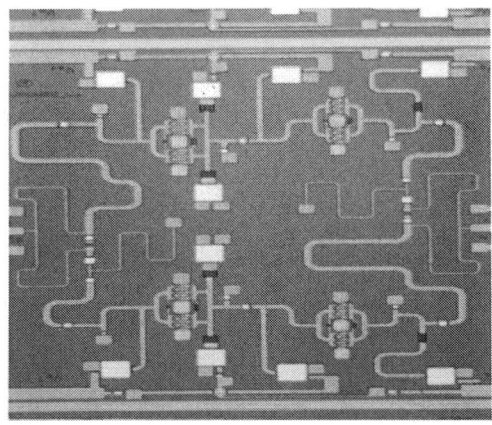

Figure 3: A 0.5 micron ion implanted MMIC amplifier

The direction toward ion implantation limited the device design space to carrier type and doping level and while this produced some very compelling device capabilities it ignored both major weaknesses and major advantages inherent in compound semiconductors. The first major weakness was the lack of an insulator/semiconductor interface without parasitic interface states that made the fabrication of a device similar to the silicon MOSFET possible. The second major weakness was the very low hole mobility which made truly complimentary devices difficult if not impossible to make practical. The major strength that was being ignored was the use of composition as a degree of freedom for the device designer. This was simply because of the focus of effort in preparing compound semiconductor active layers by ion implantation.

Evolution of the Current Industry

The earliest reference to the use of composition in compound semiconductors for RF devices goes back to the late 1969 when Esaki and Tsu postulated that the mobility of a two dimensional electron gas should be enhanced over traditional semiconductor transport. In 1978, R. Dingle et. al. reported device results for an enhanced mobility FET[5]. Over the next few years a variety of hetro-junction FETs were reported and called a variety of names including

MODFET, TEGFET, SDHT and more generally high electron mobility transistor (HEMT). This direction involved multiple compositions of compound semiconductor layers to create devices with performance that was enhanced over that which could be produced with a single material. Ion implantation proved useful however, but was relegated for use in bombardment implants to provide isolation for devices formed on epitaxially prepared active layers. In the decade of the 90s cellular communication was gathering steam. Cellular telephones emerged with the power amplifier a major user of battery energy. Battery systems were moving to higher energy density lithium ion systems with available voltage at or below 4.0 volts. Compound semiconductor devices and in particular those that utilized multiple compositions for operation, were uniquely suited to high performance at voltages below 4.0 volts. Initially these applications were in both the receivers and transmitter/power amplifier chains although more recently, integration of the transceiver function have moved compound semiconductors to the front end.

The dominant device being used for PA application is the hetro-junction bipolar transistor (HBT). This NPN emitter up device is constructed using a high band gap emitter of either AlGaAs or InGaP over a GaAs base and collector. The predominant device utilized for antenna switch components in handsets is the pseudomorphic high electron mobility transistor (pHEMT). This device involves a GaAs contact layer on top of a high bandgap layer, typically AlGaAs over a low bandgap layer, usually InGaAs. The word pseudomorphic comes from the fact that the InGaAs layer is not lattice matched to the substrate.

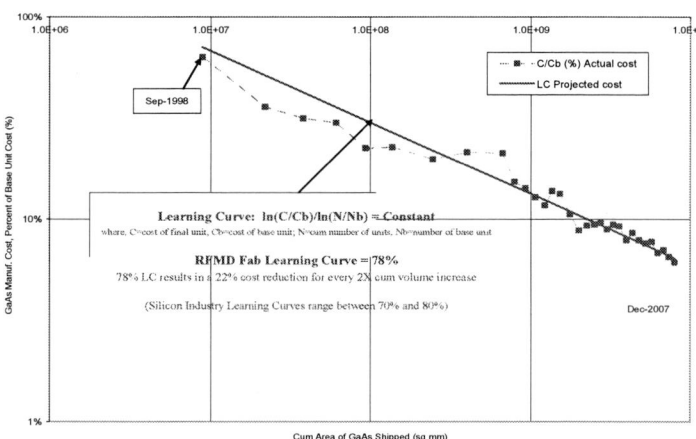

Figure 4: Relative cost versus cumulative area shipped for GaAs based cell phone components at RFMD

The market for RF compound semiconductors is in excess of $3.6 billion as of August 2008[6]. RFMD has reached shipment rates of 2.0 million handset components per day containing both HBT and pHEMT devices. Figure 4 is a plot of relative unit cost versus cumulative area shipped for GaAs based cell phone components. This cumulative area includes both HBT and pHEMT devices for cell phone and other contemporary RF applications. RFMD recently marked the shipment of 2.0 billion components to a single customer. The 22% learning curve compares favorably with that of silicon and demonstrates the maturity this industry has achieved. The current compound semiconductor industry while very competitive and has more than lived up to the hype received in earlier periods.

Future Directions

Development of existing and new device technologies in compound semiconductor benefits from the richness available by combining a universe of compositions together to obtain desired device performance envelopes. InGaP HBTs changed the band gap offset over AlGaAs and improved the temperature performance of current gain. Non-alloyed emitter contacts necessary for improved reliability were only possible through the use of strained N+ InGaAs contact layers. Super lattice buffer layers improved sub-threshold device performance in pHEMT structures. Fabrications schemes benefit from etch stop layers inserted into epitaxial layer designs making high capability processing possible through selective etching. It is clear that future technical progress in device performance is coming through the creative use of composition as the primary degree of freedom.

Group III nitrides are emerging as the next compound semiconductor platform. GaN HEMT devices based on an AlGaN/GaN hetrojunction can achieve very high currents and breakdown voltage compared with alternatives. RFMD has released a generation of GaN HEMT on silicon carbide substrates capable of operation at 48 volts with power densities between 7-10 watts/mm of device periphery[7]. These devices are produced in the same fabrication facility with the high volume GaAs based devices and enjoy the associated economies of scale. GaN based RF electronics will strongly compliment those of GaAs as already application consist of driver stages in GaAs with high power stages in GaN.

Conclusion

RF compound semiconductors began as a set of niche diode technologies where the performance commanded sale prices that offset the costly fabrication. Developments including SEMI standard wafers and planar processing led to a period of substantial investment and gave rise to a hype cycle that drove the industry along a silicon like model. The creative use of combinations of various compositions of III-V compounds has resulted in an extremely vibrant and mature industry. The major advantage that compound semiconductors have over competing technologies is the rich pool of available alloys that can produce ever more useful device performance. Pursuit of this avenue has in many ways, compensated for the lack of a true MOSFET analog.

Acknowledgments

I would like to acknowledge the members of the wafer fab team at RFMD.

References

1. W. Liu, *Handbook of III-V Hetrojunction Bipolar Transistors* 1998 p. 58
2. J.L. Heaton, R.S. Posner, T.B. Ramachandran, R.E. Walline, *International Electronic Devices Meeting*, 1978
3. E.D. Cohen, Digest *of Papers, Microwave and Millimeter-Wave Monolithic Circuits Symposium* 1988
4. C. Barratt, R. Boerstler, M.A. Shea, J. Heaton, *The Electrochemical Society SOTAPOCS,* 1985
5. M. Shur, *GaAs Devices and Circuits* 1986 p.513
6. A. Anwar, *GaAs Device Vendor Market Share 2007: North America*, Strategy Analytics, 2008
7. K. Krishnamurthy, J. Martin, B. Landberg, R. Vetury, M.J. Poulton, *IEEE MTT-S International Microwave Symposium Digest,* 2008

GaN technology for RF electronics
Development Status in Europe

H. Blanck, J. Splettstößer
United Monolithic Semiconductors GmbH
Ulm, Germany
herve.blanck@ums-ulm.de

D. Floriot
United Monolithic Semiconductors SAS
Orsay, France

Abstract—**GaN technology has gained a lot of attention in Europe over the last few years for various domains including RF electronics. This paper will try to summarize the current status achieved and illustrate it with a few representative examples. Due to the importance of the subject this paper will focus solely on the RF electronics related topics, which should be of major interest in the frame of this conference. Aspects covering material, devices up to circuits and module integration will be addressed.**

Keywords - GaN; RF; Electronics; Europe

I. INTRODUCTION

Europe has a long history on the development of III-V technologies. In particular, some of the GaAs pioneering work, like the co-invention of the HEMT device, was done in Europe. Over the last years the focus has strongly shifted to GaN-based devices and a large community including Academia, Industry and National research institutes is now strongly involved in the development of GaN devices and technology. Although in the last decade Japan and the US have mostly been in the limelight, in particular with the first products being put on the market, significant and innovative work has also been done in Europe reaching to the state-of-the-art. The paper will try to summarize and draw a picture of the current status of the GaN technology in Europe and the perspective for the years to come. The content will try to cover all topics from material to modules for RF-applications.

Since a few years the GaN activity in Europe has dramatically increased. This activity is solidly based on a few important cornerstones:

- A long and extensive experience of III-V semiconductors and RF applications in the Academia as well as the Industry. For most players the transition from GaAs to GaN was a logical move.

- A strong commitment from the industry that recognized that its future would be increasingly dependent on GaN technology even if GaAs remains a major player for many years to come.

- The awareness of the European institutions that the GaN technology would become a key enabling technology for many applications and that without it Europe would lose its independence and competitiveness and many industrial and strategic areas. In consequence, many projects were initiated to support and push the development forwards. These are essential not only through their financial support but also as catalyst, to bring top pan-European teams together, limit redundancy and maximize synergy. All domains are being addressed including Space, Commercial and Defense.

II. PROGRAMS

An exhaustive list of all activities and projects would be too long to include in this paper but we will show some representative examples to illustrate the status already achieved. To name only two such federating projects we could mention "Korrigan" [1,2] (supported by the European Defense Agency, or EDA), that covers all aspects from Substrates to Modules; or "GREAT²" [3] (supported by the European Space Agency, or ESA), that focuses on reliability improvement for Space applications.

The final objective is clearly to achieve a complete, independent and open food chain for the GaN Technology, as shown in Fig. 1, involving a close fabric of small to large companies addressing all aspects of the technology.

In parallel to a fully open commercial offer, some companies, mainly in the defense sector (EADS, Thales, Selex-SI, Qinetiq for instance) keep internal resources operational to address specific needs through their own R&D centers (Thales with ATL III-V Lab, Selex, etc...) or through specific partnership (EADS with IAF, etc...).

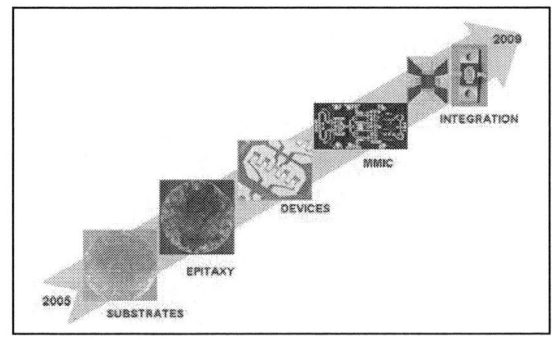

Figure 1. GaN food chain. Korrigan overview (Courtesy G. Gauthier [1]).

The mainstream is based on AlGaN/GaN HEMT on SiC substrates but the European Commission (EC) through IST calls has also contributed to the development of alternatives for nitride technologies or applications. We could mention the evaluation of InAlN/GaN HEMT for high temperature / high frequency applications (UltraGaN [4]), alternative substrates (Hyphen [5,6], MorGaN [7]).

Thermal management of high power GaN HEMT is a key. Specific transistors designs or material for assembly have been investigated through different initiatives [1, 8].

Finally, like GaAs in its time GaN technology needs a strong commercial basis to survive and expend. The base-station (BTS) market represents good opportunities, especially for future wideband standards and multi-standard products. As a major player in this field, NXP has planned the introduction of GaN RF power transistors for the next year in partnership with UMS, through a dedicated development program also involving the Fraunhofer Institute for Applied Solid-States Physics (IAF) and Chalmers University [9,10].

The table below shows an overview of the teams involved in the different topics relevant to the GaN Technology. The table does not pretend to be exhaustive but nevertheless demonstrates the importance that GaN has gained.

TABLE I.   GAN PLAYERS IN EUROPE

Topic	Players (not exhaustive)
Substrate	Soitec[a], Norstel[a]
Epitaxy	Picogiga[a], Qinetiq[a], Azzuro[a], ATL III-V Lab, IMEC, IAF, FBH, Ulm Uni., Chalmers, FORTH, EPFL, CRHEA, etc…
Device & Process	UMS[a], IAF, FBH, ATL III-V Lab, Ulm Uni., Qinetiq[a], RWTH Aachen, MicroGaN[a], NXP[a], IMEC, IEMN, IET , Selex[a], Chalmers, etc…
Circuit design	IAF, FBH, EADS[a], Thales[a], UMS[a], TESAT[a], Alcatel-Lucent[a], Qinetiq[a], SAAB[a], Ericsson[a], Selex[a], AMS[a], TNO[a], ATL III-V Lab, IEMN, etc...
Module & Systems	Thales[a], EADS[a], IAF, NXP[a], TNO[a], Qinetiq[a], Selex[a], Alcatel-Lucent[a], Astrium[a], etc...
Characterisation	Bristol Uni. CDTR, UMS[a], IAF, ATL III-V Lab, FBH, IEMN, Padova Uni., Tor Vargata Uni., IRCOM, XLIM, ISOM, DEI, MFA, IMS, etc...

a. Industrial or commercial activity

III.   MATERIAL

A.   Substrate

Promising results have been achieved regarding semi insulating Silicon Carbide (s.i. SiC). SiC substrates are available from Norstel in 2" and 3" diameter with a micropipe density down to < 2 cm^{-2}. In Fig. 2 the resistivity profiles for 2" and 3" s. i. SiC grown by High Temperature CVD are shown.

All points for 2" SiC are in the 10^{11} Ωcm range. For the 3" SiC all points are between 10^9 Ωcm and 10^{12} Ωcm.

Interesting activities are on-going on hybrids substrates in an attempt to reduce cost and increase wafer dimensions without impacting performances [10]

_____

This work was supported in part by DGA, BWB, EDA, BMBF, ESA, CNES, DLR, EC 6th and 7th Frame Work, Thales and EADS

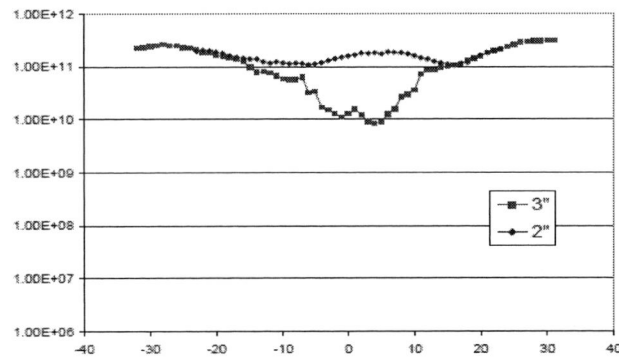

Figure 2.   Resistivity profile for 2" and 3" semi insulating SiC. Courtesy Norstel.

B.   Epitaxy

A very large expertise regarding epitaxy is available at various Research institutions as well as commercial companies (IAF, III-V Lab, IMEC, FBH, Qinetiq, Picogiga, etc…) delivering state-of-the-art quality and guarantying a European supply of "conventional" AlGaN/GaN HEMT as well as more exploratory structures (like AlInN/GaN [11]).

IV.   TECHNOLOGY AND DEVICES

RF performances have been regularly improved during the recent years. Thanks to a constant progress of the technology and a better maturity of the material, the optimization of the elementary RF transistor cell have been considerably improved. The maturity of the technology has also improved dramatically, for instance UMS intends to release its first generation in 2010.

A.   Devices

Very high power densities are being routinely achieved as well as stable operation at increased Drain bias voltage (Fig.3). Today the absolute level of performance achieved is no longer limited by the quality of the process only but also by the thermal management. Reliability and robustness also play central roles in the devices optimization.

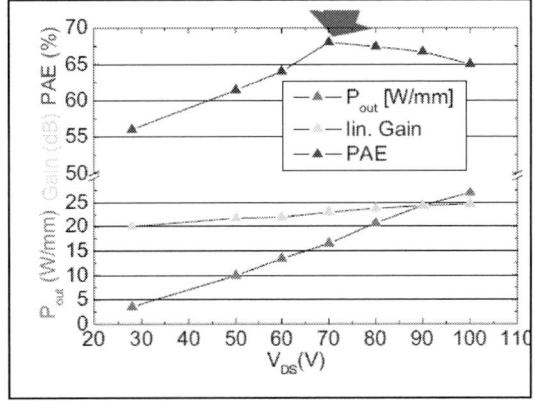

Figure 3.   2-GHz power characterisation. Courtesy IAF [11]

B. Characterisation and reliability

The transistor characterization, including reliability evaluation and analysis, are essential aspects in the development and industrialization of a new technology. These topics are being intensively addressed by several groups in Europe. For instance:

- Thermal investigations [15].

- Reliability analysis [14].

- Transient phenomena (Fig. 4, [1]).

- Transistor modeling in order to allow reliable circuit designs [13].

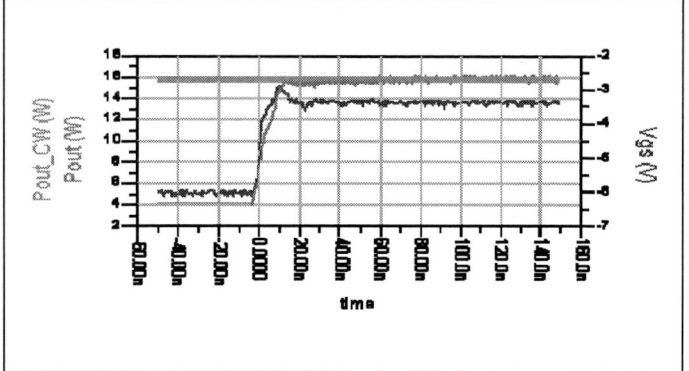

Figure 4.   Measurement of switching timeon a S-Band HPA – rising time = falling time << 50ns.Thales SA - Alcatel-Thales III-V lab.

Modeling of high power GaN transistors is a key issue. A lot of efforts are necessary to achieve a consistent non-linear electro-thermal model including trapping effects. The versatility of use (CW, pulsed through gate or drain access, bias, class of operation) increase the difficulty to integrate into only one model. Nevertheless, specific models have been developed to take into consideration the different time constants of lagging in GaN HEMTs [13].

V.   APPLICATIONS, CIRCUITS AND DEMONSTRATORS

Various applications are considered for GaN technologies: high power amplifiers (medium and wide band) for telecom and Radar applications, robust receiver including Low Noise Amplifiers (LNA), highly linear mixer, high power switch etc... Regarding these applications and the associated timeframe, the GaN technology is developed to be fully compliant in term of RF specs, reliability, electrical robustness and cost.

A. Power bars

For BTS and radar applications up to C-Band the traditional approach mostly used so far to design High Power Amplifiers (HPA) includes packaged large periphery power bars. Although this may be replaced in the future by more integrated approaches, better suited to advanced concept like switch-mode amplifiers for instance, a significant effort is being dedicated to the development and industrialization of such devices. They will probably be the first generation of devices to be included in commercial systems and excellent performances have

already been achieved comparable to the current world state-of-the-art as can be seen in Fig. 5.

The power management remains one difficult aspect limiting the possibility to leverage the potential of GaN to provide extremely high power densities. Today the limit seems to be around 4W/mm and 6W/mm dissipated power for continuous wave (CW) and long pulse (<100µs) operation respectively.

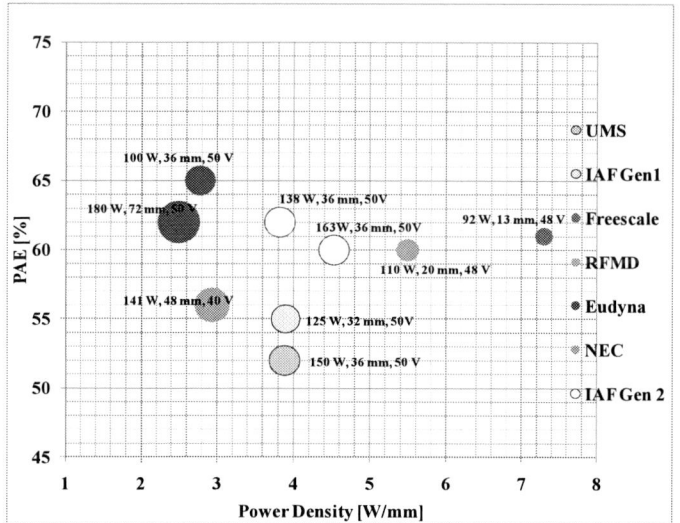

Figure 5.   2-GHz power performance of power bar devices. Power added efficiency (PAE) versus Power density (the size of the symbol is proporpotional to the gate periphery) [13].

B. HPAs and MMICs

Impressive results have been obtained in S- and X-band HPAs:

- 100-140W at 3GHz by SAAB/BAE (Fig. 5) - Korrigan project [2].

- 40-60 W at 10GHz by Selex-SI/UMS/AT III-V Lab in short pulsed mode with PAE in the range of 35% under Vds voltage ranging from 25V to 40V [13] (Fig.6) - Korrigan project [18].

- 20W with PAE around 35% in representative pulse Radar mode (100µs of pulse length – 10%) and relevant robustness validation (VSWR 4:1) by IAF – BWB project [16].

For very wide band applications (2 octaves and more) like EW or jamming system for which the PAE in CW mode is limited, the thermal management is one of the most critical parameters.

Robust X-Band LNA have also been demonstrated with adequate noise figure (2.5 dB) associated to a strong robustness, 43dBm of input level corresponding to 40W/mm (TNO) [2]. In another context, the FBH-Berlin also published recently similar results on robust X-Band LNA [8]. Wideband HPA, LNA and power switches have also demonstrated all the interest of this technology for such applications (ELT/INDRA/TNO) [2].

978-1-4244-5190-6/09 $25.00 © 2009 IEEE   14

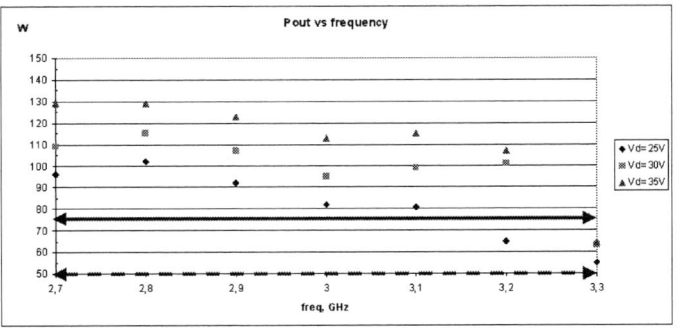

Figure 6. S-Band HPA – Pout >100W from 2.7 to 3.2 GHz – PAE > 50%. – Vds=35V -50μs pulse length – 5μ DC. Courtesy SAAB and BAE [2].

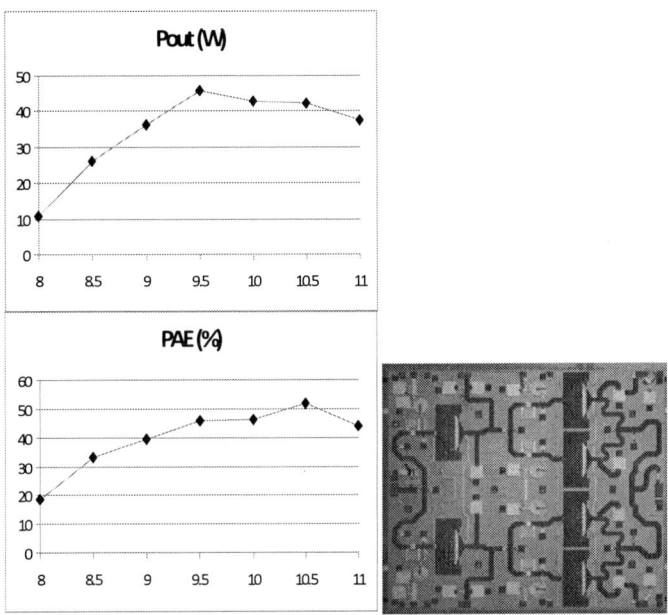

Figure 7. X-Band HPA – UMS design – Pout>40W and PAE >40% - Vds=25V – 20μs pulse length – 10% [18].

Based on the circuits developed in the running projects first transmit/receive (T/R) modules and RF transmitters have been fabricated reaching up to 800W in S-Band for instance [19].

Advanced material based on Diamond composite are being developed inside the AGAPAC project [8] aiming at improving the thermal conductivity of the base plate to roughly 500-600W/mK in comparison to 300W/mK for the current state-of-the-art. A compromise between the coefficient of thermal expansion (CTE) and the thermal conductivity (W/mK) has to be managed adequately for either small surface to height ratio (like for power bars) and large surface to edge ratio (like for MMICs).

VI. CONCLUSIONS

The GaN Technology in Europe is emerging as a key building block for academia and industry. Significant efforts are being invested in research, development and industrialization and the results obtained are very close or even at the current world state-of-the-art. Within a short time the first version of a fully European GaN technology will be industrially and commercially available to support key areas of the RF-electronics-related industry.

ACKNOWLEDGMENT

The authors would like to thank all the European GaN community for their support and contribution to this work. The authors would also like to acknowledge the financial and technical support from: EDA, EC, ESA, BMBF, DLR, BWB, DGA, CNES, WTD81, FMV, MDE, Difesa, dstl, Defensie, ANR, MiUR EPSRC.

REFERENCES

[1] G. Gauthier, Y. Mancuso and F. Murgadella, "KORRIGAN - A Comprehensive Initiative for GaN HEMT Technology in Europe", Proceedings European Microwave Week, October 2005, Paris, France.

[2] P. Duême et al., "Overview of the KORRIGAN project", Proceedings WOCSDICE 2009, Malaga, Spain.

[3] http://www.great2-project.com/index.html

[4] http://www.ultragan.eu/

[5] http://www.hyphen-eu.com/

[6] F. Zanon, F. Danesin, A. Tazzoli, G. Montanari, A. Chini,J. Thorpe, C. Gaquière, G. Meneghesso, and E. Zanoni, "High power performances of GaN HEMT on SopSiC substrate", Proceedings WOCSDICE 2008, Leuven, Belgium.

[7] http://www.morganproject.eu/

[8] http://www.agapac.eu/

[9] http://www.nxp.com/acrobat_download/literature/9397/75016518.pdf

[10] T. Rödele, "GaN – the Future Workhorse of Microwave Power?", Proceedings EuMiC 2008 – Amsterdam.

[11] N. Defrance, V. Hoel, Y. Douvry, J. C. De Jaeger, C. Gaquière, X. Tang, M. Rousseau, M. A. di Forte-Poisson, J. Thorpe, H. Lahreche and R. Langer, "AlGaN/GaN HEMT High Power Densities on SiC/SiO2/poly-SiC Substrates", IEEE Electron Device Letters, Vol. 30, Number 6, June 2009, pp596.

[12] F. Medjdoub, J.-F. Carlin, M. Gonschorek, E. Feltin, M.A. Py, D. Duccatteau, C. Gaquière, N. Grandjean, E. Kohn "Can InAlN/GaN be an alternative to high power / high temperature AlGaN/GaN devices?" Proceedings of the IEEE International Electron Device Meeting (IEDM), 2006, San Francisco, USA.

[13] An electrothermal model for AlGaN/GaN PowerHEMTs including trapping effects to improvelarge-signal simulation results on high VSWR - O. Jardel et al. – IMS 2007.

[14] E. Zanoni, G. Meneghesso, F. Rampazzo, M. Meneghini, A. Tazzoli, "Trap related instabilities and localised damages induced by reverse bias", Proceedings Workshop on GaN Microwave Component Technologies, 30-31 March 2009, Ulm, Germany.

[15] Kuball et al., CSICS 2007; Sarua et al., IEEE Trans. Electron Dev. 54, 3152 (2007).

[16] R. Quay, P. Waltereit, M. Walther, M. Schlechtweg, M. Mikulla, O. Ambacher, "AlGaN/GaN HEMTs and MMICs: Status at Fraunhofer (IAF)", Proceedings Workshop on GaN Microwave Component Technologies, 30-31 March 2009, Ulm, Germany.

[17] K. Riepe, Philugan project report, unpublished.

[18] S. Piotrowicz et al , "State of the Art 58W, 38% PAE X-Band AlGaN/GaN HEMTs microstrip MMIC Amplifiers" ,Proceedings CSIC 2008, Monterey, USA.

[19] H. Brugger, B. Adelseck, M. Oppermann, P. Schur, F. Maile, "GaN radar and EW Applications", Proceedings Workshop on GaN Microwave Component Technologies, 30-31 March 2009, Ulm, German.

A 140 GHz Heterodyne Receiver Chipset for Passive Millimeter Wave Imaging Applications

S. Koch [#1], M. Guthoerl [#1], I. Kallfass [*2], A. Leuther [*2], and S. Saito [#1]

[#1] Sony Deutschland GmbH, Hedelfinger Strasse 61, D-70327 Stuttgart, Germany

Email: koch@sony.de

[*2] Fraunhofer Institute for Applied Solid-State Physics (IAF), D-79108 Freiburg, Germany

Abstract — A heterodyne receiver chipset for 140 GHz passive millimeter wave imaging applications is presented in this paper. The chipset consists of two different millimeter wave monolithic integrated circuits (MIMICs): a voltage controlled oscillator (VCO) working in the 35 GHz frequency range and a receiver chip hosting a low noise amplifier, a down-conversion mixer, a frequency multiplier and a local oscillator buffer amplifier together with a local oscillator distribution network. Both chips presented are realized using latest 100 nm gatelength metamorphic InAlAs / InGaAs HEMT (high electron mobility transistor) technology on 50 μm thick and 4 inch diameter GaAs substrates. The chips are utilizing grounded coplanar waveguide (GCPW) technology. Within the frequency band of operation from 120 to 145 GHz the receiver is showing a noise figure of ~ 5 dB and a conversion gain between -1 and 2 dB. The voltage controlled oscillator can be tuned from 31 to 37 GHz with associated output power levels from -2 to +2 dBm. All building blocks are explained in detail and measured results are presented. Finally the overall receiver performance is given.

I. INTRODUCTION

Based on Planck's law, every object emits electro-magnetic waves, ranging from microwave to infrared. Since this emission can be blocked by materials, passive imaging sensors based on millimeter wave detection can be used for future diagnostics [1], security [2], vision-systems [3] and airborne applications [4]. Using the advantage of atmospheric windows, 94-GHz-band sensors have been reported [2] – [7]. The atmospheric window at 140 GHz is a good candidate for improving the resolution of sensors and decreasing their size. From our perspective this frequency band offers a good compromise between availability of the latest semiconductor process developments [9] and low cost production solutions.

Within this paper we present a 140 GHz heterodyne receiver chipset using a 100 nm gatelength metamorphic InAlAs / InGaAs HEMT technology. This technology is realized on 100 mm diameter and 50 μm thick wafers. Therefore this technology can offer low-cost solutions compared to pure InP substrate based designs [5]. In addition this compound semiconductor technology has the advantages in two very important parameters of passive imaging sensors: the operation frequency bandwidth (25 GHz) and the noise figure (~ 5 dB). This performance is superior to the latest presented results using CMOS technology [8].

The chipset is split into two different millimeter wave monolithic integrated circuits (MIMICs): the voltage controlled oscillator (VCO) and the heterodyne receiver chip. Splitting into two chips was motivated by the intention to have the possibility of adding different VCO MIMICs to the receiver. In addition, splitting at 35 GHz still supports low-cost conventional bond-wire technology for chip interconnection.

II. CHIPSET ARCHITECTURE

The chipset architecture is shown in Fig. 1. The received signal from the antenna port (RF$_{IN}$) is amplified within the LNA with fixed gain. After amplification the signal is down-converted to the IF by a single ended, sub-harmonic pumped resistive FET mixer.

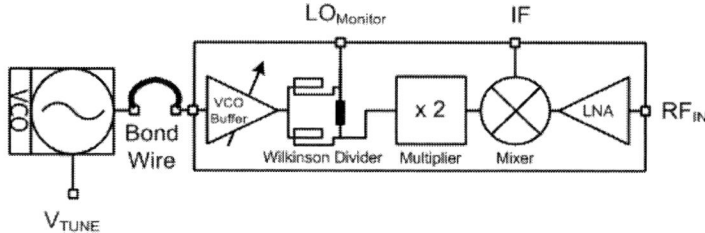

Fig. 1: Block diagram for the 140 GHz heterodyne receiver chipset.

The mixer is driven by an LO signal which is generated from the external 35 GHz VCO-MIMIC. For stabilization and tuning purposes (frequency setting) the VCO oscillation frequency is controlled by the tuning voltage (V$_{TUNE}$). The VCO is connected to the receiver chip via standard gold wire bonding technology.

To guarantee stable oscillation and to avoid load pulling effects, the VCO is decoupled with a VCO buffer amplifier. This amplifier is a variable gain amplifier to finally adjust the power level. In addition, the VCO signal is split after the buffer amplifier into two equal paths using a folded 35 GHz Wilkinson power divider [13]. One part of the signal is routed to the doubler which multiplies the 35 GHz signal to 70 GHz to drive the sub-harmonic mixer. The other part is accessible at the LO$_{MONITOR}$ output and can be used for PLL stabilization circuits.

Photographs of both chips are given in Fig. 2. The chip sizes for the VCO and receiver are 1.5 x 1.0 mm^2 and 1.0 x 4.5 mm^2 respectively. The receiver chipset is realized using a 100 nm gate length metamorphic InAlAs / InGaAs HEMT process [9] offered by Fraunhofer Institute for Applied Solid-State Physics (IAF). The process technology offers a transit frequency f$_T$ higher than 200 GHz and a maximum

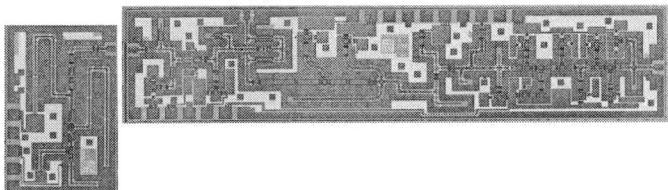

Fig. 2: Chip photographs of the VCO (left) and the 140 GHz receiver (right); Chip sizes for VCO 1.5 x 1.0 mm² and for receiver 1.0 x 4.5 mm²

oscillation frequency f_{MAX} of more than 300 GHz. Therefore, the technology is still well suited for 140 GHz applications.

III. BUILDING BLOCKS

Within this section the different building blocks used for the chipset realization are introduced and measured results are discussed.

A. Voltage Controlled Oscillator (VCO)

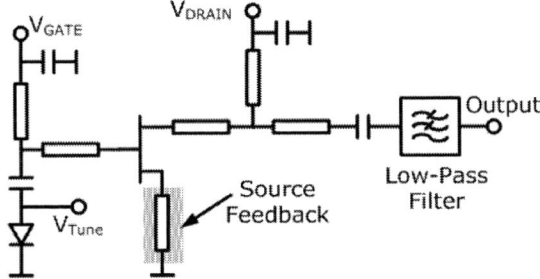

Fig. 3: Schematic diagram for the VCO circuit.

For the design of the VCO, a source feedback oscillator with a varactor tuning element connected to the gate side is realized. The varactor consists of a HEMT diode (2 x 30 µm) integrated in parallel configuration within a series-feedback resonator network. To achieve the required output power levels, a transistor size of 4 x 30 µm was chosen.

All matching networks were optimized to maximize the output signal at the output port. To suppress higher order harmonics, which could be delivered to the multiplier, a low-pass filter is added. The schematic diagram of the VCO is given in Fig. 3.

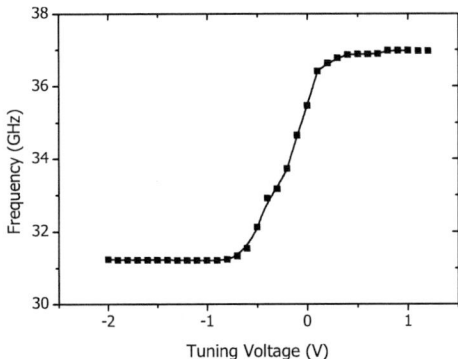

Fig. 4: Measured oscillation frequency characteristic for the VCO versus the tuning voltage.

Measured results of the oscillation frequencies versus tuning voltages with a tuning range from 31.5 to 37 GHz are shown in Fig. 4.

B. Low Noise Amplifier (LNA)

For the design of the LNA, a four stage design, with each stage using a common source transistor without inductive source feedback was chosen. The transistor size of each stage is 2 x 15 µm, yielding a total gate width of 120 µm. The amplifier is biased with a drain voltage of 1.0 V and the gate voltage is set for the maximum transconductance of the devices. A drain current draw of 35 to 40 mA is observed over wafer mapping.

In the frequency range from 115 to 150 GHz, the LNA achieves a measured small signal gain of ~ 15 dB and a noise figure between 4 and 6 dB (Fig. 5). All data were taken with a noise figure meter; therefore the impedance matching of the noise diode is influencing the flatness of the measured curves.

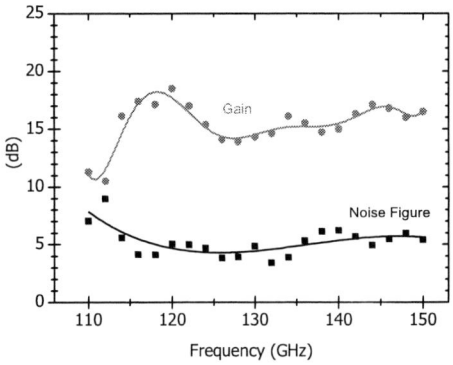

Fig. 5: Measured gain and noise figure of the LNA. Squares are for the noise figure and circles are for the gain.

C. Sub-Harmonic Resistive Mixer

For the down-conversion of the received signal a sub-harmonic pumped, single ended resistive mixer was designed. The optimum transistor size was found to be 4 x 30 µm. With this size, the compromise between down conversion gain and LO-power drive was best. More details about the design principles can be found in [10].

In Fig. 6, the measured conversion gain of the mixer is

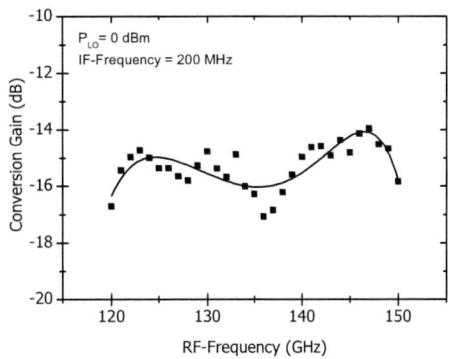

Fig. 6: Measured down-conversion gain of the sub-harmonic resistive mixer versus RF-frequency.

shown. It varies from - 14 to -17 dB in the frequency range of 120 to 150 GHz and at a local oscillator power of 0 dBm. The optimum gate bias for sub-harmonic operation was found at – 0.5 V, corresponding to a quiescent condition below threshold.

D. Frequency Multiplier (x2)

The frequency multiplier (doubler) design adopts a single-ended topology using the design techniques explained in [11] and [12], which exploits the strong generation of the second order harmonic in a transistor biased under conditions close to

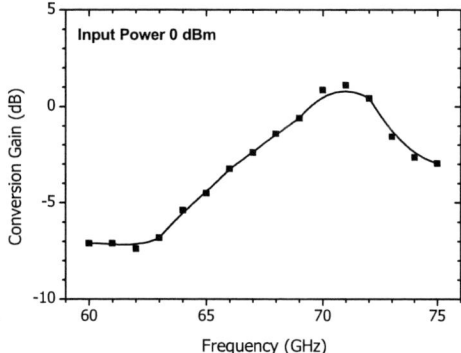

Fig. 7: Measured conversion gain versus 2 x fundamental frequency for the multiplier block. Measurement conditions are $V_D = 0.8$ V with an input power level of 0 dBm (fundamental)

class B. Besides the matching for the fundamental frequency, the input matching circuit creates a short circuit at the second fundamental frequency (70 GHz) to avoid backward reflection of the multiplied signal to the input. On the other hand, the output matching includes a short circuit for the fundamental frequency (35 GHz) which is realized by an open-ended λ/4 stub to suppress the fundamental frequency at the output. The transistor size of 4 x 30 µm was chosen for optimum conversion gain.

During measurements a current draw of 20 to 30 mA was observed using a 0.8 V drain bias. In Fig. 7, the conversion gain versus (2 x fundamental) output frequency is plotted. For the frequency range of interest (60 – 75 GHz) the gain varies

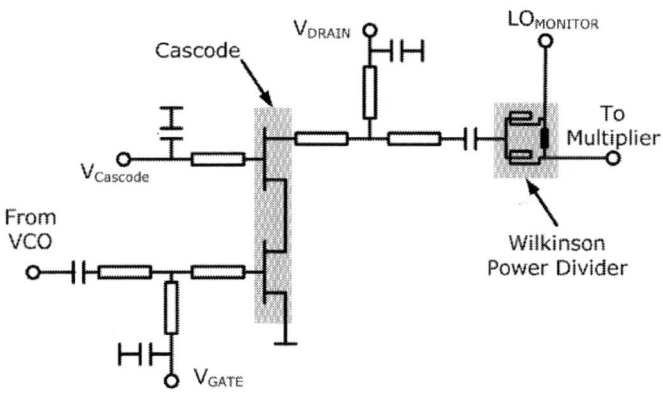

Fig. 8: Schematic diagram for the VCO buffer amplifier including the Wilkinson power divider.

between -7 and +1 dB at 0 dBm fundamental input power level.

E. VCO Buffer Amplifier

To adjust the power level for the multiplier and to decouple the VCO against the environment, a variable gain amplifier was designed as a VCO buffer. For having the possibility to lock the VCO via an external phased locked loop (PLL) circuit, a Wilkinson power divider was added to the buffer amplifier. A detailed schematic is given in Fig. 8. The buffer is realized as a single stage cascode. In the interesting frequency range from 31 to 37 GHz a maximum small signal gain of 7 dB with a reverse isolation higher than 30 dB is measured for the combination of the buffer amplifier and the folded Wilkinson power divider. The gain can be adjusted by the cascode within a >15dB range, which is sufficient for the application. The power consumption is 60 mW for a 1.8 V drain bias at maximum gain settings.

IV. OVERALL RECEIVER PERFORMANCE

The overall performance of the receiver was measured on-wafer, using GSG- (Ground-Signal-Ground) and DC-probe heads to provide biasing voltages. All drain voltages were set to 1.0 V for common source FETs. For the cascode FET configurations the drain voltages were set to 1.8 V. All gate voltages were adjusted for maximum transconductance, while for the multiplier and the mixer transistors the gate voltages were set to -0.25 V. Based on these settings, the overall power dissipation of the receiver is 120 mW (excluding the VCO). Within the measurement setup a commercial available signal generator was used for LO-generation.

These settings result in conversion gain from -1 to +2dB in the frequency range from 120 to 145 GHz (Fig. 9). Additional measurements of the conversion gain versus IF-frequency show a flat conversion gain (within +/- 1 dB) at least up to 2 GHz bandwidth.

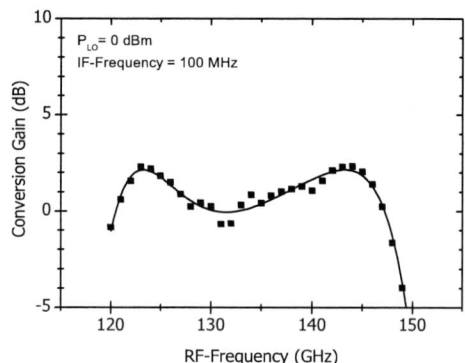

Fig. 9: Measured conversion gain of the receiver MIMIC.

V. CONCLUSIONS

A 140 GHz heterodyne receiver chipset for passive millimeter wave imaging applications has been introduced. The chipset is realized using a 100 nm gatelength

978-1-4244-5190-6/09 $25.00 © 2009 IEEE   18

metamorphic InAlAs / InGaAs HEMT technology based on 100 mm diameter and 50 μm thick wafers.

The receiver MIMIC is offering superior performance in noise figure (5 dB) and operation frequency bandwidth (25 GHz). From this, the chipset is considered to be an excellent candidate for low cost, high resolution sensor implementations operating in the 140 GHz atmospheric window.

VI. ACKNOWLEDGEMENT

The authors wish to acknowledge the contributions of J.-Y. Choi at Sony Deutschland, as well as of H. Massler from IAF for their continuous support during design and measurements.

REFERENCES

[1] K. Mizuno, H. Matono, Y. Wagatsuma, H. Warashina, H. Sato, S. Miyanaga, and Y. Yamanaka, "New applications of millimeter-wave incoherent imaging," *2005 IEEE MTT-S Int. Microwave Symp. Digest*, pp. 629-631. June 2005.

[2] D. R. Vizard, and R. Doyle, "Advances in millimeter wave imaging and radar systems for civil applications," *2006 IEEE MTT-S Int. Microwave Symp. Digest*, pp. 94-97, June 2006.

[3] Ortiz, A, "Passive millimeter wave imaging sensor enhanced vision system" *Digital Avionics Systems Conference, 1997*, Volume 1, Issue , Oct 1997, pp. 23-30.

[4] S. E. Clark, J. A. Lovberg, C. A. Martin and V. Kolinko, "Passive millimeter-wave imaging for airborne and security applications," *Proceedings of SPIE 2003*, Vol. 5077, pp. 16-21.

[5] Sato, M. Hirose, T. Ohki, T. Takahashi, T. Makiyama, K. Hara, N. Sato, H. Sawaya, K. and Mizuno, K, "InP-HEMT MMICs for passive millimeter-wave imaging sensors", Indium Phosphide and Related Materials (IPRM) 2008, pp 1-4, May 2008.

[6] Wang, H. Chang, K.W. Ton, T.N. Biedenbender, M. Chen, S.T. Lee, J. Dow, G.S., Tan, K.L. and Allen, B.R., "High-yield W-band monolithic HEMT low-noise amplifier and image rejection downconverter chips", *IEEE Microwave and Guided Wave Letters*, vol 3, issue 8, pages 281-283, August 1993.

[7] Tessmann, A., Kudszus, S., Feltgen, T., Riessle, M., Sklarczyk, C., Haydl, H.W., "Compact single-chip W-band FMCW radar modules for commercial high-resolution sensor applications", *IEEE Transactions on Microwave Theory and Techniques* 50 (12), 2002, 2995 – 3001.

[8] S. T. Nicolson1, A. Tomkins1, K. W. Tang1, A. Cathelin2, D. Belot2, and S. P. Voinigescu, "A 1.2V, 140GHz Receiver with On-Die Antenna in 65nm CMOS", *2008 IEEE Radio Frequency Integrated Circuits Symposium*, pp. 229-232, June 2008.

[9] A. Leuther, A. Tessmann, M. Dammann, W. Reinert, M. Schlechtweg, M. Mikulla, M. Walther, and G. Weimann, "70 nm low noise metamorphic HEMT technology on 4 inch GaAs wafers," in *Proc. Indium Phosphide Related Materials Conf. (IPRM)*, May 2003, pp. 215–21.

[10] N. A. Rahman, B. Y. Majlis, A. Ariffin, " A 28 GHz PHEMT GaAs MMIC Single-ended Resistive Mixer", *ICSE 2002 Proc. 2002*, Penang, Malaysia, pp. 511 – 513.

[11] Nishikawa, K.; Toyoda, I.; Tsunekawa, K.; Enoki, T. and Sugitani, S." Low-voltage and broadband V-band InP HEMT frequency doubler MMIC", ," *2007 IEEE MTT-S Int. Microwave Symp. Digest*, pp. 46-48, June 2007

[12] Chen, S.-W.; Ho, T.; Phelleps, F.R.; Singer, J.L.; Pande, K.; Rice, P.; Adair, J. and Ghahremani, M. "A high-performance 94-GHz MMIC doubler", *IEEE Microwave and Guided Wave Letters*, vol 3, issue 6, pages 167-169, June 1993.

[13] S. Koch, I. Kallfass, A. Leuther, M. Schlechtweg, S. Saito, M. Uno, "A Four-Antenna Transceiver MIMIC for 60 GHz Wireless Multimedia Applications", *2008 EuMC European Microwave Conference. Dig.*, pp. 1529-1532.

Ultra Low Power 60 GHz ASK SiGe Receiver with 3-6 GBPS Capabilities

Woorim Shin, Mehmet Uzunkol and Gabriel M. Rebeiz

Electrical and Computer Engineering

The University of California, San Diego

woshin@ucsd.edu, muzunkol@ucsd.edu, rebeiz@ece.ucsd.edu

Abstract— **This paper presents an ultra low power 60 GHz SiGe ASK receiver capable of 3-6 Gbps communications. The receiver is based on a 4-stage low-noise amplifier and an ASK detector, all in a 0.12 um SiGe transistor technology. The LNA and ASK detector were designed for wide bandwidth in order to result in a very high data rate. The LNA+Detector chip consumes 11 mW and is capable of 3-6 Gbps communications with a BER < 10^{-12} for an input power of -36 dBm. At an input power of -44 dBm, the system can maintain a 3 Gbps link with a BER <10^{-9}. This translates to 3.3-1.6 pJ/bit for the LNA+Detector, and to our knowledge, is the state of the art. The chip was tested up to 105°C and maintained > 3 Gbps with a BER < 10^{-12} over the entire temperature range.**

Index Terms— **60 GHz, amplitude detector, LNA, ASK, OOK, communication.**

I. INTRODUCTION

HIGH data rate 60 GHz communication systems have been demonstrated by several groups using ASK, BPSK and QPSK transceivers [1-9]. Most of these systems are based on complex modulation schemes which consume a lot of power per bit, but result in a wide dynamic range and immunity to interferers. As is well known, the ASK (also known as OOK) system results in the lowest power consumption, but has limited dynamic range and virtually no immunity to interferers. However, there are applications such as low-cost short distance point-to-point links where a 60 GHz ASK system is highly advantageous. This is especially applicable if the ASK system is operated with directional antennas (G > 10 dB) and over 0.1-3 meters.

This paper presents a SiGe LNA+Detector which have been optimized to result in the lowest power consumption while still being capable of delivering a 3-6 Gbps link at low input power levels (Fig. 1). The ASK detector has a very high NF (> 35 dB) and therefore, an LNA with > 25 dB is used to greatly improve the system NF and

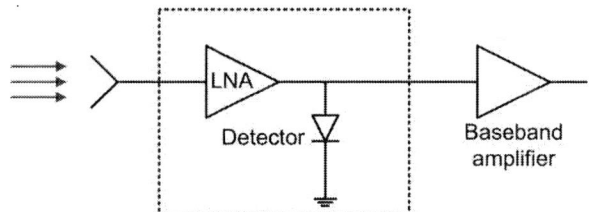

Fig. 1: ASK receiver system consisting of a SiGe LNA and a SiGe transistor detector.

sensitivity. For reference purposes, the baseband amplifier is not integrated on-chip and the quoted power is only for the LNA+Detector. Even though the communication link was established using waveguide horn antennas over variable distances, the power is quoted at the receiver input port so as to normalize out the antenna gain and free space loss.

II. LNA DESIGN AND MEASUREMENTS

Fig. 2a presents a 4-stage SiGe LNA designed for low noise and very high gain. The first stage is designed for low noise with an emitter degeneration inductor and 1 mA of current, while stages 2-4 are designed for high gain and consume 2 mA each. The bias circuits (not shown) are implemented using standard current mirrors and are not a PTAT (proportional to absolute temperature) design. The simulated gain is > 25 dB from 52-61 GHz with a peak of 26.3 dB at 56 GHz. The simulated S_{11} and S_{22} are < -10 dB and < -8 dB, respectively, over this frequency range. The simulated NF is 5.7-5.8 dB at 55-60 GHz. The amplifier consumes 10.5 mW (7 mA from a 1.5 V source).

The amplifier is designed to have an input P1dB of -32 dBm (output P1dB of around -7 dBm). As will be seen in Section III, the detector saturates in a gradual fashion and can withstand -10 dBm of input power. The simulated input IIP3 is -24 dBm for the LNA.

The LNA+Detector chip is implemented in the IBM

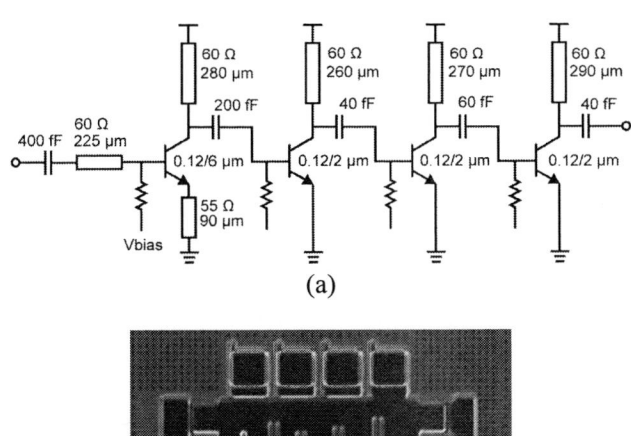

(a)

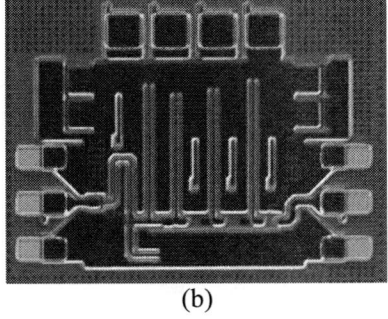

(b)

Fig. 2: Schematic (a) and microphotograph (b) of the 4-stage SiGe amplifier. (0.76 x 0.57 mm²)

8HP process with 7 metal layers. This process has 0.12μm SiGe transistors (f_T of 180-200 GHz) and 0.12μm CMOS transistors (f_T of 90-100 GHz). Grounded CPW transmission lines are used with dimensions of 11/12/11 μm for 50 Ω and with an estimated loss of 0.55 dB/mm at 60 GHz. All inductors are implemented using shorted stubs with an estimated transmission-line Q of 11-12 at 60 GHz. Standard IBM transistor cells and MIM capacitor models are used, and full electromagnetic modeling is done on the transmission-lines and stubs using Sonnet[1]. The fabricated LNA occupies 0.42 mm² including pads (Fig. 2b).

The measured S-parameters of 10 different LNA chips are shown in Fig. 3. The peak response occurs at 51 GHz with a gain of 25.7-27.3 dB. The simulated S_{11} and S_{22} agree well with measurements. A small shift in frequency is observed and this is currently being investigated. It is possible that this is due to a small additional inductance incurred in the small metal-oxide-metal (40-60 fF) inter-stage capacitors. The NF was measured on 3 different LNAs at 50 GHz (limited by our measurement set-up) and was 6.2-6.7 dB. This compares well with simulations at 50 GHz (6.1 dB). Measured output P1dB is -10.1 dBm, which is enough to drive the subsequent diode detector. (Fig. 4)

The temperature performance of an LNA chip is shown in Fig. 5. The gain of LNA gradually decreases by 4.9 dB as temperature increases to 105°C. The gain response does not shift in frequency.

Fig. 3: Measured S-parameters of 10 LNA chips: (a) S_{21}, and (b) S_{11} and S_{22}.

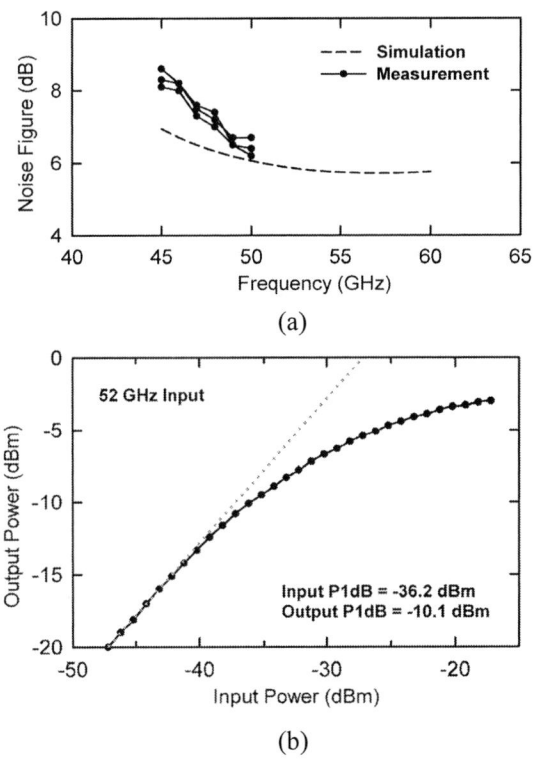

(a)

(b)

Fig. 4: Measured (a) noise figure and (b) P1dB of LNA

[1] Sonnet, ver. 11.52, Sonnet Software Inc., Syracuse, NY, 1986-2007

978-1-4244-5190-6/09 $25.00 © 2009 IEEE

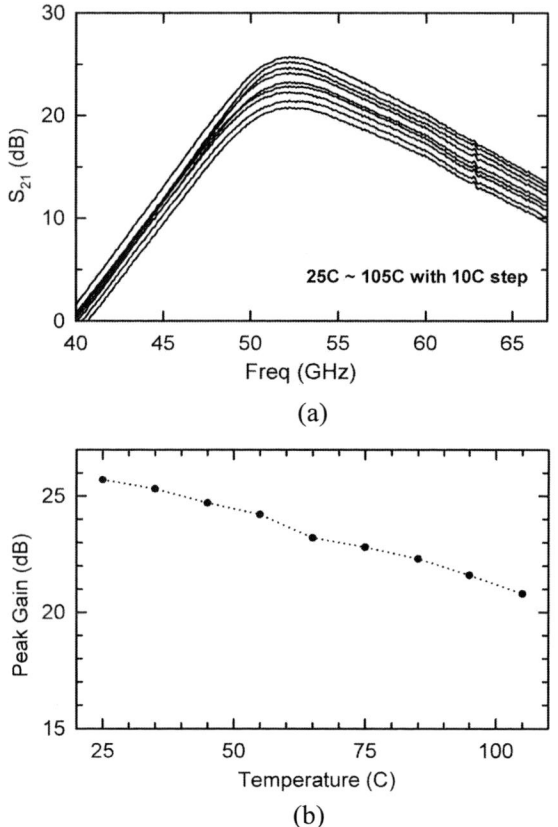

(a)

(b)

Fig. 5: Measured gain vs. temperature: (a) S_{21}, (b) peak gain.

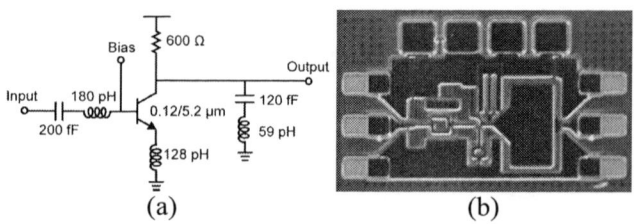

(a)   (b)

Fig. 6: Schematic (a) and microphotograph (b) of the SiGe ASK detector. (0.42×0.57 mm^2)

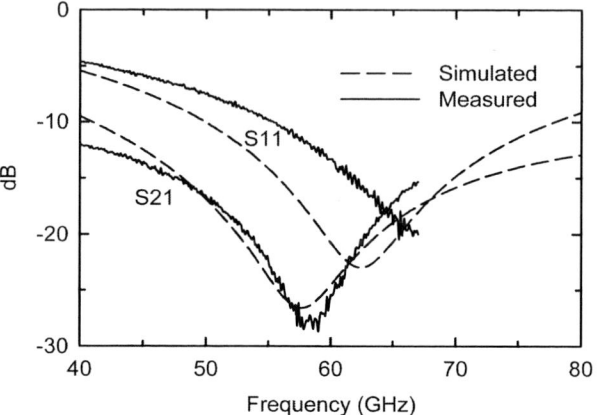

Fig. 7: Measured and simulated S-parameters of the ASK detector.

III. DETECTOR DESIGN AND MEASUREMENTS

The SiGe ASK detector is based on a bipolar amplifier biased in a near class C region with a bias current of 0.35 mA (Fig. 6). Lumped-element emitter degeneration and a base inductor are used to match the detector to 50 Ω at the input port. A 60 GHz LC notch filter (L=59 pH, C= 120 fF) is placed at the collector to filter out the 60 GHz component, and the collector impedance is set to 600 Ω for DC biasing purposes. The fabricated detector occupies 0.24 mm^2 including the pads. In most measurements, a 50 Ω load is attached to the output port. The simulated responsivity at 60 GHz is 9100 V/W into an open circuit and 700 V/W into a 50 Ω load. Noise-equivalent power (NEP) is defined as the output noise voltage (2-3 nV/Hz$^{1/2}$ at > 500 MHz) divided by the responsivity, and is simulated as 3-4.3 pW/Hz$^{1/2}$ at > 500 MHz. The NEP is the same for an open circuit or for a 50 Ω load since both the output noise and the responsivity drop by the same amount when a 50 Ω load is connected to the detector output port.

The measured S_{11} and S_{21} of the SiGe detector are presented in Fig. 7. The detector is well matched over a wide bandwidth and the notch filter response is clearly seen in S_{21}. The measured responsivity is ~635 V/W at 60 GHz with a P1dB of -15 dBm, and the measured output noise is 2.5-3 nV/Hz$^{1/2}$ > 500 MHz. The measured output noise includes the noise of 1-1000 MHz Miteq AM1533 amplifier (NF<2 dB, V_n<0.7 nV/Hz$^{1/2}$) but the gain of the amplifier was normalized out of the measurement. This results in an NEP of 4-4.8 pW/Hz$^{1/2}$ for >500 MHz and agrees well with simulations. The output noise is high at low frequencies due to the noise coupled from the biasing resistor and it is multiplied by the low frequency gain of the detector. The output noise voltage of the detector could be improved by using a smaller bias resistor.

IV. LNA+DETECTOR MEASUREMENTS

The LNA+Detector chip is shown in Fig. 8 and is based on combining the previously described LNA and ASK detector into a single circuit (area is 0.57×1.1 mm^2 including pads). From the NEP measurements and assuming an RF input bandwidth of 10 GHz, the estimated NF of the detector is ~40-41 dB. The system NF is therefore 15-16 dB with the LNA, and is completely dominated by the ASK detector. The measured responsivity of the LNA+Detector chip at 55 GHz is 180,000 V/W into a 50 Ω load with a P1dB of -36 dBm.

978-1-4244-5190-6/09 $25.00 © 2009 IEEE

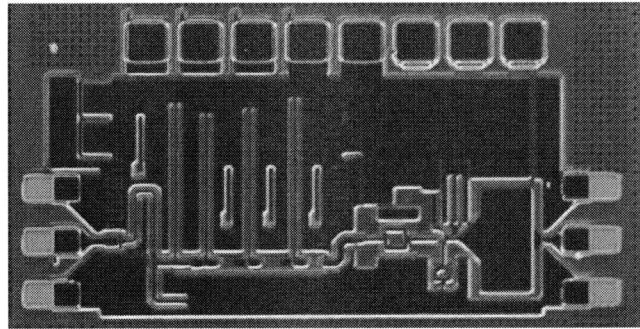

Fig. 8: Microphotograph of LNA+detecter. (1.1 x 0.5 mm^2)

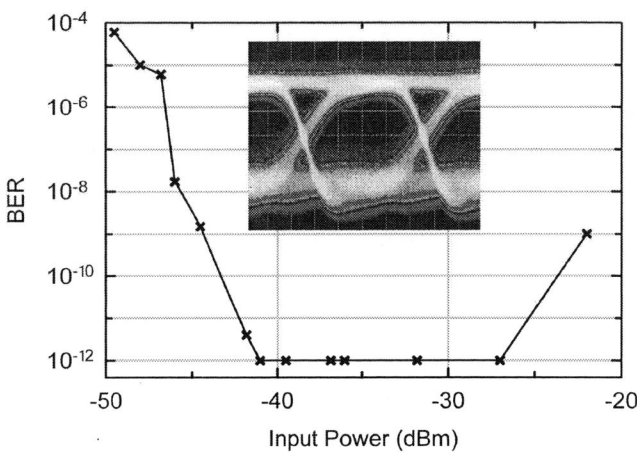

Fig. 9: Measured BER vs input power at 55 GHz. A dynamic range of 23 dB is observed for a BER < 10^{-9}.

At this power level, the LNA output power (assuming an internal 50 Ω load) is -13 dBm and agrees well with the measured stand-alone detector P1dB.

An ASK modulated system was set up with an Agilent N4903A BER tester sending 2^{31}-1 pseudo-random binary sequence (PRBS), a double-balanced mixer and a 55 GHz signal source. A center frequency of 55 GHz was chosen since it results in the widest amplifier bandwidth for double sideband operation. The ASK modulated signal was transmitted and received using horn antennas over variable distances, but this has been normalized out of the experiment and all power are referred to the available input power at the chip. Coaxial 50 Ω wideband amplifiers were used after the LNA+Detector chip and their power consumption is not included in the calculations (they consume ~150 mA to drive 50 Ω over 1 MHz to > 6 GHz). The measured BER vs. input power level for a 3 GBPS link is shown in Fig. 9. The ASK detector shows a dynamic range of 14-23 dB with a BER of less than 10^{-12} to 10^{-9}. At low powers, the system is limited by the NF, while at high powers, it is limited by the detector compression. For an input power of -36 dBm, we were able to obtain 6 Gbps with a BER of 10^{-12}.

The input power level can also be deduced from the system NF of 15 dB. For a double side-band 3 Gbps signal, the system bandwidth is around 2*(6 GHz)=12 GHz, and the available noise at the input is -174 + 15 + 101 = -58 dBm. A S/N of 15-20 dB is required for excellent BER performance in a simple ASK detector, and thus the -41 dBm minimum signal level for a BER of 10^{-12}. The system operated up to 105°C in a 3 Gbps link, and this will be presented at the conference.

ACKNOWLEDGMENT

This work was funded by Intel Corp. and the UC-Discovery Program, Drs. Ian Young and Jad Rizk, Program Monitors. We also thank Prof. James Buckwalter at UCSD for his help in the BER measurements.

REFERENCES

[1] S. Sarkar and J. Laskar, "A Single-Chip 25pJ/bit Multi-Gigabit 60GHz Receiver Module," *IEEE Int. Microwave Symposium Digest*, pp. 475-478, June 2007.

[2] J. Lee et al., "A Low-Power Fully Integrated 60 GHz Transceiver System with OOK Modulation and On-Board Antenna Assembly," *IEEE ISSCC Dig. Tech. Papers*, pp. 316-317, Feb. 2009

[3] M. Ko, "A CMOS-Compatible Schottky-Barrier Diode Detector for 60-GHz Amplitude-Shift Keying (ASK) Systems," *IEEE Int. Microwave Symposium Digest*, pp. 1557-1560, June 2008.

[4] A. Oncu et al., "60 GHz-Pulse Detector Based on CMOS Nonlinear Amplifier," *IEEE Topical Meeting on Silicon Monolithic IC in RF Systems*, pp.1-4, Jan. 2009

[5] A. Tomkins et al., "A Zero-IF 60 GHz Transceiver in 65nm CMOS with > 3.5Gb/s Links," *IEEE Custom Integrated Circuits Conference*, pp. 471-474, Sept. 2008

[6] C. Marcu et al., "A 90nm CMOS Low-Power 60 GHz Transceiver with Integrated Baseband Circuitry," *IEEE ISSCC Dig. Tech. Papers*, pp. 314-315, Feb. 2009

[7] S. Reynolds et al., "A Silicon 60 GHz Receiver and Transmitter Chipset for Broadband Communications," *IEEE J. Solid-State Circuits*, vol. 41, no. 12, pp. 2820-2831, Dec. 2006.

[8] M. Tanomura et al., "TX and RX Front-Ends for 60 GHz Band in 90nm Standard Bulk CMOS," *IEEE ISSCC Dig. Tech. Papers*, pp.558-559, Feb. 2008.

[9] T. Mitomo et al., "A 60 GHz CMOS Receiver Front-End With Frequency Synthesizer," *IEEE J. Solid-State Circuits*, vol. 43, no. 4, Apr. 2008.

A Fully Integrated, Compound Transceiver MIMIC utilizing Six Antenna Ports for 60 GHz Wireless Applications

S. Koch [1], I. Kallfass [2], R. Weber [2], A. Leuther [2], M. Schlechtweg [2], and S. Saito [1]

Sony Deutschland GmbH, Hedelfinger Strasse 61, D-70327 Stuttgart, Germany [1]

Fraunhofer Institute for Applied Solid-State Physics (IAF), D-79108 Freiburg, Germany [2]

Email: koch@sony.de

Abstract — **A fully integrated transceiver MIMIC (millimeter wave monolithic integrated circuit) with six antenna port functionality for 55 to 65 GHz wireless applications has been developed. The chip has been realized using 100 nm gatelength metamorphic InAlAs / InGaAs HEMT (high electron mobility transistor) technology on GaAs substrates together with CPW (coplanar waveguide) technology. The novel transceiver topology consists of switches, amplifiers, mixers, a voltage controlled oscillator and a frequency divider. The receiver chain shows noise figure < 2.6 dB on the low noise amplifier level and smaller than 7 dB including the antenna switching network. The medium output power amplifier delivers the saturated power level of + 14 dBm. This is the highest integration level for a 60 GHz compound semiconductor transceiver chip reported to-date.**

I. INTRODUCTION

Many different publications within the last years have been focusing towards the integration of 60 GHz transceivers. The papers showing results for receivers or building blocks were dedicated to the new IEEE 802.15.3c or Wireless HD standards. The customer demand for transmitting HDMI type of contents is increasing and thus the demand for a data-throughput beyond the gigabit / second regime is growing as well. To fulfil the strong requirements for transmission quality and to overcome propagation channel distortion and system gain margins, the presented MIMIC will support a multi-antenna communication system.

Many publications are dealing with 60 GHz integrated circuit designs [1] - [8]. Even the latest highly integrated results [15] do not take the multi-antenna aspect into account. By adding a switching capability to the device the system can support multiple sectors for the antenna covering areas with, if necessary, different antenna half-power beam widths and respective different gains.

The MIMIC outlined in this paper is a final enhancement of the MIMICs presented in [9] and [13] towards a novel, fully integrated transceiver. This MIMIC consist of, besides for compound semiconductor integration very familiar amplifiers and mixers, a 2:6 switch, a 30 GHz VCO with a LO-distribution amplifier and a 4-stage static frequency divider.

This MIMIC is, to the knowledge of the authors, the highest fully integrated transceiver MMIC reported for the 60 GHz frequency band using compound semiconductor technology.

By using compound semiconductors, the MIMIC demonstrates superior performance in output power levels of ~ 14 dBm at the medium power amplifier output port, very low noise figure performance (< 2.6 dB) at the low noise amplifier input and high output power switching capability for the antenna port switches. The antenna port switches can handle more than 16 dBm input power drive without degradation of the performance. The achieved output power levels are about 5-6 dB higher compared to the latest reported CMOS data [15] resulting in a longer distance data link or higher throughput communication. Finally gain flatness of + /- 1dB for the complete receive- or transmit-chain has been achieved over the complete frequency range from 55 to 65 GHz.

II. TRANSCEIVER TOPOLOGY

The transceiver block architecture of the MIMIC is given in Fig. 1. Six antenna signals (ANTi, i = 1, 2...6) are routed, by using a novel 2:6 antenna switch, to either a low-noise amplifier (LNA) or a medium power amplifier (MPA) depending on a desired Rx- or Tx-operation, respectively. The I- and Q-baseband signals are down- or up-converted using a sub-harmonic pumped resistive IQ-mixer.

A 30 GHz voltage controlled oscillator (VCO) is buffered by a local oscillator (LO)-buffer amplifier, which is delivering the requested LO-power. For stabilization of the VCO

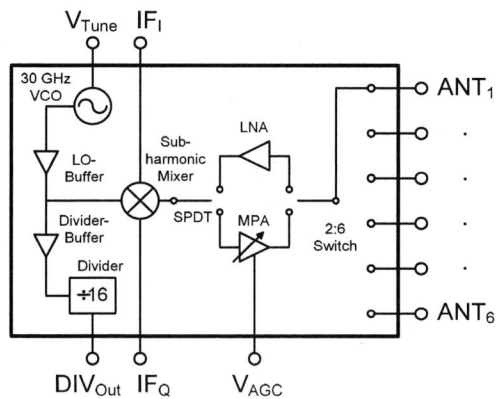

Fig. 1: MIMIC architecture for the 6-antenna-port transceiver.

(associating with an external PLL), a small amount of the LO-signal is coupled towards a divider-buffer amplifier which de-couples and amplifies the LO-signal to drive a divider-by-16 circuit. The divider is realized by four equivalent cascaded stages, each having a division factor by 2.

The chip photograph is given in Fig. 2. The chip has a size of 2.5 x 5.5 mm² and is therefore ~30 % smaller in size compared to the MIMIC presented in [9] while it still offers complete transceiver functionality. Because the chip size is not optimized for the smallest chip area, further reduction can be achieved from the authors' perspective.

All measurements presented in this paper were carried out on-wafer using coplanar GSG (ground-signal-ground) probes.

The transceiver is realized using the 100 nm gatelength

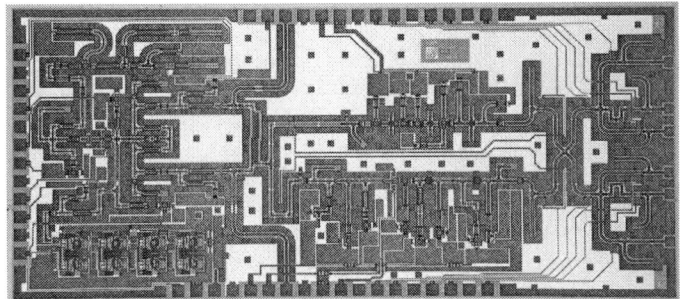

Fig. 2: Chip photograph of the MIMIC-transceiver; chip size 2.5 x 5.5 mm²

metamorphic high-electron mobility transistor (HEMT) process [10] by Fraunhofer Institute for Applied Solid-State Physics (IAF). The HEMT devices are grown on 4-inch semi-insulating GaAs substrates by Molecular Beam Epitaxy (MBE).

The T-shaped gate definition is performed using electron beam lithography in a three layer (PMMA) resist process. A Pt-Ti-Pt-Au layer sequence is used for the gate metallization. With an indium content of 65% in the channel, an extrinsic transit frequency of 220 GHz and a maximum oscillation frequency of 300 GHz have been achieved.

III. TRANSCEIVER BUILDING BLOCKS

The design of the MIMIC is carried out using grounded coplanar waveguide (GCPW) technology with a substrate thickness of 50 μm. The design library is offering passive models for the transmission lines as well as linear and non-linear models for the transistors which were validated > 100 GHz.

A. Amplifiers

Three different amplifiers are used on the transceiver MIMIC: a low noise amplifier (LNA) for the receive path, a medium power amplifier (MPA) for the transmitter and a local oscillator distribution amplifier.

The design for the LNA was not changed from [9]. The performance of the 3-stage common source design was excellent and demonstrates a small signal gain $|S_{21}|$ of ~ 21 dB and a noise figure NF of < 2.6 dB with a bias voltage of 1.0 V at maximum transconductance ($g_{M,MAX}$) in the complete 55 – 65 GHz frequency range, having a power consumption of 65 mW.

The MPA is a 4-stage variable gain amplifier to adopt for different TX input power levels. The MPA has 27 dB gain from 45 to 75 GHz which can be controlled by incorporating a cascode element at the second stage. The achieved gain control was higher than 20 dB. The MPA offers an output power level of ~ 14 dBm for saturation and P1dB of ~ 10 dBm in the frequency band of interest. The power consumption of the MPA is 200 mW. More details can be found in [13].

Finally, the local oscillator distribution amplifier is distributing the LO-signal (27.5 GHz – 32.5 GHz) to the sub-harmonic pumped IQ-mixer, as well as to the divider circuit. Therefore, a unique design with two de-coupled outputs has been derived. The schematic is given in Fig. 3. The LO-signal is amplified by the first cascode stage, offering a gain control to the amplifier (better than 15 dB). The output signal is delivered to the mixer (main part). A small part of the signal is coupled to a second, common source, one stage amplifier which is delivering the requested output power (> 0 dBm) to the divider circuit.

Small signal gains of ~14 dB for the VCO-mixer path and 11 dB for the VCO-divider path were measured. For the intended input power level of 0 dBm, output power levels of +10 dBm for the mixer and +2 dBm for the divider are realized. The graph in Fig. 4 shows the output power performance versus the input power drive for both outputs as the comparison of simulations and measurements at 27 GHz operation frequency.

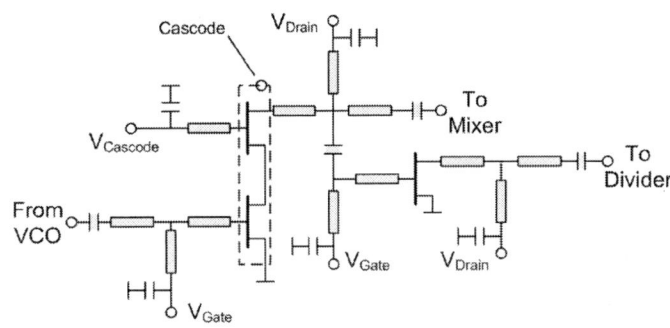

Fig. 3: Schematic of the LO-distribution amplifier.

B. Switches

Two different kinds of switches, a single pole double throw (SPDT) and a "2:6 switch", were designed and are implemented on the MIMIC. The design principles of the switches and detailed results are given in [13].

The SPDT is utilized to switch between the IQ-mixer and the Rx-Tx path. It shows a 1-dB-bandwidth of the insertion loss from 43 to 80 GHz. Minimum insertion loss is 1.4 dB at 60 GHz. The SPDT switch exhibits an isolation of 33 dB and return loss greater than 9 dB between 55 to 65 GHz.

For switching between the antenna ports and transmit or receive path, the "2:6" switch is used. The switch demonstrates an insertion loss of ~ 4 dB with again the minimum at 60 GHz and isolation values of ~ 20 dB.

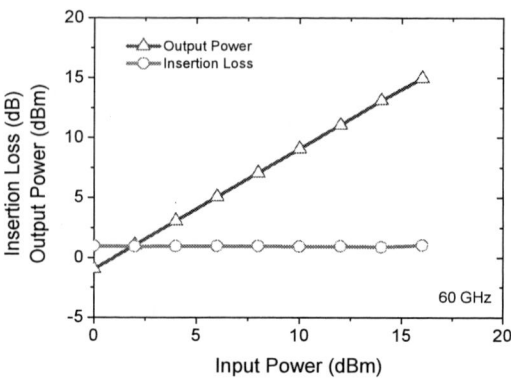

Fig. 5: Measured output power and insertion loss versus input power drive for the SPDT switch at 60 GHz.

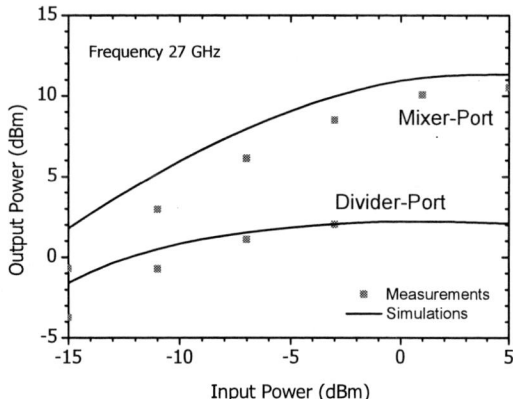

Fig. 4: Output power for the mixer and divider port versus the input power level for the LO-monitor amplifier at 27 GHz.

As the switches have to handle the output power levels in the order of 15 dBm, the power handling capability was tested for the SPDT switch. No degradation in insertion loss and output power compression could be recognized up to input power levels of 16 dBm (Fig. 5). The input power level of 16 dBm was the highest level which could be generated in the measurement setup at 60 GHz.

C. Sub-harmonic IQ-Mixer

The mixer design was first introduced in [13]. The IQ-mixer combines two single sub-harmonic pumped resistive mixer cells together with a Wilkinson power divider and a 90-degree Lange-coupler on the LO- and RF-side respectively. As the LO frequency range is in the 30 GHz regime, the chip area for the Wilkinson power divider is quite big.
Therefore we introduced a "bridged" Wilkinson power divider which is folded inside the IQ-mixer area and therefore reduces the mixer area to ~ 1 mm². A chip photograph of the IQ-mixer using the bridge Wilkinson power divider is given in Fig. 6.

The IQ-mixer shows 15 dB conversion loss for up- and down-conversion within the frequency range from 50 to 65 GHz with an LO-drive of 10 dBm. The LO-isolation is higher than 25 dB between 55 and 65 GHz.

D. Voltage Controlled Oscillator & Divider

For the frequency generation, a 30 GHz VCO was designed. The oscillator is realized as a single stage oscillator using a negative source feedback. For frequency tuning a HEMT

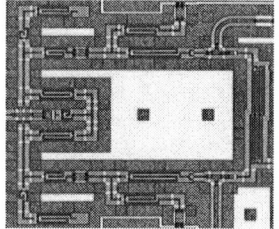

Fig 6: Chip photograph of the IQ mixer using the "bridged" Wilkinson power divider.

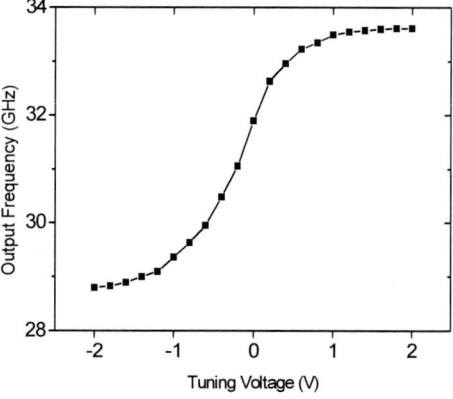

Fig 7: Measured oscillation frequency vs. tuning voltage for the VCO.

diode is added to the series feedback circuit at the gate side. The output matching (drain-side) is designed to maximize the output power level [16].

The measured oscillation frequency versus the tuning voltage is given in Fig. 7. It varies between 28.8 and 33.6 GHz which has a frequency offset of 1.3 GHz from the requested mid frequency of 30 GHz with 5 GHz bandwidth. The measured output power levels at the oscillator output are varying between -3 to 0 dBm over the tuning range. By stabilization the VCO with an external PLL (ADF4108) circuit, a phase noise of -72 dBc / Hz at 1 MHz offset could be realized. The DC power consumption of the VCO is 50 mW.

To control the VCO with external PLL circuitry, a divider with a division ratio of 16 was implemented on the chip, which consists of 4 sub-dividers, each having the division ratio of 2. The design is a static divider based on D-flip-flop design. More details about the design procedure can be found in [12].

IV. TRANSCEIVER PERFORMANCE

The overall performance of the transmit (TX) and receiver (Rx) chain was measured in the IQ frequency ranges from 1 MHz to 2 GHz. Over the complete IQ-frequency range conversion gains were flat within 1 dB. The up-conversion gain was measured 3 dB [13] and the down-conversion gain is 0 dB. To adjust power levels for A/D or D/A converters, additional amplifiers can be added at low frequency level on lowest cost base. The power consumption is 300 mW for up-conversion and 200 mW for down-conversion (excluding divider power consumption of 1 W).

V. CONCLUSIONS

A fully integrated transceiver MIMIC with six antenna port functionality for the frequency range from 55 to 65 GHz has been presented. With the six antenna ports, the MIMIC supports directly multi-antenna systems. No external switching network needs to be implemented. The chip has been realized using state-of the art 100 nm gatelength metamorphic InAlAs / InGaAs HEMT technology on 4-inch GaAs substrates.

All building blocks have been discussed and measured results presented. The receiver chain shows the noise figure smaller than 2.6 dB on the low noise amplifier level and smaller than 7 dB including the 2:6 switching network. The medium output power amplifier delivers the saturated power level of + 14 dBm. To the authors' knowledge, this is the highest integration level for a 60 GHz compound semiconductor transceiver chip reported to-date.

ACKNOWLEDGEMENT

The authors wish to acknowledge the contributions of M. Guthoerl and J.-Y. Choi at Sony Deutschland, as well as for H. Massler and A. Tessmann at IAF.

REFERENCES

[1] S. Gunnarsson, C. Kärnfelt, H. Zirath, R. Kozhuharov, D. Kuylenstierna, A. Alping, and C. Fager, "Highly Integrated 60 GHz Transmitter and Receiver MMICs in a GaAs pHEMT Technology," *IEEE Journal of Solid-State Circuits.*, vol. 40, no. 11, pp. 2174-2186, November 2005.

[2] K. Fujii, M. Adamski, P. Bianco, D. Gunyan, J. Hall, R. Kishimura, D. Lesko, M. Schefer, S. Hessel, H. Morkner, and A.

Niedzwiecki, "A 60GHz, MMIC Chipset for 1-Gbit/s Wireless Links", *2002 IEEE MTT-S Int. Microwave Symp. Dig.*, pp. 1725-1728.

[3] O. Vaudescal, B. Lefebvre, V. Lehoué, P. Quentin, "A Highly Integrated MMIC Chipset for 60 GHz Broadband Wireless Applications," *2002 IEEE MTT-S Int. Microwave Symp. Dig.*, pp. 1729-1732.

[4] Y. Mimino, K. Nakamura, Y. Hasegawa, Y. Aoki, S. Kuroda, and T. Tokumitsu, "A 60 GHz Millimeter-wave MMIC Chipset for Broadband Wireless Access System Front-end", *2002 IEEE MTT-S Int. Microwave Symp. Dig.*, pp. 1721-1724.

[5] K. Ohata, K. Maruhashi, M. Ito, S. Kishimoto, K. Ikuina, T. Hashiguchi, N. Takahashi, S. Iwanagam, "Wireless 1.25Gh/s Transceiver Module at 60GHz-Band", *2002 IEEE International Solid-State Circuits Conference Dig.*, February 2002.

[6] B. Floyd, S. Reynolds, U. Pfeiffer, T. Beukema, J. Grzyb, and C. Haymes, "A Silicon 60 GHz Receiver and Transmitter Chipset for Broadband Communications", *2006 IEEE International Solid-State Circuits Conference Dig.*, vol. 49, pp. 130-131, February 2006.

[7] C-H. Wang, Y-H. Cho, C-S. Lin, H. Wanf, C-H. Chen, D-C. Niu, J. Yeh, C-Y. Lee, and J. Chern, "A 60 GHz Transmitter with Integrated Antenna in 0.18µm SiGe BiCMOS Technology", *2006 IEEE International Solid-State Circuits Conference Dig.*, vol. 49, pp. 132-133, February 2006

[8] B. Gaucher, T. Beukema, S. Reynolds, B. Gloyd, T. Zwick, U. Pfeiffer, D. Lin, J. Cressler, "MM-Wave Transceivers Using SiGe HBT Technology", *2004 Topical Meeting on Silicon Integrated Circuits in RF Systems*, pp. 81 – 84.

[9] S. Koch, I. Kallfass, R. Weber, A. Leuther, M. Schlechtweg, M. Uno, " An Analogue, 4:2 MUX/DEMUX Front-End MIMIC for Wireless 60 GHz Multiple Antenna Transceivers", *2007 IEEE MTT-S Int. Microwave Symp. Dig.*, pp. 1121-1124.

[10] A. Leuther, A. Tessmann, M. Dammann, W. Reinert, M. Schlechtweg, M. Mikulla, M. Walther, and G. Weimann, "70 nm low noise metamorphic HEMT technology on 4 inch GaAs wafers," in *Proc. Indium Phosphide Related Materials Conf. (IPRM)*, May 2003, pp. 215–21

[11] N. A. Rahman, B. Y. Majlis, A. Ariffin, " A 28 GHz PHEMT GaAs MMIC Single-ended Resistive Mixer", ICSE7.002 Proc. 2002, Penang, Malaysia, pp. 511–513.

[12] M. Lang, A. Leuther, W. Benz, U. Nowotny, O. Kappeler, and M. Schlechtweg, "66 GHz 2:1 static frequency divider using 100 nm metamorphic enhancement HEMT technology", *Electron. Lett.*, vol. 38, no. 14, pp. 716-717, July 2002.

[13] S. Koch, I. Kallfass, A. Leuther, M. Schlechtweg, S. Saito, M. Uno, "A Four-Antenna Transceiver MIMIC for 60 GHz Wireless Multimedia Applications", *2008 EuMC European Microwave Conference. Dig.*, pp. 1529-1532.

[14] Kallfass, I.; Diebold, S.; Massler, H.; Koch, S.; Seelmann-Eggebert, M.; Leuther, A, "Multiple-Throw Millimeter-Wave FET Switches for Frequencies from 60 up to 120 GHz", *2008 EuMC European Microwave Conference. Dig.*, pp. 1453-1456.

[15] Sarkar S., Sen P. Perumana B., Yeh, D., Dawn, D., Pinel, S., Laskar, J., "60 GHz Single-Chip 90nm CMOS Radio with Integrated Signal Processor", *2008 IEEE MTT-S Int. Microwave Symp. Dig*, pp. 1167-1170.

[16] R. Weber, M. Kuri, M. Lang, A. Tessmann, M. Seelmann-Eggebert, and A. Leuther., "A PLL-Stabilized W-Band MHEMT Push-Push VCO with Integrated Frequency Divider Circuit", *2007 MTT-S Int. Microwave Symp. Dig*, pp. 653-656.

A 12-Gb/s, Direct QPSK Modulation SiGe BiCMOS Transceiver for Last Mile Links in the 70-80 GHz Band

I. Sarkas[1], S.T. Nicolson[1], A. Tomkins[1], E. Laskin[1], P. Chevalier[2], B. Sautreuil[2], and S. P. Voinigescu[1]

1) Edward S. Rogers Sr. Dept. of ECE, University of Toronto, Toronto, ON M5S 3G4, Canada
2) STMicroelectronics, 850 rue Jean Monnet, F-38926 Crolles, France

Abstract — **This paper describes a novel single-chip W-band wireless transceiver utilizing a direct mm-wave QPSK modulator. The transceiver was fabricated in a 130 nm SiGe BiCMOS technology and can operate at data rates in excess of 10 Gb/s. The Zero-IF receiver peak gain is 50 dB, the noise figure is 7 dB while the 3-dB IF bandwidth extends over 6 GHz. The differential transmitter achieves a maximum output power of +9 dBm, while the transceiver occupies 1.9 mm × 1.1 mm. The total power consumption, including the 4 × 20 Gb/s PRBS generator, is 1.2W from 1.5, 2.5 and 3.3V power supplies.**

I. INTRODUCTION

The mm-wave frequency spectrum offers plenty of bandwidth rendering it ideal for multi-gigabit wireless communications. Specifically, the 70-80 GHz frequency band is of particular interest for outdoor point-to-point, multi-gigabit links because of its relatively low atmospheric attenuation.

In this paper, a novel, large power, direct QPSK modulator is proposed which achieves bit rates higher than 10 Gb/s and an output power of +9 dBm. The modulator was integrated in a zero-IF transceiver along with a multi-lane 4×20Gb/s PRBS generator to facilitate a 12-Gb/s link demo.

II. DESIGN

The transceiver architecture, illustrated in Fig. 1, consists of: (i) a direct QPSK modulator with +9 dBm output power, (ii) a 4-channel 2^7-1 PRBS generator capable of transmitting at 20 Gb/s per channel [1], (iii) I-Q transmit and receive LO distribution networks, (iv) an LNA with active 4-way signal splitting at its output stage, (v) transformer-coupled double-balanced Gilbert cell I-Q mixers, and (vi) two variable gain baseband amplifiers with 50-Ω output buffers [2]. An 80-GHz VCO and PLL were also fabricated as a separate chip, but will be discussed elsewhere.

The novel QPSK modulator is realized using two MOS-HBT doubly-balanced Gilbert cells (or direct BPSK modulators) depicted in Fig. 2, driven in quadrature by the LO signal. The 130-nm n-MOSFETs in the modulator are

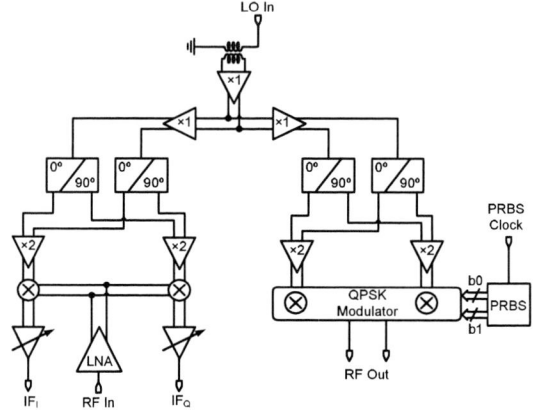

Fig. 1: Transceiver top-level block diagram.

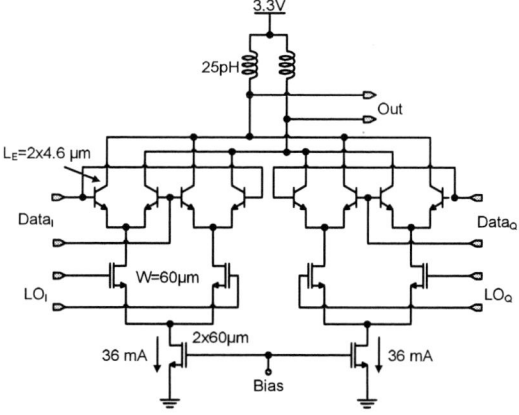

Fig. 2: Simplified schematic of the QPSK modulator. All emitter widths and gate lengths are 0.13μm.

biased at 0.3 mA/μm and placed at the LO inputs for improved linearity at high LO signal levels. The mixing quads are realized with SiGe HBTs to minimize the CML voltage swing applied at their bases for increased data rate and to maximize the output voltage swing and output power of the transmitter. Recently, a 6-Gb/s link was demonstrated at 80 GHz over 300 m using a transmit power of only -6 dBm and transmit and receive antennas with 40dBi gain [3].

978-1-4244-5190-6/09 $25.00 © 2009 IEEE

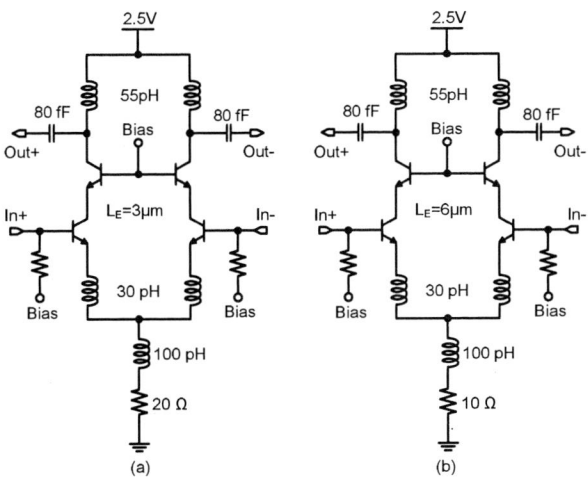

Fig. 3: Schematics of the LO buffers at (a) the input of the 90-degree hybrids (×1) and (b) at the output (×2).

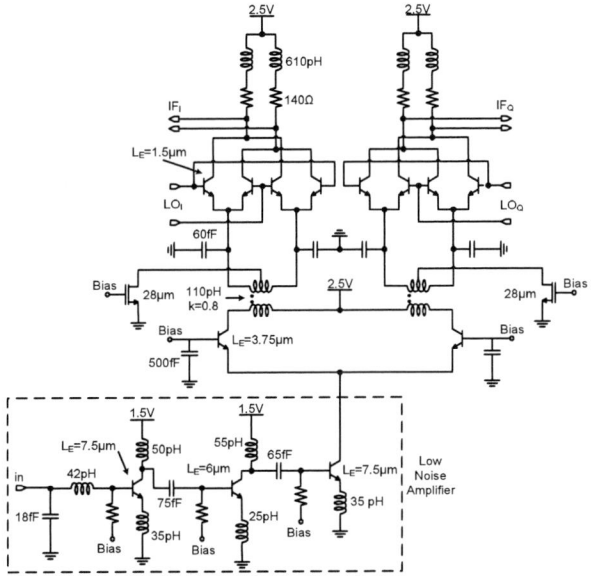

Fig. 4: Schematic of the receiver (excluding the VGAs). All emitter widths and gate lengths are 0.13μm.

Since the QPSK modulator can generate 9 dBm differentially in 50-Ω loads, no upconverter and no power amplifier are required, thus improving the EVM and power added efficiency of the entire transmitter. The I and Q LO inputs of the modulator are buffered for improved isolation between the I and the Q paths. The LO (×1 and ×2) buffers are illustrated in Fig. 3. The transformer-based 90-degree hybrid employed to generate the I and Q LO signals is identical to the one described in [4]. The simulated phase error of the hybrid is less than 2° in the 70-90 GHz range. It should be noted that this direct modulation transmitter is a modern replica of the first diode or GaAs MESFET based solid-state mm-wave QPSK transmitters [5] and modulators [6] of the 1970's.

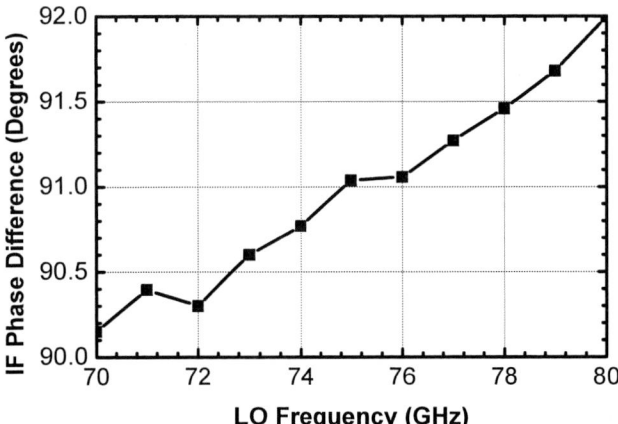

Fig. 5: Simulated phase difference of the two IF outputs of the receiver versus the LO frequency.

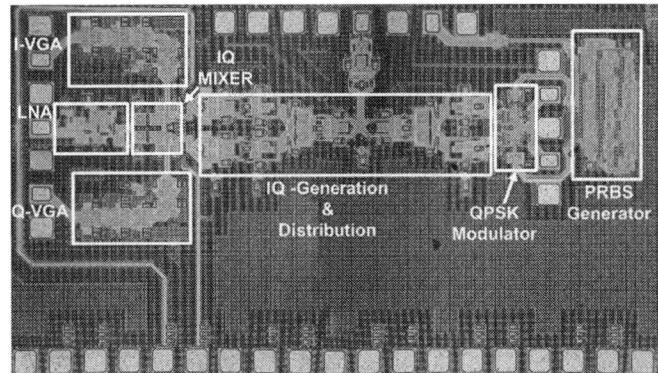

Fig. 6: Transceiver micro-photograph.

In the zero-IF receiver, the LO and RF signals are converted to differential mode with the use of transformers, before driving the two double-balanced Gilbert cell mixers, which are shown in Fig. 4. Each mixer is followed by a 5-stage differential baseband amplifier with a peak gain of 30 dB and over 35 dB of gain control. The LNA employs two CE stages and includes a 2-step, 4-output active power splitter realized with a cascode stage with two common-base HBTs and two transformers. As illustrated in Fig. 4, each transformer acts as a 2-way power splitter. The simulated LNA gain is 17 dB at 77 GHz. The simulated receiver down-conversion gain, DSB NF and P_{1dB} for an LO of 77 GHz and maximum baseband amplifier gain setting, are 51 dB, 6.8 dB and -51 dBm, respectively. Fig. 5 reproduces the simulated phase mismatch between the I and Q baseband outputs of the receiver with respect to the LO frequency.

III. FABRICATION AND MEASUREMENT RESULTS

The circuit was fabricated in a SiGe BiCMOS process with a dedicated mm-wave back-end, 130-nm MOSFETs, and HBT f_T and f_{MAX} of 230 GHz and 280 GHz, respectively [7]. The back-end features 6 copper layers with the top two being 3-μm thick.

978-1-4244-5190-6/09 $25.00 © 2009 IEEE

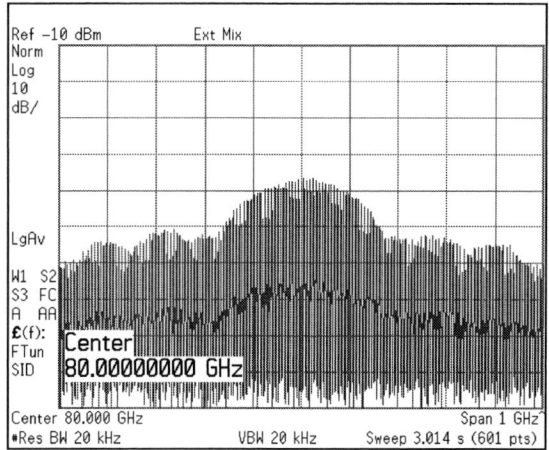

Fig. 7: Measured 400 Mb/s QPSK spectrum at the 80 GHz output of the transmitter.

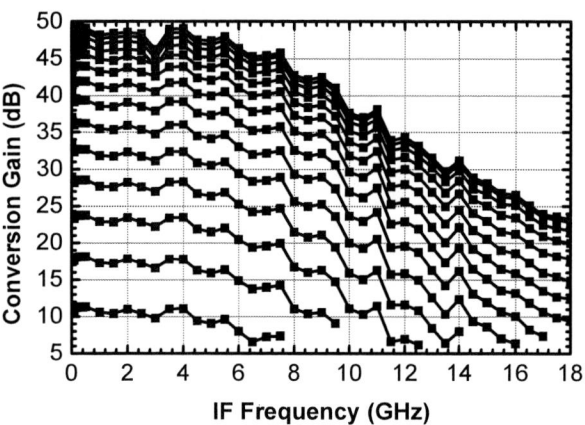

Fig. 8 Measured down-conversion gain of the receiver with different gain settings in the baseband amplifier.

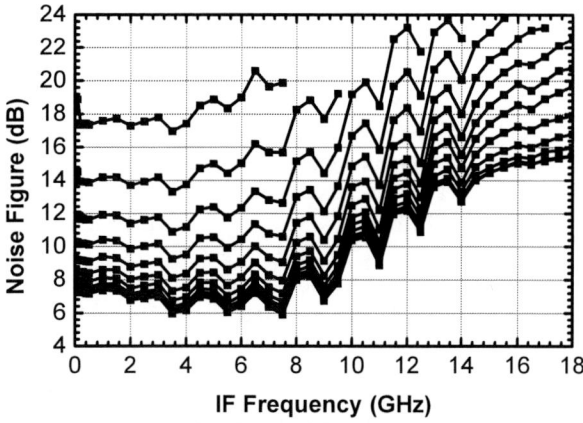

Fig. 9: Measured noise figure of the receiver with different gain settings in the baseband amplifier.

The die photo of the transceiver chip is shown in Fig. 6. It occupies an area of 1.9 mm × 1.1 mm, including all pads.

The transmitter output power was measured first with an un-modulated carrier. An output power of +6 dBm per side was observed in the 60-70 GHz range. Higher output power may be possible but the external LO signal was not large enough to saturate the transmitter. The measured transmitter output spectrum for an LO signal at 80 GHz is illustrated with 400 Mb/s aggregate QPSK data rate in Fig. 7. The lobes

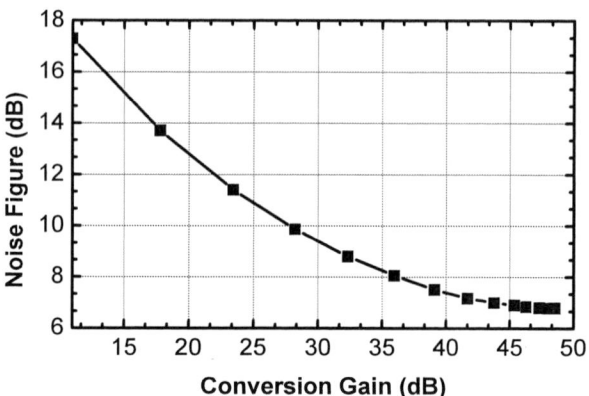

Fig. 10: Measured receiver noise figure at LO=76 GHz and IF=2 GHz vs. receiver gain.

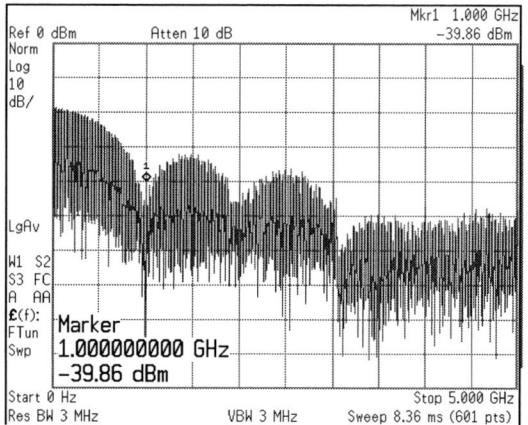

Fig. 11: Baseband spectra at the I output of the receiver for 2 Gb/s QPSK data rate at 76 GHz.

of the QPSK modulation spectrum are clearly visible at low bit rates. As the bit rate increases, image folding from the spectrum analyzer's external harmonic mixer makes the spectrum more difficult to interpret and verify. The receiver baseband spectrum and baseband eye diagrams were employed to verify correct operation at higher rates.

The peak measured down-conversion gain of 50 dB, and the 7 dB, 50-Ω noise figure are shown versus IF frequency in Fig. 8 and 9, respectively, for an LO frequency of 76 GHz. The receiver gain is adjustable between 10 and 50 dB. Fig. 10 shows the measured receiver noise figure as a function of receiver gain.

Finally, a loop back test was conducted with the transmitter connected to the receiver through a high loss V-cable and monitoring the I and Q baseband outputs simultaneously with an oscilloscope and with the PSA. The received spectra at 2 Gb/s and 12 Gb/s QPSK data rates are illustrated in Fig. 11 and 12. The location of the null of the main lobe of the spectrum, as well as the tone spacing within the main lobe, clearly indicate that the received signals are correct 2^7-1 PRBS sequences. The received time domain signals at 4 Gb/s, and 12 Gb/s are depicted in Fig. 13 and 14, respectively. The I and Q outputs are almost in perfect quadrature, although the mismatch in probes and test cables has not been calibrated out and no attempt was made to compensate the offset of the baseband amplifiers. The captured 6-Gb/s PRBS data sequence at the Q-channel

978-1-4244-5190-6/09 $25.00 © 2009 IEEE

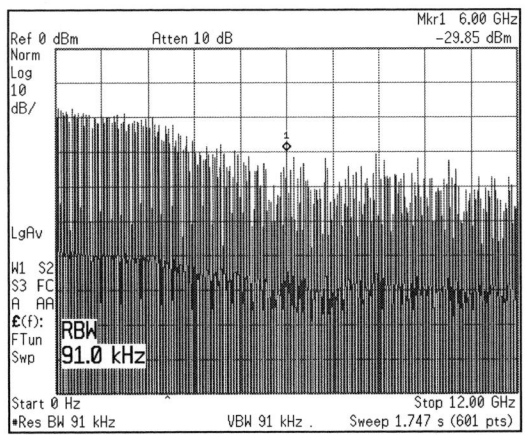

Fig. 12: Measured 12 Gb/s spectrum at the I baseband output of the receiver.

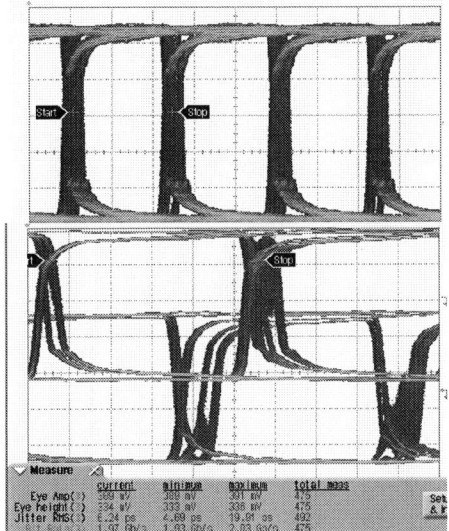

Fig. 13: Measured eye diagrams at I and both I and Q outputs corresponding to 4 Gb/s QPSK transmission, (2 Gb/s per channel). The baseband amplifier offset voltage was not corrected.

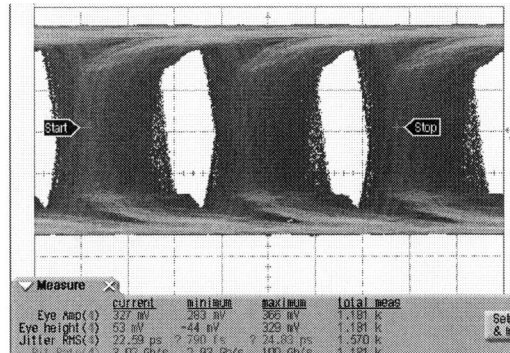

Fig. 14: Measured 6 Gb/s eye diagram at the Q baseband output corresponding to 12 Gb/s QPSK transmission.

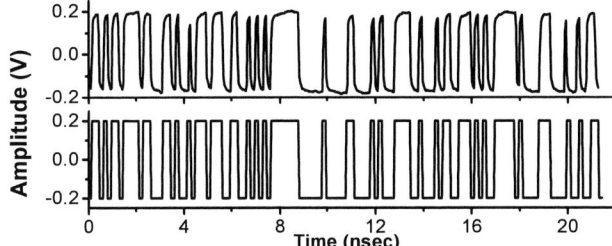

Fig. 15: Captured 6 Gb/s data sequence at the I output (for 12 Gb/s transmission) versus an ideal 2^7-1 bit PRBS sequence.

baseband output, corresponding to an aggregate 12-Gb/s data rate, is compared to an ideal 2^7-1 PRBS in Fig. 15. Correct transmission up to 12 Gb/s was verified for both I and Q channels.

At bit rates beyond 12 Gb/s, the received signal spectrum presents the correct tone spacing but errors start appearing in the received PRBS sequence. This indicates that, although the QPSK modulator is capable of properly modulating the carrier at higher bit rates, the demodulated signal is affected by the channel delay and the frequency response of the baseband amplifiers. Some form of equalization will be necessary for operation at higher rates than 12 Gb/s.

The transmitter, receiver and LO distribution network consume 630 mW, 400 mW and 173 mW, respectively, from 3.3 V, 2.5 V and 1.5 V supplies. The power consumption is distributed as follows: 100 mW in LNA and IQ mixers, 300 mW in the baseband amplifiers and 50 Ω drivers, 238 mW in the modulator and 380 mW in the PRBS generator and data buffers.

ACKNOWLEDGEMENT

This work was funded by NSERC. Fabrication was provided by STMicroelectronics, equipment by NSERC, CFI, OIF and ECTI. CAD support from CMC and Jaro Pristupa are also acknowledged.

REFERENCES

[1] E. Laskin et al., "A 60 mW per Lane, 4 × 23-Gb/s 2^7-1 PRBS Generator," *IEEE J. Solid-State Circuits*, vol. 41, no. 10, pp. 2198-2208, Oct. 2006.

[2] E. Laskin, *et al.* "A 140-GHz Double-Sideband Transceiver with Amplitude and Frequency Modulation Operating over a few Meters," submitted to *IEEE BCTM 2009*.

[3] V. Dyadyuk, *et al.* "A Multigigabit Millimeter-Wave Communication System with Improved Spectral Efficiency" *IEEE Trans. Microwave Theory & Tech.*, vol. 55, no. 12, pp. 2813-2821, Dec. 2007.

[4] I. Sarkas, *et al.* "W-band 65-nm CMOS and SiGe BiCMOS Transmitter and Receiver with Lumped I-Q Phase Shifters," *IEEE RFIC Symp. Dig.*, pp.441-444, Jun. 2009.

[5] Y.-W. Chang, *et al.*, "High Data-Rate Solid-State Millimeter-Wave Transmitter Module," *IEEE Trans. Microwave Theory & Tech.*, vol. 23, no. 6, pp. 470–477, Jun. 1975.

[6] C.L. Cuccia *et al.*, "PSK and QPSK Modulators for Gigabit Data Rates," *IEEE MTT Symp. Dig*, pp. 208-211, Jun. 1977.

[7] G. Avenier, *et al.*, "0.13µm SiGe BiCMOS Technology for mm-Wave Applications," *IEEE BCTM* , pp. 89-92, Oct. 2008.

Normally-Off Operation GaN Based MOSFETs for Power Electronics

Yuki Niiyama, Shinya Ootomo, Hiroshi Kambayashi, Nariaki Ikeda, Takehiko Nomura, and Sadahiro Kato

Yokohama R&D
The Furukawa Electric, Co., Ltd.
2-4-3 Okano, Nishi-ku, Yokohama, 220-0073 Japan
niyama.yuki@furukawa.co.jp

Abstract— **GaN is promising for high-power, high-temperature devices due to some of its large bandgap, high critical electric field, and high saturation velocity compared with Si and SiC. The MOSFET structure can be operated at a positive threshold voltage in the normally-off mode, which is preferable for power transistors in terms of fail-safe operation. In order to realize MOSFET operation, low resistance in the n^+-contact layer and good interface quality at SiO_2/GaN are strongly required. We could reduce the interface state density at SiO_2/GaN by annealing at 900°C for 30 min. Furthermore, we successfully realized the formation of the n+ contact layer by annealing at 1260°C for 30 s. Finally, we have fabricated GaN MOSFETs and have achieved more than 2.5 A operation in the normally-off mode at more than 250°C. The breakdown voltage was more than 1550 V.**

Keywords-GaN; normally-off mode; power devices; ion implantation; interface state density; MOSFETs

I. INTRODUCTION

Gallium nitride (GaN) has good characteristics for power devices because of a high saturation velocity and a high critical electric field compared to those of SiC and Si. Figure 1 shows the calculated specific on resistance against the breakdown voltage of Si, SiC, and GaN [1,2]. The specific on resistance of SiC and GaN was calculated to be less than Si due to its large band gap and a high breakdown field. The on state resistance of GaN was less than 1/1000 of that of Si. Therefore, GaN is expected to be useful for high-performance power transistors. Previously, many researchers have studied AlGaN/GaN heterostructure field effect transistors (HFETs), since a large electron density is easy to be obtained due to two-dimensional electron gas (2DEG) generated by the piezo effect at the AlGaN/GaN interface [3,4]. However, the AlGaN/GaN HFETs were intrinsically normally-on mode operation, which was not suitable for power switching application in terms of fail-safe operation. On the other hand, GaN-based metal-oxide-semiconductor (MOS) field effect transistor (MOSFET) is one of the candidates for normally-off devices. Consequently, the GaN-based MOSFETs are expected for power switching application.

The technique of ion implantation is attractive for forming an n-type region, since it can introduce well-defined impurity concentrations in selected regions. In the case of selective area growth, it seems to be difficult to re-grow on the etched GaN surface because many voids existed at the boundary of the grown and re-grown GaN [5]. A thermal-diffusion technique used as a doping degraded the surface morphology, so that the resistance of the layer was still high, and the processing time was quite long. On the other hand, the ion implantation technique is attractive in terms of the low resistance, short processing time, and easy processing, and so on. Previously, it had been shown that Si implants can be used to convert un-doped and p-type GaN into n-type GaN [6-8]. However, the characteristics of Si implanted Mg-doped GaN after activation annealing was unclear.

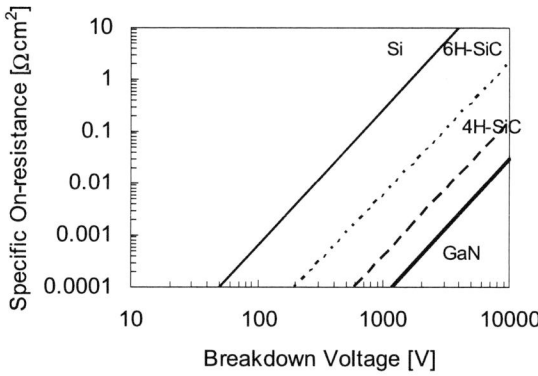

Figure 1. Comparison with trade-off characteristics of GaN, Si, and SiC.

An SiO_2 is a good candidate as a gate oxide because of a high critical electric field, robustness at high temperature, large bandgap, and so on. Therefore, the inversion electrons could be easily confined at the interface between an oxide and a semiconductor. The interface quality at SiO_2/GaN is an important issue. Up to now, it was reported that the interface state density was 3×10^{11} $cm^{-2}eV^{-1}$ on the PECVD SiO_2/GaN by cleaning with organic solvents, NH_4OH and N_2 plasma treatment [9].

We previously demonstrated the 2 A operation of GaN MOSFETs on sapphire substrate, although the breakdown voltage was approximately 40 V [10]. RPI reported on a lateral GaN reduced surface field (RESURF) MOSFET resulting in a breakdown voltage of 940 V [11]. However, the drain current in on-state operation of the GaN MOSFETs was not so large due to the insufficient the ion-implanted activation in the ion

978-1-4244-5190-6/09 $25.00 © 2009 IEEE

implanted RESURF zone. Actually, the sheet resistance of the RESURF zone was more than 100 kΩ/sq.

In this paper, we review the developments and characterizations of unit process steps for GaN MOSFETs. We also describe the fabrication and the 1550 V/2.5 A operation of GaN MOSFETs.

II. SI ION IMPLANTATION

We have studied annealing conditions for activation of an n^+ contact layer [12]. The Si ion as an n-type dopant was selected for conventional material. A 2-μm-thick un-doped and Mg-doped ([Mg]~1×10^{17} cm^{-3}) GaN was grown on a sapphire (0001)c substrate by metal organic vapor phase epitaxy (MOVPE). A screen oxide of 20-nm-thick SiO$_2$ was deposited on a GaN surface in order to suppress any damage of implantation, and to adjust the depth profile of Si ions. In order to achieve the box-shaped depth profile, the Si ion was implanted in GaN at the following energies: 190 keV, 120 keV, 60 keV, and 30 keV. The implanted depth was designed to be 300 nm. Once having removed the SiO$_2$ screen oxide, SiO$_2$ as a capping layer was deposited onto the GaN surface again. A rapid thermal annealing (RTA) was used for activation annealing, because Si atoms cannot diffuse in the GaN layer. The sheet resistance was checked by Van der Pauw.

Figure 2 shows the results of Hall measurements of Si implanted GaN after activation annealing. The activation ratio was decreased while decreasing the Si doses. It is expected that the crystalline defects compensate the generated donors like poly-Si [13]. The activation ratio of the Mg-doped GaN was lower than the one of the un-doped GaN. It was speculated that the acceptors compensate the generated donors. The activation ratio of the implanted GaN annealed at 1260°C for 30 s was improved compared with the annealing at 1200°C for 10 s. In other words, the activation ratio was improved by a higher thermal budget. Consequently, we could activate the GaN implanted lower Si doses.

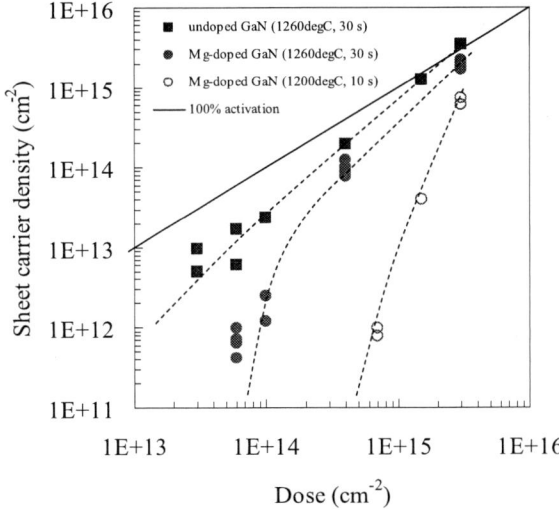

Figure 2.   Relationship between Si total doses and sheet carrier density.

III. INTERFACE STATES AT SIO$_2$/GAN

SiO$_2$/GaN MOS capacitors were fabricated for examining an interface state density. A 1-μm-thick Si doped GaN (N_d^--N_a^+~2×10^{17} cm^{-3}) was grown on a sapphire (0001)c substrate by MOVPE. GaN surfaces were cleaned using the conventional RCA method. After the deposition of 50-nm-thick SiO$_2$ by plasma enhanced chemical vapor deposition (PECVD), the samples were annealed at 800, 900, and 1000°C for 30 min in the N$_2$ ambient. Then, ohmic (Ti/AlSi/Mo) and gate (Ti/Au) electrodes were formed using sputtering methods and lift-off processes. The details of the process and the method used in the measurement are given in Ref. [14].

Figure 3 shows the density distribution of the interface state of samples annealed at several temperatures, calculated from capacitance-voltage (CV) data at 200°C by the Terman method. The interface state density (D_{it}) decreased as the annealing temperature increased. The D_{it} of a sample annealed at 900-1000°C was approximately 8×10^{10}-9×10^{10} cm^{-2}eV^{-1} at E_c-0.4 eV and 1×10^{11}-2×10^{11} cm^{-2}eV^{-1} at E_c-0.2 eV. We consider that these values are sufficiently low for MOS devices. In addition, these values were definitely lower than D_{it} of the previously reported SiC and GaAs.

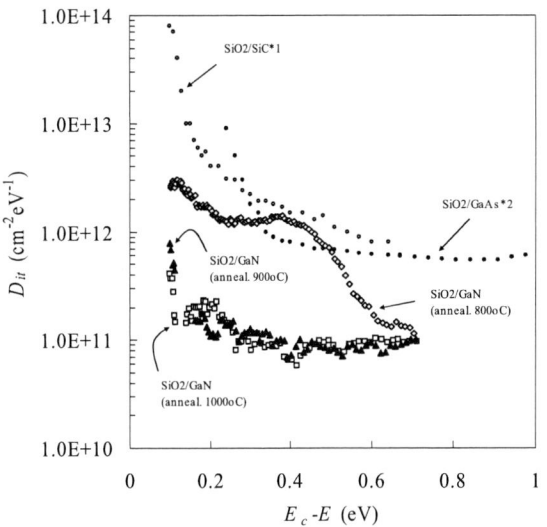

Figure 3.   Interface state density (D_{it}) of SiO$_2$/GaN, which was annealed at 800, 900, and 1000°C. (*1: ref. [15], *2: ref. [16]).

IV. GAN MOSFETs

We were able to understand the formation of an n^+ contact layer with low resistance and good interface quality at SiO$_2$/GaN. Therefore, we fabricated GaN MOSFETs [10,17,18]. At first, the Mg-doped GaN ([Mg]=1×10^{17} cm^{-3}) was epitaxially grown on a sapphire (0001)c substrate. Then, an n^+ contact layer was formed by Si ion implantation. Activation annealing was performed at 1260°C for 30 s in an Ar ambient by RTA. Next, 60-nm-thick SiO$_2$ as a gate oxide was deposited on a GaN surface. The SiO$_2$/GaN was annealed at 900°C for 30 min in the N$_2$ ambient for reducing the interface states. The Ti (25 nm) and Al (300 nm) as source and drain electrodes were evaporated on an n+ contact layer by using sputtering methods. After electrode annealing at 650°C for 5 min in N_2, the ohmic contact resistance and sheet resistance were found to be 1.1×10^{-7} Ωcm^2 and 53 Ω/sq. by transmission line matrix (TLM) measurements. Then, Ti (25 nm) and Au (200 nm) as a gate electrode were evaporated onto the SiO_2 layer using sputtering methods.

Figure 4 shows a current-voltage curve of GaN MOSFETs. The channel length (L_{ch}) and width (W_{ch}) were 4 μm and 16 mm, respectively. Apparently, we achieved normally-off operation mode GaN MOSFET at 250°C; that is, V_{th} was +3 V. To our knowledge, this operation temperature is highest in the world. The on/off ratio was more than 10^5. The subthreshold slope (S) was 328 mV/dec. The gate leakage current at gate voltage (V_g) of 0 V was less than 0.1 μA, this means, normally-off operation was certainly realized. Regarding the output characteristics, a linear and saturation region were clearly observed. The drain current (I_d) of more than 2 A was realized at a V_g of +15 V for the first time.

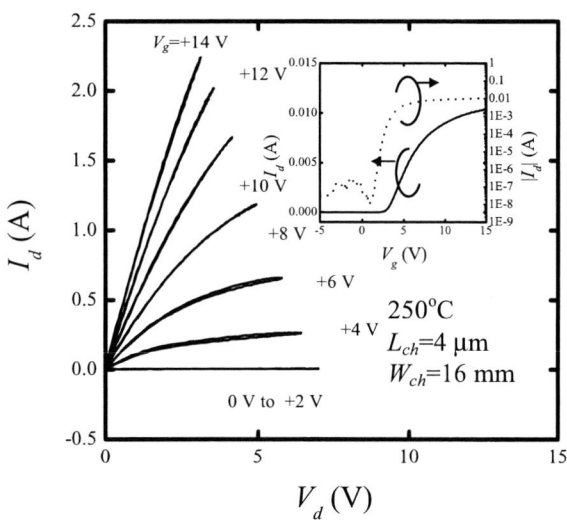

Figure 4. Output characteristics of GaN MOSFETs at 250°C. The gate voltage (V_g) was changed from 0 V to +14 V in 2 V steps. The inset of transfer characteristics at V_d=0.1 V at 250°C.

However, the breakdown voltage of this device was approximately 40 V, which was broken at the gate edge. In order to improve the blocking characteristics, we introduced a RESURF structure between the gate and the drain regions. The RESURF was also formed by Si ion implantation. The fabrication process was quite similar to that in previous work. The total Si dose was 6×10^{13} cm^{-2}. The sheet resistance and carrier density of the RESURF zone were 23 kΩ/sq. and 1.1×10^{12} cm^{-2}, respectively, after activation annealing at 1260°C for 30 s. The channel and RESURF length were 4 and 20 μm, respectively. The gate width was 150 mm.

Figure 5 shows the output characteristics of the GaN RESURF-MOSFETs. The breakdown voltage was more than 1550 V. The relationship between the voltage and the RESURF

length was approximately 80 V/μm, the value of which was never achieved by lateral Si MOSFETs. Furthermore, the operation current exceeded 2.5 A. We achieved both more than 2.5 A and 1550 V for the first time. These results are indicated that our lateral GaN-MOSFETs are able to be applied to power switching application.

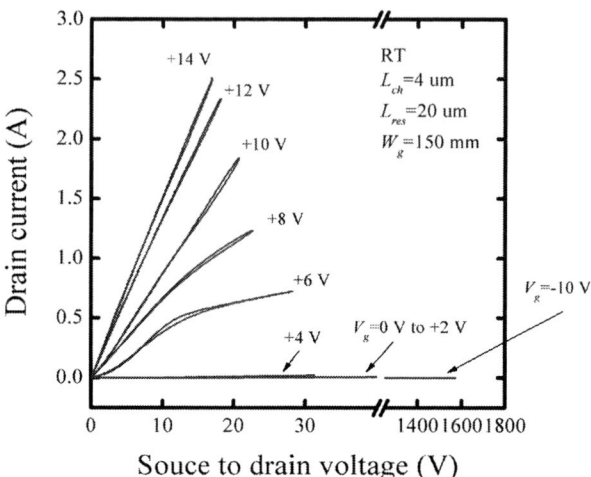

Figure 5. Output characteristics of GaN RESURF-MOSFETs at room temperature (RT). The output characteristics of GaN MOSFETs are shown at RT. The gate voltage (V_g) was changed from 0 V to +15 V in 5 V steps.

V. CONCLUSION

In order to realize the operation of GaN MOSFETs, we have studied the formation of the n+ contact layer and the reduction of interface state density. The n+ contact regions were developed with a total Si implant dose of 3×10^{15} cm^{-2}. After optimizing implant activation condition, the sheet resistance was 26 Ω/sq. obtained, with a sheet carrier density of ~3×10^{15} cm^{-2}. The SiO_2/GaN was characterized as a function of the annealing temperature. The highest-quality SiO_2/GaN interface was achieved by annealing at 900°C for 30 min in a N_2 ambient. The Terman method was used to extract an interface state density of 8×10^{10}-9×10^{10} $cm^{-2}eV^{-1}$ at 0.4 eV below the conduction band edge.

Then, GaN MOSFETs were fabricated on a sapphire substrate. The threshold voltage was +3 V. Operation at 250°C was observed. The drain current exceeded 2.5 A. In addition, GaN RESURF-MOSFETs were fabricated to improve the blocking characteristics. Finally, we achieved more than 1550 V and 2.5 A operation on GaN MOSFETs.

ACKNOWLEDGMENT

We would like to thank Professor T. P. Chow and Dr. W. Huang from Rensselear Polytechnique Institute for their expertise and constant suggestions. We would also like to acknowledge Professor T. Hashizume of Hokkaido University

978-1-4244-5190-6/09 $25.00 © 2009 IEEE

for his helpful discussions. We partially carried out the experiments at Tokyo City University. We also thank Vice President Y. Shiraki and Dr. K. Sawano from Tokyo City University for their help with the ion implantation and the RTA experiments.

REFERENCES

[1] B. J. Baliga, "Power semiconductor device figure of merit for high-frequency applications," IEEE Elect. Dev. Lett. vol. 10, pp. 455-457 Oct. 1989.

[2] S. Yoshida, J. Li, T. Wada and H. Takehara, "High-Power AlGaN/GaN HFET with a Lower On-state Resistance and a Higher Switching Time for an Inverter Circuit," Ext. Abst. 15th international symposium on power semiconductor devices and ICs (ISPSD03), Cambridge, UK, April 14-17, 2003, S3.3, p. 58-61.

[3] T. Kikkawa, "Highly Reliable 250 W GaN High Electron Mobility Transistor Power Amplifier," Jpn. J. Appl. Phys. vol. 44, pp. 4896-4901, 2005.

[4] W. Saito, T. Nitta, Y. Kakuuchi, Y. Saito, K. Tsuda, and I. Omura "On-resistance modulation of high voltage GaN HEMT on sapphire substrate under high applied voltage," IEEE Elect. Dev. Lett., vol. 28, pp. 676-678, Aug. 2007.

[5] T. Kachi, "Recent Advances on GaN Power Devices," Ext. Abst. 2006 International Conference on Solid State Devices and Materials (SSDM), E-10-1, Yokohama, 2006, pp. 970-971.

[6] C. J. Pan, G. C. Chi, B. J. Pong, J. K. Sheu and J. Y. Chen, "Si diffusion in p-GaN," J. Vac. Sci. Tech., B, vol. 22, pp. 1727-1730, July 2004.

[7] J. K. Sheu, C. J. Tun, M. S. Tsai, C. C. Lee, G. C. Chi, S. J. Chang and Y. K. Su, "n$^+$-GaN formed by Si implantation into p-GaN," J. Appl. Phys., vol. 91, pp. 1845-1848, Feb. 2002.

[8] S. J. Pearton, C. B. Vartuli, J. C. Zolper, C. Yuan and R. A. Stall, "Ion implantation doping and isolation of GaN," Appl. Phys. Lett. vol. 67, pp. 1435-1437, 1995.

[9] R. Nakasaki, T. Hashizume, and H. Hasegawa, "Insulator-GaN interface structures formed by plasma-assisted chemical vapor deposition," Physica E, vol. 7, pp. 953–957, 2000.

[10] Y. Niiyama, H. Kambayashi, S. Ootomo, T. Nomura, S. Yoshida and T. P. Chow, "Over 2 A Operation at 250°C of GaN Metal-Oxide-Semiconductor Field Effect Transistors on Sapphire Substrates," Jpn. J. Appl. Phys., vol. 47 pp. 7128-7130, 2008.

[11] W. Huang, T. Khan and T. P. Chow, "Enhancement-Mode n-Channel GaN MOSFETs on p and n⁻ GaN/Sapphire substrate" Ext. Abst. 18th international symposium on power semiconductor devices and ICs (ISPSD06), Italy, June 4-8, 2006, p. 10-1.

[12] Y. Niiyama, S. Ootomo, H. Kambayashi, T. Nomura and S. Yoshida, "High activation of Si implanted un and Mg-doped GaN on sapphire substrate by using rapid thermal annealing," Ext. Abst. 15th International Conference on Crystal Growth (ICCG-15), Salt Lake City, Utah, August 12-17, 2007, wg-40.

[13] S. M. Sze, Physics of Semiconductor Devices (Wiley, New York, 1981), 2nd ed., p. 390.

[14] Y. Niiyama, T. Shinagawa, S. Ootomo, H. Kambayashi, T. Nomura and S. Yoshida, "High-quality SiO$_2$/GaN interface for enhanced operation field-effect transistors," phys. stat. solid. (a), vol. 204, pp. 2032-2036, 2007.

[15] K. Fukuda, M. Kato, K. Kojima, and J. Senzaki, "Effect of gate oxidation method on electrical properties of metal-oxide-semiconductor field-effect transistors fabricated on 4H-SiC C(000-1) face," Appl. Phys. Lett., vol. 84, pp. 2088-2090, 2004.

[16] J. S. Herman and F. L. Terry, "Hydrogen sulfide plasma passivation of gallium arsenide," Appl. Phys. Lett., vol. 60, pp. 716-717, 1992.

[17] Y. Niiyama, S. Ootomo, J. Li, H. Kambayashi, T. Nomura, S. Yoshida, K. Sawano and Y. Shiraki, "Si Ion Implantation into Mg-Doped GaN for Fabrication of Reduced Surface Field Metal–Oxide–Semiconductor Field-Effect Transistors" Jpn. J. Appl. Phys., vol. 47 pp. 5409-5416, 2008.

[18] Y. Niiyama, H. Kambayashi, S. Ootomo, T. Nomura, and S. Kato, "Over 1500 V/2 A operation of GaN RESURF-MOSFETs on sapphire substrate," Electron. Lett., vol. 45, pp. 379-380, 2009.

On-Wafer Seamless Integration of GaN and Si (100) Electronics

Jin Wook Chung, Bin Lu and Tomás Palacios[*]

Department of Electrical Engineering and Computer Science and Microsystems Technology Laboratories
Massachusetts Institute of Technology, 77 Massachusetts Ave. Bldg. 39-567B, Cambridge, MA 02139, USA

[*]*Corresponding author; phone: +1-617-324-2395; FAX: +1-617-258-7393; email: tpalacios@mit.edu*

Abstract

The high thermal stability of nitride semiconductors allows for the on-wafer integration of (001) Si CMOS electronics and electronic devices based on these semiconductors. This paper describes the technology developed at MIT to seamlessly integrate GaN and Si transistors in very close proximity (<5 μm). This integration, the first of any III-V field effect transistor with (001) Si electronics, enables tremendous new possibilities to circuit and system designers. For example, we will study the use of hybrid GaN-Si circuits to improve the power distribution networks in Si microprocessors.

Keywords: GaN, high electron mobility transistors, nitride electronics, high frequency, heterogeneous integration.

1. INTRODUCTION

Moore's law has been one of the main drivers behind the unprecedented development of semiconductors in the last forty years. However, this economical and technological paradigm that has helped to create modern Si electronics is now jeopardizing its future. Traditional Si scaling is not only becoming unaffordable, but the performance improvement due to scaling is diminishing.

Our group is working on an approach different from Moore's law to increase the performance of electronics: the heterogeneous integration of different semiconductor materials on the same wafer. In this paper, we describe our work on the seamless integration of GaN-based devices and Si electronics. While Si electronics has shown unsurpassed levels of scaling and circuit complexity, nitride semiconductors offer excellent optoelectronics and high frequency/power electronic properties. The ability to combine these two material systems in the same chip and in extremely close proximity would allow unprecedented flexibility for advanced applications.

2. HETEROGENEOUS INTEGRATION WITH SILICON DIGITAL ELECTRONICS

The unique properties of AlGaN/GaN High Electron Mobility Transistors (HEMTs) have made them the best option for many RF amplifiers. The unsurpassed high current levels possible in these devices [1], in combination with their very high breakdown voltage allow almost 10 times higher maximum power density than GaAs amplifiers [2]. In addition, their high frequency performance, an f_{max} of 300 GHz has recently been demonstrated [3], enables extremely high gain and power added efficiencies. Also, the high output resistance resulting of the small device width significantly simplifies the design of the matching networks in RF amplifiers. Finally, the recent demonstration of device lifetimes in excess of 10^6 hours at a channel temperature of 175C make this technology one of the most reliable semiconductor technologies [4].

978-1-4244-5190-6/09 $25.00 © 2009 IEEE

In spite of the excellent performance demonstrated by nitride transistors, these devices cannot compete with Si MOSFETs in terms of scalability and level of integration. Modern microprocessors, for example, have more than one billion Si transistors on a single chip [5]. In spite of this unsurpassed scalability, traditional Si electronics is facing tremendous challenges to continue its scaling and performance improvement due to short channel effects and power dissipation. The on-chip integration of nitride and Si technology would enable new flexibility in the circuit and device design to increase the system performance.

Previously, several authors have reported heterogeneous integration of Si and GaAs devices (i.e. field effect transistors, light emitting diodes), by the low-temperature selective epitaxial growth of GaAs on a *miscut* Si(100) substrate [6-8]. With similar technology, several groups have reported the growth of GaN structures on *miscut* Si(100) or Si(110) substrates by molecular beam epitaxy

(MBE) [9] and metalorganic vapor phase epitaxy (MOVPE) [10]. However, this approach is challenging because of the difficulty of growing high quality wurtzite GaN on (100)-oriented cubic Si substrates [10]. Moreover, the use of miscut substrates increases the density of surface states in the Si material, degrading the performance of Si electronics designed therein.

The technology to integrate GaN and Si electronics in the same wafer starts by fabricating a virtual Si (001) / GaN / Si (001) substrate by wafer bonding with a SiO_2 bonding interlayer (Fig. 1) [11]. Due to the high thermal stability of GaN, Si CMOS electronics can then be processed in these new substrates without affecting the nitride layers underneath the surface. After the Si devices are fabricated, the Si material is removed from the regions where nitride devices are needed. Then, the nitride devices (transistors, LEDs, lasers or sensors) are processed and, finally, an interconnection layer forms the final hybrid circuits (Figs. 2 and 3).

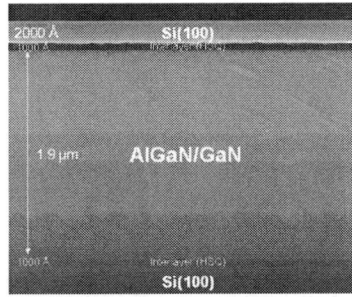

Fig. 1. Scanning electron micrograph of the cross-section of a Si/nitride/Si wafer.

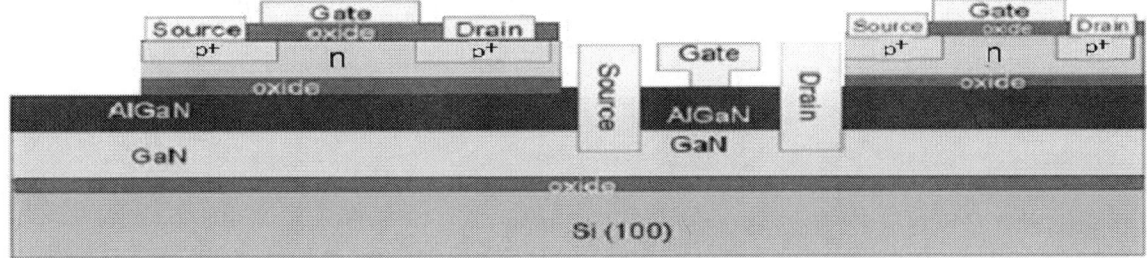

Fig. 2. Diagram showing the cross-section of a virtual wafer with Si pMOS transistors and Nitride HEMTs are fabricated in very close proximity.

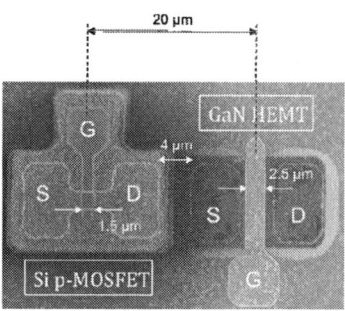

Fig. 3. Scanning electron micrograph of a GaN HEMT and a Si p-MOSFET fabricated side by side in a virtual wafer.

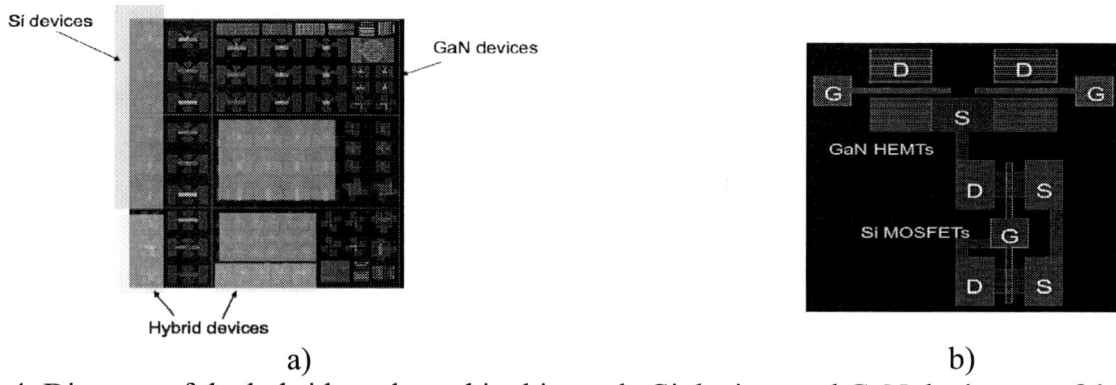

a)   b)

Fig. 4. Diagram of the hybrid mask used in this work. Si devices and GaN devices are fabricated at the same time. b) shows the structure of a high power differential amplifier where the GaN devices form the differential pair and the Si MOSFETs act as a current source.

Using this new technology several hybrid circuits are currently being developed, including high power differential amplifiers, normally-off power transistors (Fig. 4) and highly compact DC-DC power converters for advanced power distribution in Si microprocessors.

As an example of the new hybrid GaN-Si circuits currently under developing in our group, Fig. 5 shows the schematic of a GaN-Si hybrid power converter. In this hybrid circuit, switches M1 and M2 see the highest voltage stress and should be implemented with GaN transistors. In the other devices, the voltage stresses are much lower, which allow their implementation with Si electronics. This circuit is currently under fabrication and Fig. 6 shows the simulation results of the circuit described in Fig. 5. The operating frequency was 300 MHz to minimize circuit area. Simulation verifies the excellent behavior of

the new design, although additional circuit optimization is needed to improve the efficiency.

3. CONCLUSIONS

Nitride transistors have shown an outstanding improvement in performance during the last few years, which has allowed them to become the first option for power amplification below 10 GHz. However, the range of applications where nitride devices can revolutionize electronics does not stop in power amplifiers. Their high thermal stability allows their seamless integration with Si (100) devices, which enables numerous applications that take advantage of the unsurpassed integration density of Si electronics and the high breakdown and operating frequency of nitrides. This paper has demonstrated the first on-wafer integration of GaN and Si(100) devices and some of the new hybrid circuits allowed by this integration.

978-1-4244-5190-6/09 $25.00 © 2009 IEEE

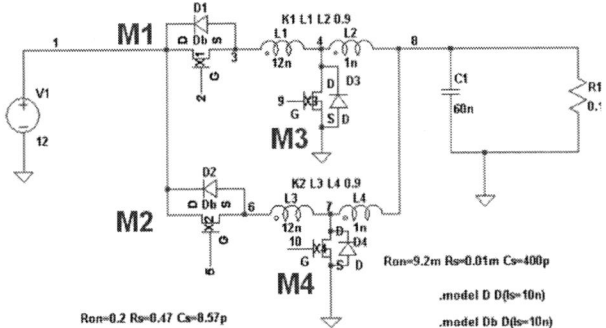

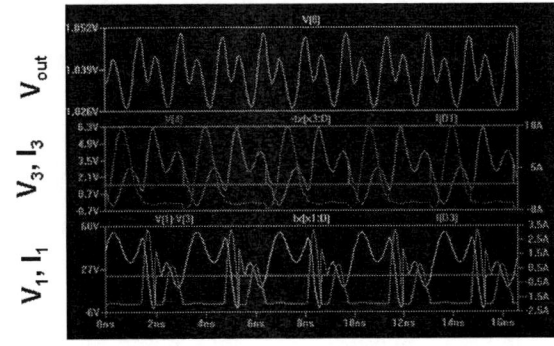

Fig. 5. Circuit schematic of the new GaN-Si hybrid power converter. M1 and M2 are GaN-based power transistors while M3 and M4 are Si MOSFETs.

Fig. 6. Simulation of the current and voltage waveforms in the 12:1 V hybrid voltage regulator studied in this project.

4. ACKNOWLEDGEMENTS

The results described in this paper have been partially funded by the DARPA Young Investigator Award (Dr. Mark Rosker) and the FCRP IFC project of the Semiconductor Research Corporation.

5. REFERENCES

[1] Y. Cao, D. Deen, J. Simon, J. Bean, N. Su, J. Zhang, P. Fay, H. Xing, and D. Jena, "Ultrathin MBE-Grown AlN/GaN HEMTs with Record High Current Densities," *Proc. Of the 2007 Int. Semiconductor Device Research Symp.*, College Park, MD, 2007.

[2] U. K. Mishra, L. Shen, T. E. Kazior, and Y.-F. Wu, *"GaN-Based RF Power Devices and Amplifiers,"* Proc. of the IEEE 96, pp. 287-305, 2008 .

[3] J. W. Chung, O. Saadat, and T. Palacios, "Gate-recessed AlGaN/GaN HEMT with a Record fmax of 300 GHz," *Proc. Of International Conference on Nitride Semiconductors*, Jeju Island, South Korea, 18-23 October 2009.

[4] K. V. Smith, "GaN Reliability Through the Decade," *Proc. Of the Meeting of the Materials Research Society, Boston, MA, 2008.*

[5] Intel Press Release, "Intel Demonstrates Industry's First 32 nm Chip and Next-

Generation Nehalem Microprocessor Architechture," Sept. 18, 2007.

[6] R. Fischer, T. Henderson, J. Klem, W. Kopp, C. K. Peng, and H. Morkoc, "Monolithic integration of GaAs/AlGaAs modulation-doped field-effect transistors and N-metal-oxide-semiconductor silicon circuits," *Appl. Phys. Lett.*, vol. 47, pp. 983-985, Nov. 1985.

[7] R. N. Ghosh, B. Griffing, and J. M. Ballantyne, "Monolithic integration of GaAs light-emitting diodes and Si metal-oxide-semiconductor field-effect transistors," *Appl. Phys. Lett.*, vol. 48, pp. 370-371, Feb. 1986.

[8] H. K. Choi, G. W. Turner, and B-Y. Tsaur, "Monolithic integration of Si MOSFET's and GaAs MESFET's," *IEEE Electron Device Lett.*, vol. 7, no. 4, pp. 241-243, Apr. 1986.

[9] S. Joblot, F. Semond, Y. Cordier, P. Lorenzini, and J. Massies, "High-electron-mobility AlGaN/GaN heterostructures grown on Si(001) by molecular-beam epitaxy," *Appl. Phys. Lett.*, vol. 87, 133505, Sep. 2005.

[10] F. Schulze, O. Kisel, A. Dadgar, A. Krtschil, J. Blasing, M. Kunze, I. Daumiller, T. Hempel, A. Diez, R. Clos, J. Christen, and A. Krost, "Crystallographic and electric properties of MOVPE-grown AlGaN/GaN-based FETs on Si(001) substrates," *J. Crystal Growth*, vol. 299, pp. 399-403, Feb. 2007.

[11] J.W. Chung, J.-K. Lee, E. L. Piner, and T. Palacios, "Seamless On-Wafer Integration of Si (100) MOSFETs and

GaN HEMTs," *IEEE Electron Dev. Letts.*, in press, 2009.

978-1-4244-5190-6/09 $25.00 © 2009 IEEE

High Linearity AlGaAs/InGaAs pseudomorphic HEMT Driver Amplifier using Tunable Field-plate Voltage Technology

Chia-Shih Cheng , Shao-Wei Lin, Jeffrey S. Fu *IEEE Senior Member*, Hsien-Chin Chiu *IEEE Member*

Abstract—A high-linearity AlGaAs/InGaAs pseudomorphic HEMT RF driver amplifier was developed using a tunable field-plate (FP) bias voltage technology in this study. In order to improve the circuit linearity performance, an FP device was employed at the output stage to provide an additional mechanism to suppress the power of the second and third-order harmonics in a two stage 5.2GHz driver amplifier. A standard Class AB driver amplifier without using FP technology was also implemented for comparison under the identical power consumption. The circuit with an FP device biased at V_{FP}= -4V in the output stage demonstrated at least 2dB improvement on the third-order intercept point at input (IIP$_3$) performance over the standard one within the useful power range in two tone measurement.

Index Terms—field-plate, linearity, driver amplifier, pHEMT

I. INTRODUCTION

Recently, microwave power devices play an important role in wireless communication systems. Both the output power density and the device linearity are important factors in increasing the dynamic range, and satisfying the requirements of new-generation communication systems. The electric field profile across the channel of a pHEMT is a key consideration for device breakdown voltage which dominates the device output power density. Besides, the electric-field modification by field plate (FP) has resulted in dramatic improvement in the large-signal performance of GaAs-based microwave pHEMTs. Therefore, several investigations have demonstrated marked improvements in the breakdown voltage (V_{BR}) of pHEMT by using the field plate (FP) technology [1–3].

Device linearity is also an important factor in order to meet the requirements of new generation communication systems. Previous work on GaAs-based HEMTs proposed increasing device trasconductance as a viable solution to improve device linearity [4]. The advantages of increasing device transconductance have also been demonstrated for AlGaN–GaN HEMTs in [5]. A method proved effective in reducing the dispersion phenomena is the introduction of a field-plated gate structure. This method was successfully applied on AlGaAs–GaAs and AlGaN–GaN HFETs, resulting in a significant improvement in device performance. However, in many investigations, the field plate (FP) of pHEMT is connected to the gate terminal to facilitate the fabrication of devices. The FP-induced depletion region of gate-terminated field plate pHEMTs (FP-G pHEMT) is also modulated by the input power signal of gate terminal. Accordingly, the nonlinearity caused by the FP modulation further affects the linearity of FP-G pHEMT [6].

The authors are with the Department of Electronic Engineering, Chang Gung University, No. 259 Wen-Hwa 1st Road, Kwei-Shan, Tao-Yuan, Taiwan, R.O.C. (phone: 886-3-211-8800 ext. 3350; fax: 886-3-211-8507; e-mail: hcchiu@mail.cgu.edu.tw)

In this study, we connected the field-plate metals of the pHEMT to a single pad and evaluated its gate leakage current, RF and power performance from a V_{FP} of +4 V to −10 V. Field-plate metals improve dc-to-RF dispersion of HEMT, which is caused by surface states. By performing a negative bias on the field-plate metal, the FP-induced surface depletion region is thicker and the carriers between drain and gate terminals are distant from the surface. Therefore, the device dc-to-RF dispersion and linearity of pHEMT can be improved by applying negative V_{FP}. Additionally, the performance of FP pHEMT can be adjusted by applying various V_{FP} without any dc power consumption. Through the above advantages, we developed a two stage 5.2 GHz RF driver amplifier by employing an FP device at the output stage to provide an additional mechanism to suppress the power of the second and third-order harmonics. And a standard driver amplifier without using FP technology was also implemented for comparison under the identical power consumption in this work.

II. DEVICE FABRICATION AND MEASURED RESULTS

For device fabrication, the epitaxial structure adopted in this study was a double recess design for high breakdown voltage consideration. A 12 nm undoped InGaAs channel layer was sandwiched between two Si planar δ-doping layers for high current and high power consideration. A 28 nm-thick n⁻ AlGaAs layer was grown on an intrinsic AlGaAs spacer layer as a Schottky layer, which improves parallel conduction at high gate voltage. Finally, a 20 nm n⁻-GaAs and a 25 nm n⁺-GaAs cap layer were grown to improve the resistivities of the ohmic contacts. The etching stop layer between cap layers and Schottky layer was 1.5 nm AlAs. For device fabrication, the devices were processed by optical stepper lithography and lift-off technology. Ohmic contacts were realized by using Au/Ge/Ni/Au alloy followed by a 430°C, 15 seconds rapid thermal annealing (RTA) alloy. Ion-implant isolation technology was used for mesa isolation to prevent the flow of any side-wall gate leakage current. The sub-micron gate-length exposure was performed using electron beam lithographic system, and a bi-layer resist profile was used. After the highly selective succinic acid chemical gate recess process, 5000 Å-thick Ti/Pt/Au-gates were deposited by lift-off process. At this stage, the Schottky layer and channel layer beneath the gate recess region were easily oxidized by moisture and generated surface states. Therefore, before the deposition of FP metal, a 1500 Å SiN$_x$ was deposited by plasma enhance chemical vapor deposition (PECVD) at 280 °C for passivation and insulator between gate and FP metals. The 11000 Å-thick FP metal (Ti/Au =3000Å/8000Å) was then deposited on the SiN$_x$ passivation layer using an electron-beam evaporator. The FP

978-1-4244-5190-6/09 $25.00 © 2009 IEEE

metal extension is identical to the gate recess region because the channel is close to the surface after recess process. Therefore, FP induced depletion region is used to keep the carrier far away from surface states. Finally, a dense SiN_x layer was deposited as a protection layer for improving device reliability. Fig.1 shows the SEM cross-sectional view photograph of the field-plate device.

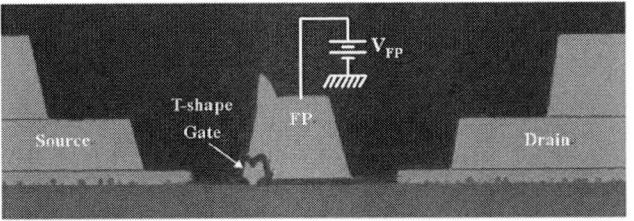

Fig.1 The field-plate device SEM cross-sectional view photograph.

The microwave power characteristics were evaluated by a load-pull system with automatic tuners, which provides conjugate-matched input and load impedances simultaneously for the maximum output power. Microwave load-pull power performance was conducted at 5.2 GHz under a drain bias of 3 V with various V_{FP}. The field-plate voltages were supplied by a single dc probe and its ground was connected with the device source terminal to guarantee the exact field-plate biases. The gate bias was chosen at a class AB operation. The output power density degraded following trends of I_{ds} at various V_{FP}. However, this is not the case for the device linearity characteristics. The third-order inter-modulation (IM3) product from the device output spectra versus the input RF power, which is an important index of the device linearity, was determined by injecting two-tone frequencies, 5.200 GHz (f_1) and 5.201 GHz (f_2). These two adjacent signals generated IM3 output power ($2f_1 - f_2$ and $2f_2 - f_1$) owing to device intrinsic nonlinearity. Fig.2 presents the measured third-order output interrupt points (OIP$_3$) as a function of V_{FP} for a single device. The OIP$_3$ was determined by extracting the curves intersection point of fundamental and IM3 output versus input power. The OIP$_3$ is 22.8 dBm at a V_{FP} of +4V operation and 25.5 dBm at a V_{FP} of −6V operation. Furthermore, the IM3 output power of the device with a V_{FP} of −6 V (−30.5 dBm) outperforms the device with a V_{FP} of +4V (−23.9 dBm) at an input power of −10 dBm. The device performed a lower output power density at negative V_{FP} which results in a small dynamic range. However, there are two mechanisms to explain the IM3 improvement at a negative V_{FP} operation. First, the gate leakage current (I_g) of a field-plate device with a negative V_{FP} is lower than a standard one. Fig.3 shows the gate leakage current characteristics between a fabricated field-plate device and a standard device under the identical drain current (I_d). Based on a previous investigation [7], the higher I_g will generate extra harmonics power and degrade the device linearity. For the observation of the gate leakage current characteristics in the field-plate structure depending on the V_{FP}, the gate leakage current was measured at a bias V_{FP} = -4 V. The measured gate leakage current level for the case of a field-plate device biased at V_{FP} = -4 V is 3.9 mA at P_{in}=10dBm, which is 17 % lower than the reference case of a standard device. The gate leakage current level is found to be improved through the decreasing of field-plate voltage, which can lead to improve breakdown voltages. Second, the carrier transportation path between drain and terminals was suppressed to a deeper channel at a negative V_{FP} and carriers were distant from the surface traps which resulted in small harmonic power and lower dc–RF dispersion. The tunable field-plate voltage technique is effective in adjusting the device power and linearity performance in a single device. On the other hand, during a high-power amplifier module development, the tunable V_{FP} can be applied to obtain a higher output power density or linearity instead of the complicated pHEMT epitaxy structure modification and process-related regulation (the depth or width of the gate recess region). Therefore, this technique exhibits a high potential for retrenching the fine-tuned procedure of the commercial microwave circuit modules without extra dc consumption.

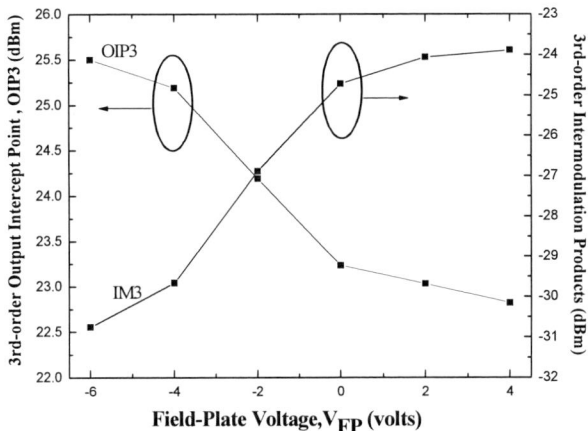

Fig.2 IM3 and OIP3 versus various FP voltages of a FP device at 5.2GHz.

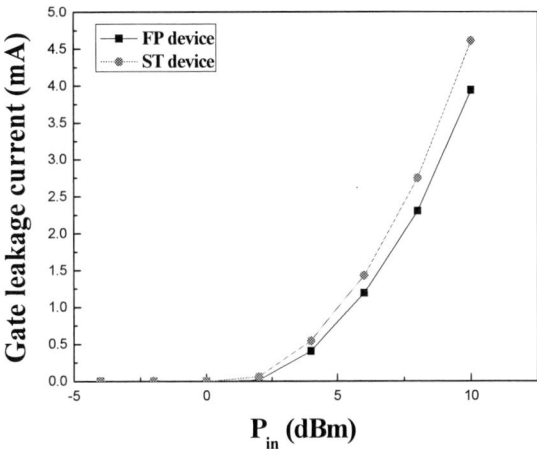

Fig.3 Gate leakage current characteristics between a fabricated field-plate device and a standard device under the identical drain current (I_d).

On-wafer microwave S-parameters evaluation of a $W_g =$ 150 μ m device was carried out in a common-source configuration by Agilent E8364B PNA network analyser from 0.05 GHz to 40 GHz. By extracting the equivalent circuit model from device S-parameters at various V_{FP} shown in Fig.4, both the gate-to-drain capacitance (C_{gd}) and intrinsic transconductance (g_{mi}) increased with V_{FP}. However, the gate-to-source capacitance (C_{gs}) was almost identical at different V_{FP}. C_{gd} and C_{ds} were strongly influenced by the thickness of the FP-induced depletion region which is controlled by V_{FP} and this phenomenon also shows the same trend with dc measurement results. Based on the C_{gs} measured results, the gate-to-source depletion region which is the most important factor to determine the device microwave performance was less influenced by V_{FP}. In other words, the tuneable FP voltage technology can adjust and modulate the device dc and RF characteristics with few sacrifice of device gain-bandwidth product. Besides, the intrinsic gm also increased from 53 mS to 62 mS which is attributed by the drain current improvement from V_{FP}=−10 to V_{FP} = +4.

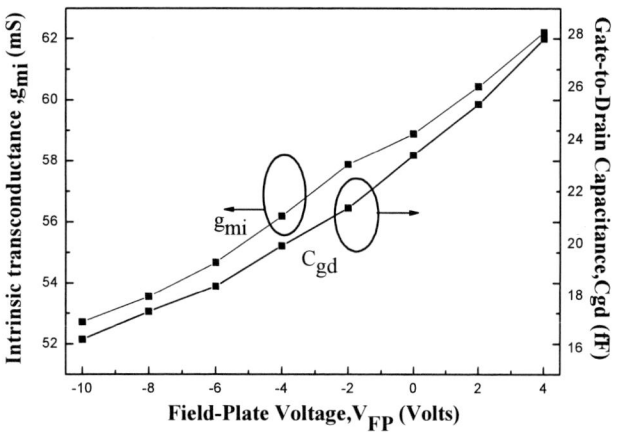

Fig.4 Relation between g_{mi}, C_{gd} and various V_{FP}.

According to a previous study [8], using nonlinear current method, Volterra model is extracted from a standard GaN HEMT to predict IMD3 highlighting various distortion mechanisms, by which, the dominant distortion behavior are efficiently localized. Among others, the 3rd-order nonlinearity arising from gate-drain feedback capacitance C_{gd} has major influence on total IMD3L. Taking the benefit of FP structure, which increasing the C_{gd}, it was seen through simulation that by scaling C_{gd} to 150%, the device linearity improved by around 3 dB. Therefore, C_{gd} can be effectively controlled to suppress distortion by means of the field-plate (FP) technology, originally intended to improve current collapse and breakdown characteristics.

III. DRIVER AMPLIFIER DESIGN AND MEASURED RESULTS

The FP driver amplifier circuit diagram and the corresponding photograph of the fabricated MMIC for the completed amplifier were shown in Fig.5 and Fig.6,

respectively. The chip dimension of the MMIC is 1.5 mm x 1 mm. A standard amplifier had also been fabricated for comparison under the identical design approach and the difference between these two amplifiers was the selection of a standard device or a FP device at output stage. For both designs, the gate biases were chosen at a class AB operation with a current of 35 mA for first stage and a current of 47 mA for the output stage, which is compromised by considering device output power (P_{out}). The first matching network for a PA is the output matching network which is designed to transfer maximum output power from the FET to a 50-Ohm system. According to the Cripps technique, the required optimized large-signal load impedance $Z_{opt,Q2}$ is composed of $R_{opt,Q2}$ and $C_{ds,Q2}$. The output power matching network was designed to transfer the $Z_{opt,Q2}$ to 50 Ohm which comprised a series spiral inductance and a series blocking capacitance. The inter-stage matching network was designed to transfer $Gamma_{IN,2}$ to the $Z_{opt,Q1}$ in order to minimize the mismatch loss. A gate resistor is used in each stage in order to achieve a lossy match and to improve circuit stability. Then the circuit stability is further confirmed by the simulation of stable factor. Finally, the input network was designed to smooth the small signal gain and to improve impedance match for a better input return loss.

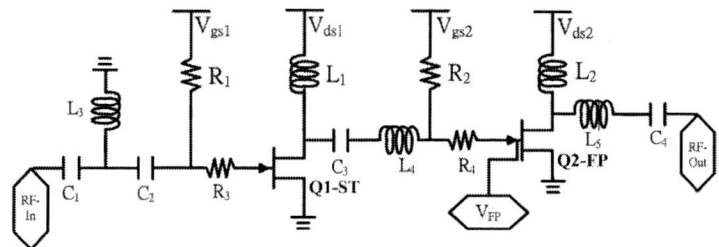

Fig.5 The circuit topology of proposed FP driver amplifier.

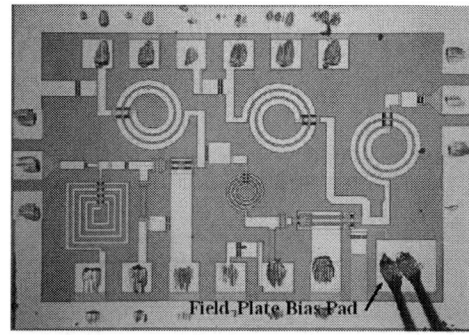

Fig.6 The photograph of proposed FP driver amplifier.

While considering the FP device power density degraded following trends of I_{ds} with the decreased field-plate voltage V_{FP} at the output stage, the gate bias of a FP device should be given a little higher to generate identical power density with respect to a standard device. Under the bias condition of V_{ds1}= V_{ds2}= 3 V, V_{gs1}= -0.5V, V_{gs2}=-0.45V and V_{FP}=-4V, the small-signal gain of the field-plate amplifier was measured with 31 dB at 5.2 GHz (see Fig.7).

Under the identical power consumption, the comparisons between a standard and a field-plate amplifier in output power, associated power gain as a function of input power at 5.2 GHz

978-1-4244-5190-6/09 $25.00 © 2009 IEEE

were shown in Fig.8. The 1-dB gain compression power (P_{1dB}) of the FP amplifier was 15.1 dBm with respect to the standard one was 14.2 dBm. The maximum output power of the FP amplifier was 17.8 dBm with respect to the standard one was 17 dBm. The measured third-order intercept point at input (IIP$_3$) was -2 dBm of FP amplifier and this value was -4 dBm of standard amplifier under the identical third-order intercept point at output (OIP$_3$) 20dBm (see Fig.9). Through the measurements, a high linearity driver amplifier using a simple tunable field-plate voltage technique to adjust circuit output power and linearity has been developed successfully.

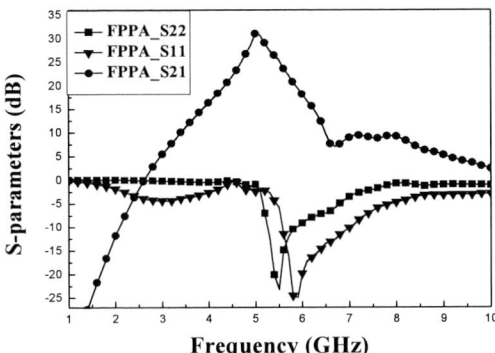

Fig.7 The measured S-parameters of proposed FP driver amplifier.

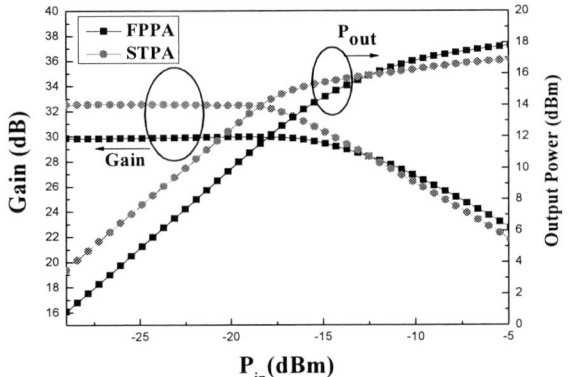

Fig.8 The comparisons between a standard and a field-plate amplifier in output power, power gain as a function of input power at 5.2 GHz.

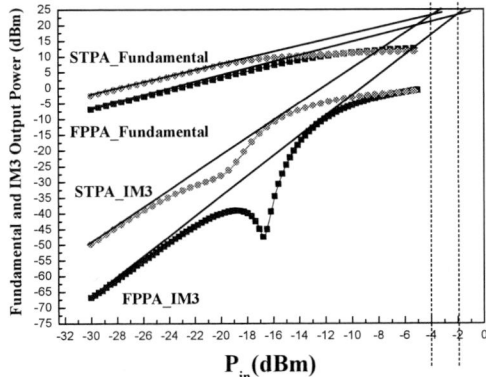

Fig.9 The third-order inter-modulation product from circuit output spectra versus the input RF powers between standard and FP driver amplifiers.

IV. Conclusion

Large-signal performance of AlGaAs/InGaAs pHEMT was remarkably improved through implementing field-plate structure. Taking the benefit of FP structure rather than standard process, which increasing the feedback gate-to–drain capacitance, the device linearity can be improved. By performing a negative bias on the field-plate metal, the FP-induced surface depletion region is thicker and the carriers between drain and gate terminals are distant from the surface. Therefore, the device dc-to-RF dispersion and linearity of pHEMT can also be improved by applying negative V_{FP}. Through the suitable bias selection of V_{FP}, field-plate technology had also been demonstrated in a 5.2GHz driver amplifier, showing great promise of the circuit linearity improvement.

Acknowledgment

The authors are grateful to WIN Semiconductors Corporation for device fabrication.

References

[1] Yunju S and Eastman L F 2005 Large-signal performance of deep sub-micrometer AlGaN/AlN/GaN HEMTs with a field-modulating plate *IEEE Trans. Electron Devices* **52** 1689–92

[2] Chini A, Buttari D, Coffie R, Shen L, Heikman S, Chakraborty A, Keller S and Mishra U K 2004 Power and linearity characteristics of field-plated recessed-gate AlGaN-GaN HEMTs *IEEE Electron Device Lett.* **25** 229–31

[3] Huili X, Dora Y, Chini A, Heikman S, Keller S and Mishra U K 2004 High breakdown voltage AlGaN-GaN HEMTs achieved by multiple field plates *IEEE Electron Device Lett.* **25** 161–3

[4] Y. Nakasha, M. Nagahara, Y. Tateno, H. Takahashi, T. Igarashi, K.Joshin, J. Fukaja, and M. Takikawa, "A low-distortion high-efficiency E-mode GaAs power fet based on a new method to improve device linearity focused on g$_m$ value," in *IEDM Tech. Dig.*, Dec. 1999, pp.405–408.

[5] A. Chini, D. Buttari, R. Coffie, L. Shen, S. Heikman, S. Keller, and U. K. Mishra. Effect of gate recessing on linearity characteristics of AlGaN–GaN HEMTs. *Proc. 2004 Device Research Conf.*

[6] Wu Y F, Moore M, Wisleder T, Chavarkar P M, Mishra U K and Parikh P 2004 High-gain microwave GaN HEMTs with source-terminated field-plates *IEEE Int. Electron Devices Meeting Digest* pp 1078–9

[7] Chiu H C, Yang S C, Chien F T and Chan Y J 2002 Improved device linearity of AlGaAs/InGaAs HFETs by a second mesa etching *IEEE Electron Device Lett.* **23** 1–3

[8] Embar R. Srinidhi and Günter Kompa 2007 Investigation of IMD3 in GaN HEMT Based on Extended Volterra Series Analysis *Proc. 2007 2nd European Microwave Integrated Circuits Conf.*

High Aspect Ratio CPW Fabricated Using a Micromachining Process Combining DRIE, Thermal Oxidation, Electroplating, and Planarization

Shane T. Todd[#1], Xiaojun T. Huang[*2], John E. Bowers[#3], and Noel C. MacDonald[*4]

[#]*Department of Electrical and Computer Engineering and* [*]*Department of Mechanical Engineering*
University of California, Santa Barbara
Santa Barbara, CA, USA 93106

[1]email: stodd@ece.ucsb.edu, phone: +1.805.893.5341, fax: +1.805.893.7990
[2]email: huang@engineering.ucsb.edu, phone: +1.805.893.7902, fax: +1.805.893.8651
[3]email: bowers@ece.ucsb.edu, phone: +1.805.893.8447, fax: +1.805.893.7990
[4]email: nmacd@engineering.ucsb.edu, phone: +1.805.893.5118, fax: +1.805.893.8651

Abstract— A micromachining process has been developed to fabricate high aspect ratio CPW. The tall conductor sidewall created from the high aspect ratio process reduces the resistance per length of the transmission line which lowers the attenuation. The micromachining process combines Si DRIE, thermal oxidation, electroplating, and planarization to create tall CPW with Au conductors and SiO_2 dielectrics. Transmission lines with characteristic impedances of 16 - 21 Ω have been fabricated on high resistivity Si. Transmission line characteristics were measured from 1 - 50 GHz and showed propagation loss of 1.1 - 1.3 dB/cm at 10 GHz.

I. INTRODUCTION

Conventional coplanar waveguide (CPW) can suffer from high conductor loss due to the high current density that exists along the surface of the conductor sidewalls. An approach to reducing conductor loss is to fabricate high aspect ratio CPW (HARC) with tall conductors to reduce the line resistance per length. The tall conductor walls also improve field confinement which increases the isolation between adjacent transmission lines and allows for an increased packing density on the surface of the substrate. Micromachining methods have been used to fabricate HARC with tall conductors of thickness ~ 10 - 200 μm [1-3].

Willke and Gearhart fabricated high aspect ratio HARC using a micromachining method that combined LIGA and electroplating [1]. This work demonstrated transmission lines and filters with low insertion loss and a wide range of achievable impedances. Sun *et al.* demonstrated a similar LIGA micromachining method for creating interdigital CPW with tall conductor walls [2]. Minimum insertion loss of 0.5 dB/cm was measured at 10 GHz. Problems with LIGA based methods include the expensive fabrication costs of LIGA and the difficulty fabricating transmission lines with a planar surface. It is difficult to integrate devices (such as micromachined switches and active microwave circuits) on top of the surface of non-planar transmission lines.

Huang *et al.* demonstrated a Ti based micromachining method to achieve HARCPW with a planar surface [3]. In this process, tall conductor mesas were etched into a Ti substrate to create the CPW conductors and the trenches between mesas

were filled with BCB polymer dielectric. Backside planarization isolated the signal and ground lines to create the CPW structure. Au was coated on the sidewall surfaces to reduce conductor loss. Limitations of this method were the mechanical instability of BCB and the conductor loss of Ti.

In this paper we present a novel micromachining method that creates HARC with a planar surface. The method combines Si DRIE, thermal oxidation, electroplating, and planarization (DTOEP) to fabricate HARC with Au conductors and SiO_2 dielectrics. The HARC is fabricated on a Si substrate which exploits the cost advantages of Si and makes the process potentially compatible with CMOS technology.

II. DEVICE PRINCIPLE

The advantage of using tall conductor walls in CPW is easily understood by considering the attenuation of a parallel plate waveguide shown in Fig. 1(a). CPW is comparable to two parallel plate waveguides attached in parallel as shown in Fig. 1(a). For two parallel plate waveguides of impedance $2Z_0$ connected in parallel with conductor height, h, much greater than the separation distance between conductors, d, the attenuation due to conductor loss is given by [4]

$$\alpha_{c,pp} = \frac{R_S}{2Z_0 h}, \ for \ h \gg d \qquad (1)$$

where R_S is the surface resistivity of the conductor and Z_0 is the total characteristic impedance of the waveguides attached in parallel. If the characteristic impedance of the transmission line is kept constant, the attenuation can be reduced by increasing the height of the conductor. As the conductor height increases, the separation distance must also increase to keep the characteristic impedance constant. Increasing the conductor height spreads the current over a larger surface area, which lowers attenuation for a given total current. This is equivalent to lowering the line resistance per length.

The parallel plate attenuation is not a good approximation of CPW, shown in Fig. 1(b), because fringing fields exist at the top and bottom of the waveguide. This is especially true for thin conductors where h is not much greater than d.

978-1-4244-5190-6/09 $25.00 © 2009 IEEE

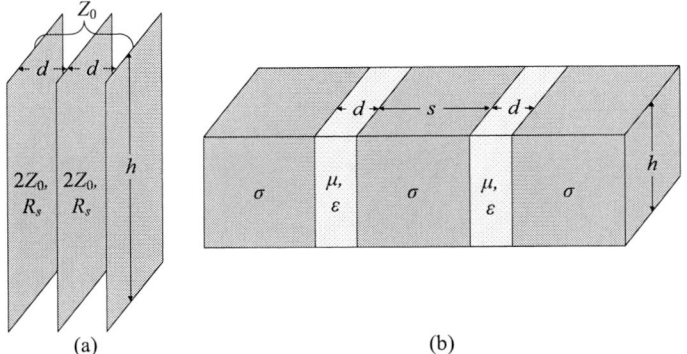

Fig. 1 (a) Two parallel plate waveguides attached in parrallel (b) Coplanar waveguide.

However, CPW will follow a similar trend in that the attenuation will decrease as the conductor height increases. This trend is demonstrated in Fig. 2 which shows HFSS simulations of CPW attenuation versus conductor thickness compared to the parallel plate attenuation predicted by (1) at 10 GHz for CPWs with signal line widths of 5, 50, and 100 µm. Both the simulated and calculated transmission lines contain Au conductors and the simulated CPW contains SiO_2 dielectrics between the Au conductors. Notice that the simulations match the parallel plate prediction reasonably for small values of s and for larger values of h. Thus the parallel plate attenuation in (1) provides a reasonable approximation of HARC for small s and large h.

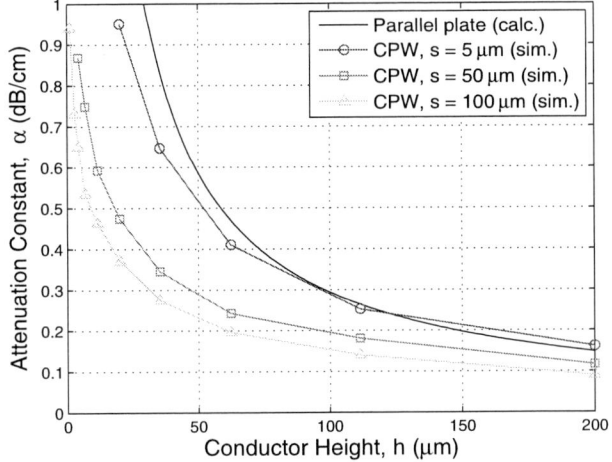

Fig. 2 Simulated CPW and parallel plate attenuation constant versus conductor thickness for 50 Ω transmission lines at 10 GHz.

III. FABRICATION

The HARC is fabricated on high resistivity Si in a process that combines DRIE, thermal oxidation, electroplating, and planarization (DTOEP). The fabrication method, shown in Fig. 3, starts with Si DRIE to create mesas and trenches which define the topology of the HARC.

Next, wet thermal oxidation at 1050 °C transforms the Si mesas into thermal oxide, creating a low-loss dielectric for the transmission lines. A Ti/Au seed layer is then deposited using

a sputtering system with good step coverage. Good step coverage of the seed layer is needed to cover the sidewalls of the SiO_2 mesas with conductor. The conductor coated sidewalls electrically connect the ground plane and signal lines that are defined by the SiO_2 mesa topology. This allows Au electroplating to fill both the ground and signal trenches with conductor in the next step. The electroplated Au grows conformally from all sides of the SiO_2 mesas. Next, the top of the structure is planarized using lapping and CMP to electrically isolate the signal and ground lines.

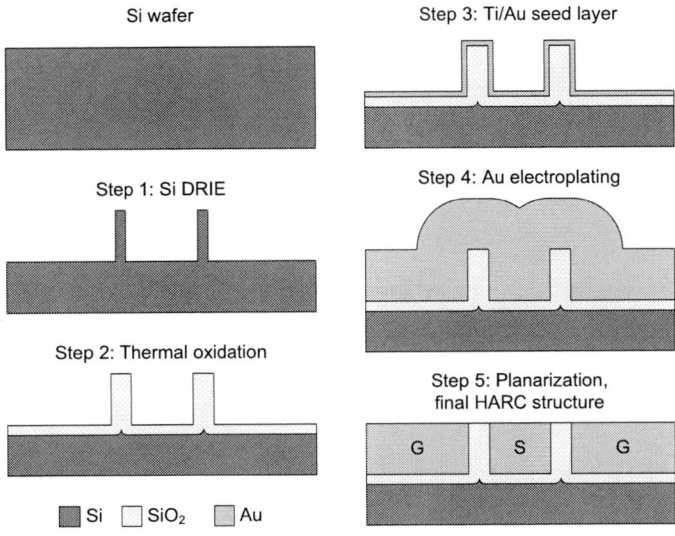

Fig. 3 The DTOEP fabrication process flow for creating HARC.

A final backside DRIE of the Si underneath the CPW would be desirable to improve microwave performance by isolating the fields from the lossy Si substrate. This step has been attempted but has not yielded working devices at this point. The reason this step failed is because intrinsic stress in the electroplated Au caused the transmission lines to deform and the dielectric material to crack after the substrate was removed underneath. The intrinsic stress of the electroplated Au was measured to be ~ 150 MPa. This is a relatively high value of stress for electroplated Au compared to values reported by Margesin *et al.* [5]. Margesin *et al.* showed that zero stress electroplated Au can be achieved at certain temperatures and current densities. Further research is being pursued to create a process with near zero stress for the electroplated Au.

The thermal oxidation step is a critical part of the fabrication sequence. Similar fabrication methods to obtain solid SiO_2 mesas using DRIE combined with thermal oxidation have been reported in literature [6-7]. The main issues with the thermal oxidation step are the process time, mesa deformation, and stress development. For thick oxide, the oxidation thickness is proportional to the square root of the oxidation time [8], so it is time consuming to create wide SiO_2 mesas. The mesa width expands during oxidation to make the final SiO_2 mesa width approximately 2.27 times as wide as the

978-1-4244-5190-6/09 $25.00 © 2009 IEEE

initial Si width [6]. This must be accounted for in the design layout.

Another concern is the stress that evolves in the SiO_2 during thermal oxidation. A somewhat surprising result is that the stress present in the SiO_2 mesas is very small. This is because the mesas grow both laterally and vertically which allows most of the strain developed from SiO_2 growth to be relaxed. There will be stress in the SiO_2 at the bottom of the trenches, however, that is very close to the stress expected from thermal oxidation on a planar surface.

Stress-dependent thermal oxidation simulations have been conducted using Silvaco to demonstrate the stress that evolves in the mesa structures. Fig. 4 shows a simulation of wet thermal oxidation growth and stress evolution at 1050 °C in a mesa with an initial Si width and thickness of 3 μm and 50 μm respectively. The maximum stress in the mesa is less than 10 MPa after the mesa is completely oxidized. The stress of the oxide at the bottom of the trench is ~ 100 MPa, close to a planar surface value.

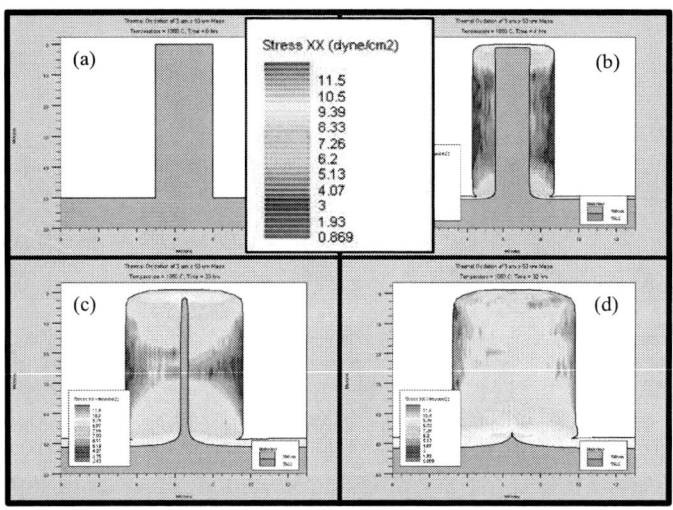

Fig. 4 Wet thermal oxidation simulation at 1050 °C of a 3 μm × 50 μm Si mesa at (a) 0 hours, (b) 4 hours, (c) 20 hours, and (d) 32 hours. The stress contour is in log scale with units of dynes/cm^2. Note: the dimension scales are different for the x and y axes.

IV. EXPERIMENTAL RESULTS

Three HARC arrays with different characteristic impedances have been successfully fabricated with the DTOEP method. Before thermal oxidation, the width of the Si mesas are 3, 4, and 5 μm for the different HARC arrays. The trench depth is 25 μm. Wet thermal oxidation was performed at 1050 °C for 91 hours to grow 6 μm of thermal SiO_2 to ensure complete oxidation of the dielectric mesas. The mesa widths are 6.8, 9.1, and 11.4 μm after thermal oxidation.

After Au electroplating, lapping and CMP was carefully monitored so that the surface was planarized to the top of the mesas without removing significant SiO_2, yielding an Au conductor thickness of ~ 25 μm. The Au resistivity was measured to be 28.5 nΩ·m. Each HARC array consists of a thru line, a short line, and line lengths of 100, 200, 400, 800, and 1600 μm for thru-reflect-line (TRL) calibration. Fig. 5

shows SEM micrographs of one thru line after each of the main steps of the DTOEP process. Fig. 5 (d) shows the transmission line after planarization and probe pad markings were etched into the Au surface.

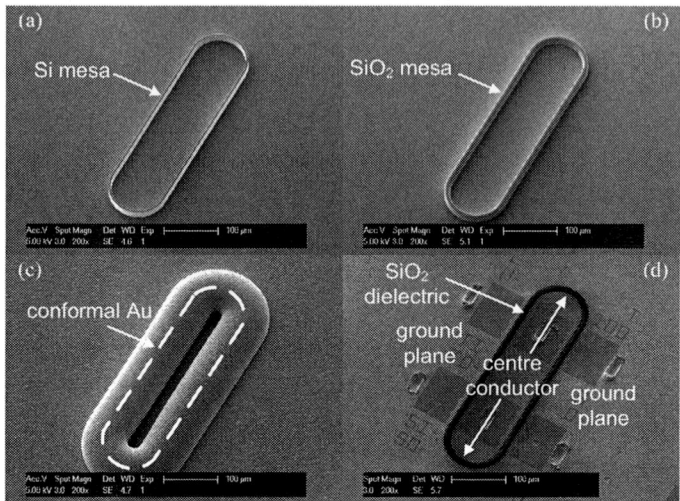

Fig. 5 SEM micrographs of a thru line after (a) Si DRIE, (b) thermal oxidation, (c) electroplating, and (d) planarization.

Microwave measurements were taken using an Agilent E8364A network analyser and simulations were conducted in HFSS. The characteristic impedances of the transmission lines were measured by first performing a short-open-load-thru (SOLT) calibrations using a 50 Ω calibration substrate from 1 - 50 GHz. Next, S parameters were measured from the HARC lines and converted into ABCD parameters using the 50 Ω reference impedance. The characteristic impedance of each line was calculated using $Z_0 = \mathrm{Re}\left(\sqrt{B/C}\right)$ [4]. The measured and simulated characteristic impedances extracted from 200 μm long lines (averaged over the frequency range) are shown in Table I. Fig. 6 shows measurements and simulations of characteristic impedance versus frequency of 200 μm long lines in each HARC array.

TABLE I
HARC TRANSMISSION LINE PARAMETERS

SiO_2 width	Z_0 (Ω)	ε_{eff}	α @ 10 GHz (dB/cm)
6.8 μm	16.3 (meas.)	4.2 (meas.)	1.3 (meas.)
	15.6 (sim.)	4.2 (sim.)	1.6 (sim.)
9.1 μm	18.8 (meas.)	4.2 (meas.)	1.1 (meas.)
	18.6 (sim.)	4.2 (sim.)	1.2 (sim.)
11.7 μm	20.5 (meas.)	4.2 (meas.)	1.1 (meas.)
	21.0 (sim.)	4.2 (sim.)	1.0 (sim.)

TRL calibrations of the HARC transmission lines were performed to extract the phase constant and propagation loss from 1 - 50 GHz. After calibration, S parameters were measured in each line and converted to ABCD parameters, and the propagation constant was calculated using $\alpha + j\beta = (1/L)\cosh^{-1}(A)$ where α is the propagation loss, β is the phase constant, and L is the line length [4].

Fig. 7 shows measurements and simulations of phase constant versus frequency of 1600 μm long lines. Notice that at around 7 GHz the phase constants become unstable. This is because the short phase limits (~ 20°) of the 1600 μm long lines are reached at this frequency. Similarly, the phase constants of the 1600 μm lines become unstable at around 44 GHz because the long phase limits (~ 160°) are reached. The effective dielectric constants, ε_{eff}, of the 1600 μm lines in each HARCPW array were calculated by performing linear regression of the phase constants and are included in Table I. Fig. 8 shows measurements and simulations of propagation loss versus frequency of 1600 μm long lines. The attenuation constant was averaged using 21 data points over a span of 5 GHz for each value shown in Fig. 8. The error bars in Fig. 8 represent the range of one standard deviation greater and less than each point on the plot. The attenuation constants at 10 GHz of the 1600 μm lines in each HARC array are shown in Table I.

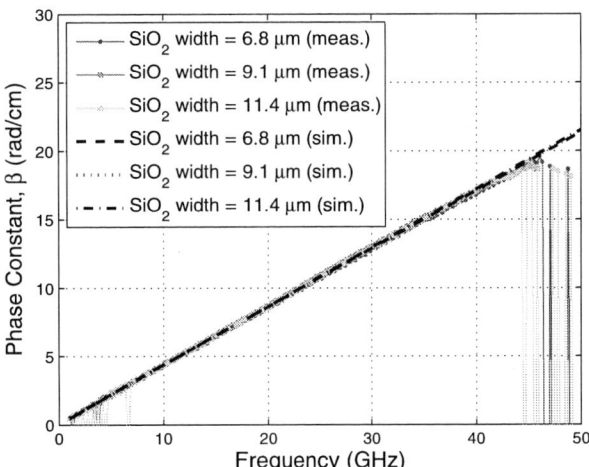

Fig. 7 Phase constant versus frequency for 1600 μm long lines.

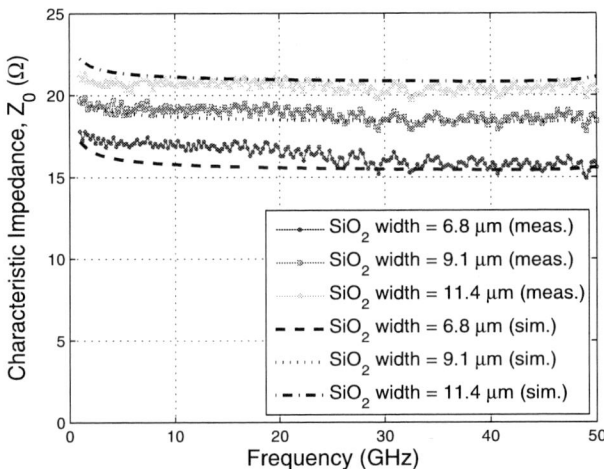

Fig. 6 Characteristic impedance versus frequency for 200 μm long lines.

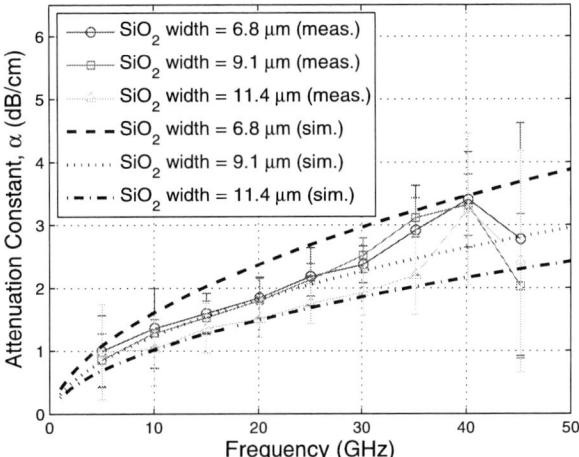

Fig. 8 Propagation loss versus frequency for 1600 μm long lines.

V. CONCLUSION

A novel micromachining process has been developed to create high aspect ratio coplanar waveguides using a Si substrate. The process combines DRIE, thermal oxidation, electroplating, and planarization to create HARC with a planar surface. The tall conductors of the HARC allow for lower propagation loss because of the lower line resistance per length. HARC arrays with characteristic impedances between 16 - 21 Ω were successfully fabricated. The effective dielectric constant was measured to be 4.2 for all transmission lines and attenuation constants were measured to be 1.9 -1.4 dB/cm at 10 GHz.

ACKNOWLEDGMENTS

We thank Tony Bosch for help with electroplating and Hui-Wen Chen for help with microwave measurements. This research was supported by the Kavli Foundation.

REFERENCES

[1] T. L. Willke and S. S. Gearhart, "LIGA Micromachined Planar Transmission Lines and Filters," *IEEE Trans. Micro. Theory Tech.*, vol. 45, no. 10, pp. 1681-1688, Oct. 1997.

[2] X. Sun, G. Ding, D. Gu, B. Li, and M. Shen, "New Micromachined Interdigital Coplanar Waveguide," *Micro. and Opt. Tech. Let.*, Vol. 49, No. 5, pp. 1007-1010, May 2007.

[3] X. T. Huang, S. T. Todd, C. Ding and N. C. MacDonald "Bulk-titanium waveguide, a new platform for planar microwave circuits", *Nanotech 2008*, Boston, Massachusetts, USA, June 1-5, 2008.

[4] D. M. Pozar, *Microwave Engineering*, 2nd Edition, Wiley, New York, 1998.

[5] B. Margesin, A. Bagolini, V. Guarnieri, F. Giacomozzi, A. Faes, R. Pal, and M. Decarli, "Stress Characterization of Electroplated Au Layers for Low Temperature Surface Micromachining," *Proc. DTIP 2003*, pp. 402-405, Cannes-Mandelieu, Canada, May 2003.

[6] M. Rais-Zadeh and F. Ayazi, "Characterization of high-Q Spiral Inductors on Thick Insulator-On-Silicon," *J. Micromech. Microeng.*, vol. 15 (2005), pp. 2105-2112.

[7] C. Zhang and K. Najafi, "Fabrication of Thick Silicon Dioxide Layers for Thermal Isolation," *J. Micromech. Microeng.*, vol. 14 (2004), pp. 769-7.

[8] E. Kobeda and E. A. Irene, "Intrinsic SiO₂ film stress measurements on thermally oxidized Si," *J. Vac. Sci. Tech. B*, vol. 5, no. 1 (1987), pp. 15-19.

Design Method for UHF Class-E Power Amplifiers

Néstor D. López, *Member, IEEE,* John Hoversten, *Student Member, IEEE,* Zoya Popović, *Fellow, IEEE*

Abstract— This paper describes a method for designing single-ended high-efficiency switched-mode class-E UHF power amplifiers. The design procedure consists of a modified load pull transistor characterization from which a power/efficiency metric is calculated. Results for four prototypes using different device technologies are presented in detail. Amplifiers with Si-LDMOS, SiC-MESFET, GaN-HEMT on a Si substrate, and GaN-HEMT on a SiC substrate produce power over 40W with power-added efficiency greater than 75% and gain between 13 dB and 17 dB.

Index Terms— UHF power amplifiers, Switching amplifiers

I. INTRODUCTION

In a high-power UHF or microwave transmitter the final stage power amplifier (PA) dominates the loss budget. Efficiency of this stage can be improved by operating it at or beyond the 1-dB compression point at the expense of linearity. Harmonic control offers an additional increase in efficiency when the device has sufficient gain at harmonics of the center frequency. The class-E mode of operation takes advantage of deep compression, low-quiescent current bias point, and harmonic control to achieve ultra-high efficiency. Linearity can be achieved using external circuits (e.g. EER, LINC [1]).

Class-E PAs have been shown to achieve 95% drain efficiency in the low-MHz range [2], 84% at UHF [3], [4], 75% at 2 GHz [5], and 70% at X-band [6]. In this mode the transistor is assumed to operate as an ideal switch and therefore must have an f_T significantly higher than the frequency of operation. Recent progress in wide band gap semiconductor technology [7], [8], [9] has significantly extended the frequency range of large periphery devices, making the class-E approximation reasonable for high-power transistors.

Nonlinear models which adequately model device operation in the low-quiescent current regime are not readily available for these devices, and load pull becomes a standard approach for obtaining empirically based device models. However, standard load pull measurement techniques do not provide the data required for class-E design. Here we demonstrate a modified load pull method applied to four transistor technologies: (1) a Si-LDMOS from Triquint (AGR09045E); (2) a SiC-MESFET from Cree (CRF24060); (3) a GaN HEMT on a Si substrate from Nitronex (NPTB00050); and (4) a GaN-HEMT on a SiC substrate from RFMD (RF3932). The paper organized as follows:

- Section 2 describes the class-E mode of operation and design procedure, including a modified load pull technique.
- Section 3 includes device and load pull characterization data for each transistor using a break-apart fixture as shown in Figure 1.
- Section 4 details design and performance of class-E PA prototypes at 370 MHz based on the four devices.

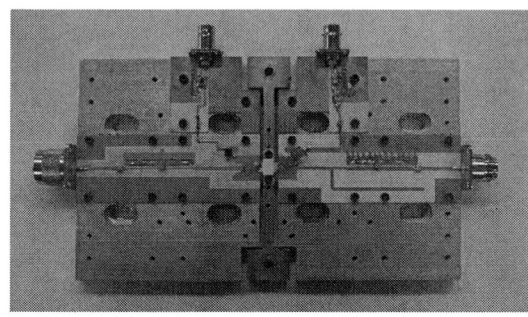

Fig. 1. Photograph of the break-apart fixture used for pre-matched harmonic-tuned load pull characterization. The second harmonic open load circuit is realized with an open stub. Bias networks are included. The active device can be replaced with TRL standards during load pull calibration.

II. CLASS-E PA DESIGN METHODOLOGY

High efficiency is obtained in the class-E mode by minimizing overlap of the current and voltage across the device output. There are three conditions which must be satisfied in the time domain [e.g. [10]]:

- $i_D = 0$ when the switch stops conducting;
- $v_{DS} = 0$ when the switch begins conducting;
- $dv_{DS}/dt = 0$ when the switch begins conducting;

These conditions are satisfied for a 50% duty cycle when the fundamental impedance is

$$Z_E = \frac{0.28}{\omega_s \cdot C_{OUT}} e^{j49°} \quad (1)$$

where C_{OUT} is the parallel combination of C_{DS} with ($C_{GS} + C_{GD}$) and ω_s is the switching rate in radians [11]. Additionally, all harmonics are presented with an open circuit at the output. Note that the ideal load impedance is a function only of frequency and the device output capacitance, which can be estimated from S-parameters.

High power transistors typically require very low load impedances due to a large number of parallel unit cells. Pre-matching networks internal to the package are often included in commercial devices in order to facilitate matching over a broad band. This introduces another impedance transformation inside the package which makes class-E matching at the fundamental and harmonic more complicated. For this reason we limit device selection to those without internal output pre-matching. A good initial approximation of input impedance is a conjugate match, also obtained from device S-parameters.

Non-sinusoidal voltage and current waveforms result when the class-E conditions listed above are satisfied. These waveforms are rich in harmonic content and have voltage and current maximum swings given by

$$V_{DS,max} = 3.56 V_{DS} \qquad I_{DS,max} = 2.86 I_{DS} \quad (2)$$

978-1-4244-5190-6/09 $25.00 © 2009 IEEE

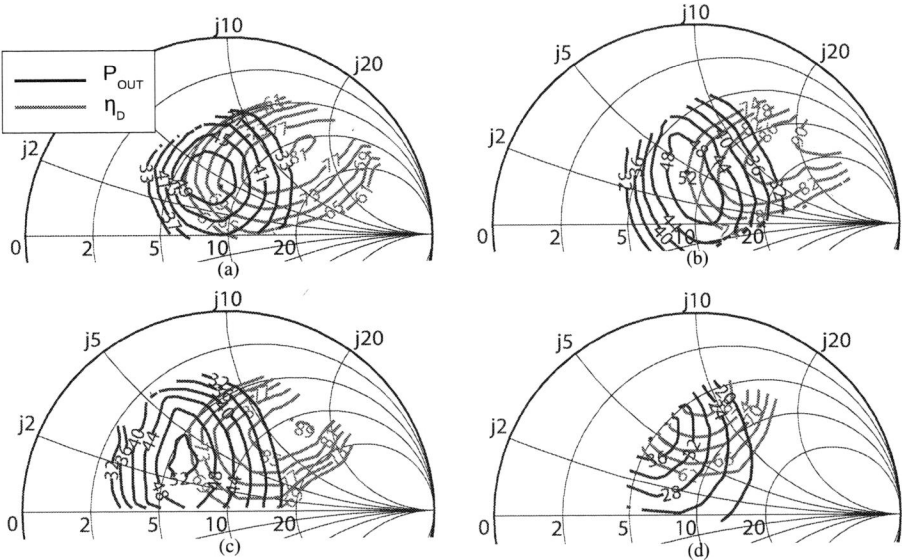

Fig. 2. P_{OUT} and η_D load-pull contours in a 10-Ω Smith Chart for each of the four different transistor technologies at a supply voltage of 28 V. (a) GaN HEMT on a Si substrate, (b) GaN HEMT on a SiC substrate, (c) SiC MESFET, and (d) Si LDMOS. Notice that optimum P_{OUT} contours do not overlap with optimum η_D contours.

C_{OUT} is dominated by C_{DS}, and very large periphery devices with many parallel cells have large nonlinear intrinsic drain capacitance. As a result of this nonlinearity the peak of the drain current waveform can be significantly higher (e.g. up to 30% for a square root nonlinearity [12]). These increased peak drain voltage and current values stress the device and must be taken into account when selecting bias conditions.

The theoretical class-E mode is capable of 100% drain efficiency but in reality is limited by intrinsic resistances inside the transistor, losses in input, output, and bias circuits, and parasitic reactance leading to drain voltage and current overlap. Important parameters when choosing a device for class-E operation therefore include C_{OUT}, packaging, f_T, breakdown voltage, maximum current, R_{ON}, and in addition maximum junction temperature as summarized in Table I.

Typical load pull uses mechanical tuners to present a constellations of impedances to the transistor source and load, yielding an empirical device model. For low impedance high-power devices, optimal device impedances are not close to the standard 50 Ω. Therefore, load pre-matching circuits are used to transform the tuner's 50 Ω constellation to one centered at the ideal class-E impedance of Equation 1. The source pre-matching circuit is similarly transformed to the conjugate match determined using S-parameters. To achieve accurate load pull results, careful calibration of the fixture and all other parts of the system is required. A modular fixture is designed in which input circuit, output circuit, and device are separate blocks. The ability to remove the device under test and insert standards allows TRL calibration, removing the effects of bias and pre-matching circuits.

In addition to transforming the fundamental impedance the pre-matching circuit also needs to transform all harmonic impedances to an open, per the class-E requirement. In practice only the second harmonic is considered because it has the largest effect. This termination is achieved using a quarter-wave open-circuit stub placed a quarter wave from the device as shown in Figure 1. Finally, the load pull test fixture includes bias circuits placed close to the device to increase low-frequency stability.

III. TRANSISTOR CHARACTERIZATION

Figure 2 shows load pull data for each of the transistors at 370 MHz when biased with a drain supply voltage of 28 V and quiescent current just above cutoff (approximately 10 mA). The source impedance for each of these measurements is chosen to obtain maximal gain. It is interesting to observe how the contours depend on C_{OUT} and device periphery, e.g. the optimal impedances for Si-LDMOS are smaller compared to the other transistor technologies due to the significantly larger device periphery and output capacitance.

TABLE I

SUMMARY OF TRANSISTOR PROPERTIES AND RESULTS OF CLASS-E LOAD PULL CHARACTERIZATION

Device	V_{DSS}	C_{OUT}	R_{ON-DC}	T_J	V_{TH}	Optimum P_{OUT}	Optimum η_D
16-mm GaN/Si HEMT	100 V	9 pF	0.23 Ω	200°C	-2.1 V	63 W, 68% @ 5+j7.5Ω	30 W, 84% @ 14+j22Ω
10-mm GaN/SiC HEMT	150 V	9 pF	0.32 Ω	250°C	-4.2 V	53 W, 75% @ 8+j4.4Ω	14 W, 93% @ 11.8+j2.9Ω
30-mm SiC-MESFET	120 V	11 pF	0.46 Ω	250°C	-10 V	53 W, 70% @ 6+j3.6Ω	20 W, 89% @ 13.5+j16Ω
120-mm Si-LDMOS	65 V	23 pF	0.35 Ω	200°C	3.5 V	40 W, 74% @ 5.2+j5.6Ω	24 W, 82% @ 4+j9.3Ω

978-1-4244-5190-6/09 $25.00 © 2009 IEEE

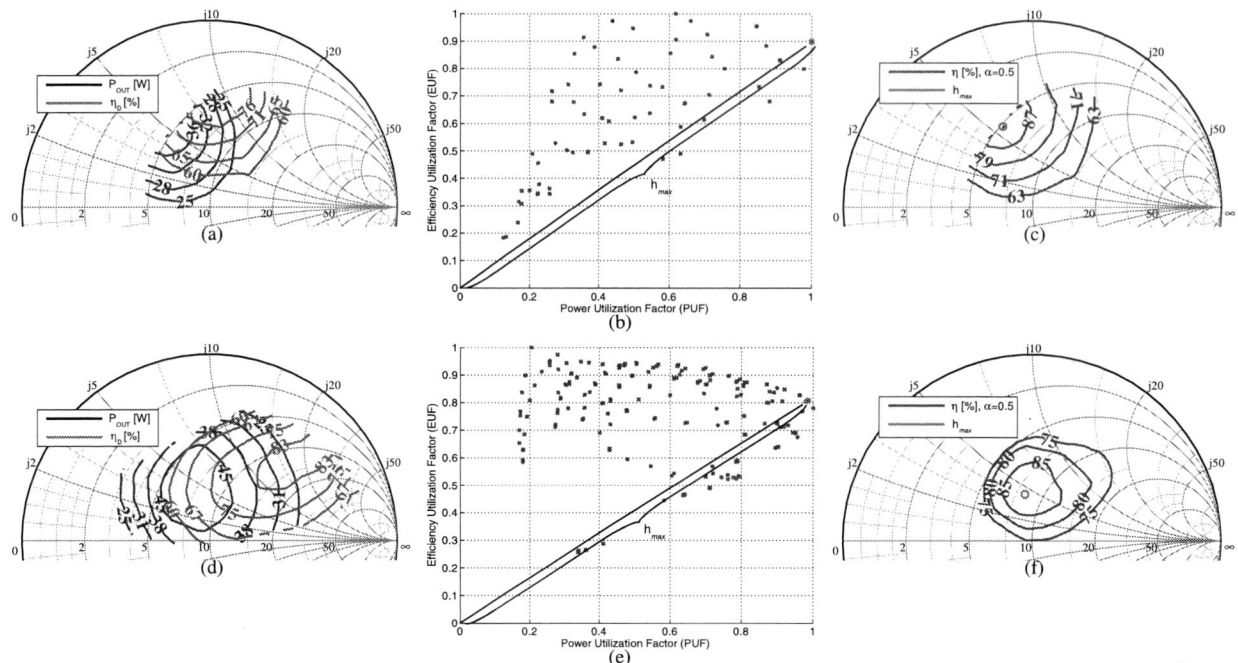

Fig. 3. (a) and (d) show load pull data for the Si-LDMOS and the GaN on Si devices with contours of output power and drain efficiency. For each impedance on the contours the normalized efficiency (EUF) is plotted vs. normalized power (PUF) in (b) and (e). For a weighting factor $\alpha = 0.5$ and based upon Equation (1) the metric h is plotted on the smith charts of (c) and (f).

These contours vary significantly from traditional load pull results obtained without explicit harmonic terminations. Notice that the degree to which regions of high output power and high efficiency overlap varies among the different technologies, and a tradeoff exists between optimum-power and optimum-efficiency designs. The results are summarized in Table I.

For a particular PA design it would be useful to have a guideline to determine the tradeoff between output power and efficiency. For each impedance of the measured load pull data in Figures 3(a) and 3(d), the measured output power is normalized and plotted versus normalized efficiency in Figures 3(b) and 3(e). A metric h is defined as a weighted Euclidean distance of each point from the origin

$$h = \sqrt{(1 - \alpha) \cdot \left(\frac{P_{\text{OUT}}}{P_{\text{OUT, max}}} \right)^2 + \alpha \cdot \left(\frac{\eta_{\text{D}}}{\eta_{\text{D, max}}} \right)^2} \quad (3)$$

This metric is then plotted on Figures 3(c) and 3(f) for a given weighting factor α for two of the devices. For example, when the output power is maximized at the expense of efficiency $\alpha = 0$. For equal weighting of both parameters $\alpha = 0.5$ and is plotted in Figure 3 for the Si-LDMOS and the GaN on Si devices.

IV. PA PROTOTYPES

Four PA prototypes were designed to achieve a minimum of 40 W output power, with the resulting weighting factors (α) in Table II. The value of α in this case is an indicator of how much power or efficiency could be improved for a given device by adjusting the matching.

TABLE II
PROTOTYPE PA POWER SWEEP PERFORMANCE

Part Number	V_{DS}	α	P_{OUT}	Gain	η_{D}	PAE
GaN/Si HEMT	28 V	0.5	43 W	17 dB	83%	81%
GaN/SiC HEMT	28 V	0.5	45 W	15 dB	85%	82%
SiC-MESFET	35 V	0.7	41 W	15 dB	80%	78%
Si-LDMOS	18 V	0.5	43 W	13 dB	80%	75%

The load pull data presented in the previous section was collected with a second harmonic termination which must be included in PA prototype designs. Measured performance at 370 MHz of the four PA prototypes is shown in figure 4 (a) and (b). Gain and PAE were measured as a function of output power with constant drain voltage. This voltage varies per prototype to prevent drain voltage breakdown as previously discussed in Equation 2. The optimal gain for each of the 370-MHz amplifiers ranged between 18 dB and 22 dB and the compressed output power is above +46 dBm for all the amplifiers. Note that heavy gain compression is required to satisfy the switching approximation and achieve high-efficiency class-E operation. This implies highly nonlinear behavior, which can be corrected with the use of supply or load modulation techniques such as EER [13] and LINC [1].

Figure 4 (c) and (d) show output power and output voltage to a 50-Ω load as a function of supply voltage. In these type of measurement the input power is fixed to a constant value. Increasing the supply voltage, effectively increases the gain of the amplifier and the output power. The slope of the V_{OUT} lines depends on impedance selection. From this figure we can conclude that the Si-LDMOS matching gives more weight

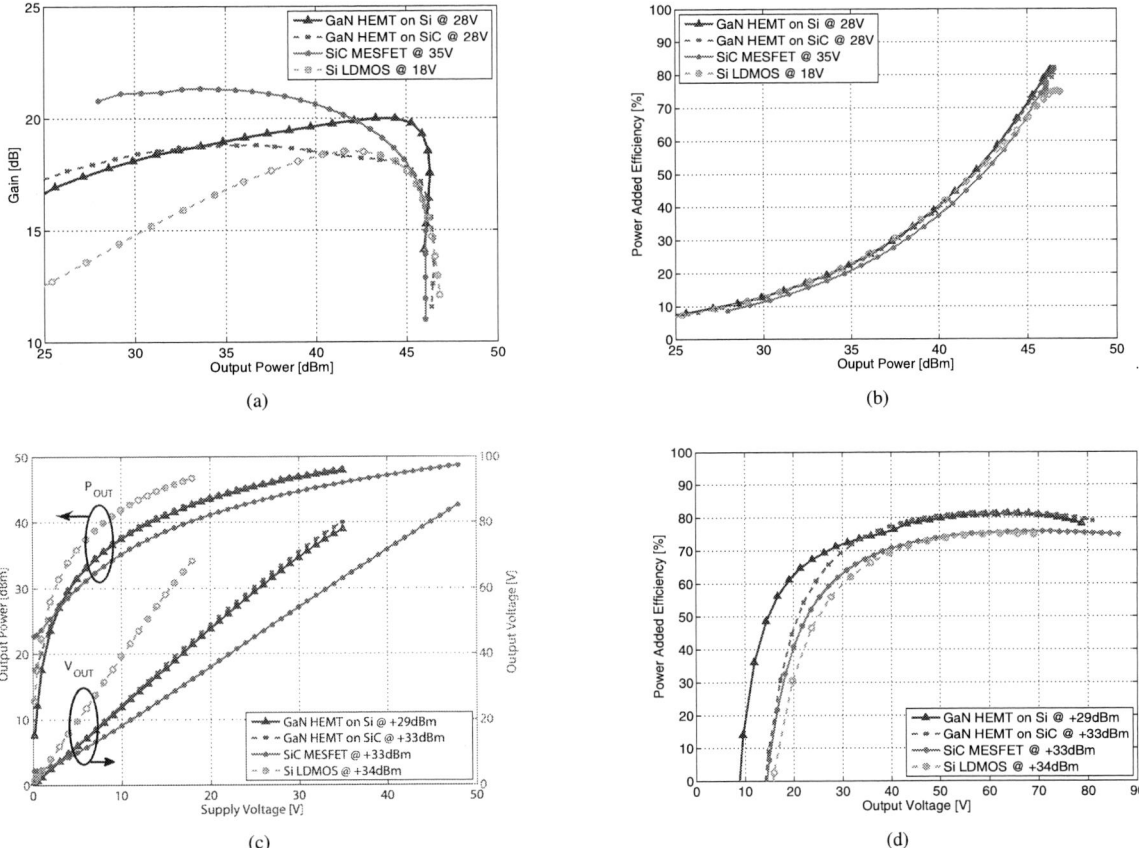

Fig. 4. Measured performance at 370 MHz of the four PA prototypes. (a) and (b) gain and PAE as a function of output power with the constant drain voltage noted in the legend. (c) output power and voltage across a 50 Ω load as a function of drain supply voltage for a fixed input power, noted in the legend. (d) PAE as a function of output voltage for a fixed input power.

towards achieving maximal output power at a lower voltage compared to the other technologies, so it has a larger slope. Figure 4(d) shows PAE vs. V_{OUT} for each of the amplifiers showing that output voltage variations can be achieved while maintaining an overall high PAE.

V. ACKNOWLEDGEMENTS

The authors acknowledge the collaboration with Matthew Poulton with RFMD and David Choi, formerly with RFMD, as well as with Todd Nichols, formerly with Nitronex. Néstor López and John Hoversten also acknowledge support from the Department of Education under a GAANN fellowship at the University of Colorado at Boulder.

REFERENCES

[1] F. H. Raab, P. Asbeck, S. Cripps, P. B. Kenington, Z. B. Popovic, N. Pothecary, J. F. Sevic, N. O. Sokal, "Power Amplifiers and Transmitters for RF and Microwave," *IEEE Trans. on Microwave Theory and Techn.*, Vol. 50, No. 3, Mar. 2002, pp. 814-826.

[2] H. Zirath, D.B. Rutledge, "LDMOS VHF class-E power amplifier using a high-Q novel variable inductor," *IEEE Trans. on Microwave Theory and Techn.*, Vol. 47, No. 12, Dec. 1999, pp. 2534-2538.

[3] N.D. Lopez, J. Hoversten, M. Poulton, and Z. Popović, "A 65-W High-Efficiency UHF GaN Power Amplifier," *IEEE MTT-S Int. Microw. Symp. Dig.*, June 2008, pp. 65-68.

[4] J. Martinetti, Al Katz, and M. Franco, "A Highly Efficient UHF Power Amplifier using GaAs FETs for Space Applications," *IEEE MTT-S Int. Microw. Symp. Dig.*, June 2007, pp. 3-6.

[5] Y-S Lee, and Y-H Jeong, "A High-Efficiency Class-E GaN HEMT Power Amplifier for WCDMA Applications," *IEEE Microw. and Wireless Letters*, Vol. 17, No. 8, Augutst 2007, pp. 622-624.

[6] S. Pajic, N. Wang, P. Watson, T. Quach, and Z. Popovic, "X-Band Two-Stage High-Efficiency Switched-Mode Power Amplifiers," *IEEE Trans. on Microwave Theory and Techn.*, vol. 53., No. 9., September 2005, 2899-2907.

[7] R. Trew, "SiC and GaN Transistors - Is There One Winner for Microwave Power Applications?," *Proceedings of the IEEE*, vol. 90., No. 6., June 2002, pp. 1032-1047.

[8] J. W. Milligan, S. Sheppard, W. Pribble, Y.-F. Wu, StG. Mueller, J. W. Palmour, "SiC and GaN Wide Bandgap Device Technology Overview," *Radar Conference, 2007 IEEE*, Apr. 2007, pp. 960-964

[9] R. Vetury, Y. Wei, D. S. Green, S. R. Gibb, T. W. Mercier, K. Leverich, P. M. Garber, M. J. Poulton, J. B. Shealy, "High Power, High Efficiency, AlGaN/GaN HEMT Technology for Wireless Base Station Applications," *IEEE MTT-S Int. Microw. Symp. Dig.*, June 2005, pp. 487-490.

[10] F.H. Raab, "Idealized operation of class-E tuned power amplifier," *IEEE Trans. Circuits Syst.* vol. CAS-24, No. 12, Dec. 1977, pp. 725-735.

[11] T.B. Mader, and Z. Popovic, "The Transmission-Line High-Efficiency Class-E Amplifier," *IEEE Microw. and Wireless Letters*, Vol. 5, No. 9, Sept. 1995, pp. 290-292.

[12] T. Mader, *Quasi-Optical Class-E Power Amplifiers*, PhD Dissertation, Department of Electrical and Computer Engineering, University of Colorado, Boulder 1995.

[13] N. Wang, N.D. Lopez, V. Yousefzadeh, J. Hoversten, D. Maksimovic and Z. Popovic, "Linearity of X-band Class-E Power Amplifiers in a Digital Polar Transmitter," *IEEE International Microwave Symposium Digest*, June 2007, pp. 1083-1086.

0.5–2.5 GHz, 10W MMIC Power Amplifier in GaN HEMT Technology

K. Krishnamurthy, D. Green, R. Vetury, M. Poulton, J. Martin

High Power Product Line, RF Micro Devices Inc., Charlotte, NC 28269, USA.

Abstract — We report a broadband lossy matched GaN on SiC HEMT power amplifier MMIC with 15dB gain, 0.5–2.5 GHz bandwidth, 9–13.6 W CW output power and 44.9–63.6% drain efficiency over the band. The amplifier operates from a 48V drain supply and is packaged in a ceramic SO8 package. These amplifiers are intended for use in wideband digital communication applications.

Index Terms — Broadband amplifiers, Gallium Nitride (GaN), High-electron-mobility transistors (HEMTs), Power Amplifiers.

I. INTRODUCTION

Next generation software defined radios under development require ultra-broadband power amplifiers (PAs) to be able to handle multiple bands and multiple operating standards. In addition high efficiency and compact size will also be required if the amplifier is employed in a handheld or mobile platform. A single amplifier capable of multiband operation will offer system cost reduction and reduced inventory requirements compared to using individual narrowband amplifiers for each of the sub-bands of interest.

The bandwidth limitation in high power amplifiers comes from the fact that large periphery devices capable of delivering high power have large input capacitances. Also they have low optimal load impedances, and impedance transformation to 50Ω further limit the bandwidth. Broadband multi-section matching networks could be used to improve bandwidth, but when used at the output they add to the loss and limits output power and efficiency. So, to obtain the wide bandwidth and high power one requires devices with low device capacitance and high optimal load impedance. GaN HEMT devices have high breakdown voltage and high power density [1], which results in smaller periphery and lower input

capacitance, as well as higher optimal load for a given output power when compared to other device technologies. This enables higher power broader band designs.

II. BACKGROUND

Distributed amplifiers offer wide bandwidth, but the inductive sections take up a lot of die area at these low frequencies. Further they have drawbacks from the reverse termination which limit maximum efficiency. Improvements like the tapered drain line have implementation issues for high power MMICs as high impedance high current lines are needed [2]. A previously reported GaN on Si MMIC distributed amplifier for the 0.1–2.2 GHz band obtained 8.7 W with the power added efficiency (PAE) rolling off from 66–30% across the band, in a 3.2 x 1.6 mm² die area [3]. Table I summarizes the results of published MMIC PAs with output power >5W covering the 0.5–2.5 GHz band.

A second publication reported power levels of 3–5 W over a 0.03–2.5 GHz bandwidth, with 20–40% PAE using a resistive feedback topology with stacked GaAs MESFET devices [4]. Resistive feedback offers broadband performance and rugged stability from the negative feedback, but the efficiency reported has been typically low.

Here we report a lossy matched amplifier incorporating a RLC all pass input matching network. This type of lossy match is optimal for its compact size, flat gain response, and excellent return loss over the bandwidth. In an earlier demonstration we had employed this in a multi-chip module using a GaN HEMT die and two GaAs passive dies for input and output matching to demonstrate a CW 8W PA operating over 0.05–2.0 GHz bandwidth with 12 dB gain and 36.7–

TABLE I
MMIC POWER AMPLIFIERS WITH OUTPUT POWER > 5W AND COVERING 0.5 – 2.5 GHZ BAND

Output power (W)	Bandwidth (GHz)	Gain (dB)	Supply Voltage (V)	PAE (%)	Circuit Topology	Implementation	Device Technology	Die Size (mm²)	Ref.
8.7	0.1 – 2.2	13	28	30 – 66	Distributed amplifier	MMIC	0.5 µm GaN on Si HEMT	3.2 x 1.6	3
3 – 5	0.03 – 2.5	21	20	20 – 40	HIFET Resistive Feedback	MMIC (2 stages)	GaAs MESFET	2.23 x 1.82	4
9 – 13.6	0.5 – 2.5	15	48	40 – 56	RLC matched	MMIC	0.5 µm GaN on SiC HEMT	2.0 x 1.0	This Work

978-1-4244-5190-6/09 $25.00 © 2009 IEEE

65.4% efficiency [5].

In this work we have a used a MMIC approach to incorporate the input match on the SiC substrate. Since this application did not require operation down to 50 MHz, we added a 4-element 4:1 impedance transformation to further boost gain by about 3 dB over the 0.5–2.5 GHz bandwidth. At the output a simple two element low loss output match consisting of a series inductor and shunt capacitor, obtains a drain efficiency of 44.9–63.6% over the band with a P3dB of 9–13.6 W.

III. GaN MMIC Technology

RFMD has developed a GaN MMIC process using our 3" GaN HEMT on SiC baseline transistor technology. The SiC substrate provides excellent thermal conductivity, with high power density capability. The GaN HEMT transistor utilizes MOCVD grown AlGaN/GaN material and a source connected field plated transistor with a gate length of 0.5 μm. The field-plate devices obtain breakdown voltages in excess of 180V. The device topology and fabrication process are detailed in an earlier publication [6].

Passive elements were implemented using PECVD silicon nitride for MIM capacitors, NiCr for thin film resistors, and a three layer Ti/Au metal interconnect for spiral inductors. The backside via (BSV) process was accomplished by thinning the SiC substrate to 100um, using an ICP process to etch the SiC vias, and subsequently depositing 4 μm of Ni/Au on the backside.

Nominal DC parameters for the device include a pinch-off voltage of about -5 V a peak current density of 0.9 A/mm and a peak trans-conductance of 250 mS/mm. Typical RF performance metrics include current and power gain cut-off frequencies (f_t and f_{max}) of 10.5 GHz and 17 GHz at 48 V drain voltage, and 44 mA bias current. Under CW operation at 2.1 GHz and class-AB bias at 48V, test devices obtained output power density of 8.2 W/mm with 69% peak PAE.

IV. Circuit Design and Implementation

A 2.2 mm device periphery was used in the design to target an output power > 12W over the band. First the output matching network was designed to present the optimum load to the device, so that the input match can later be designed to compensate for its frequency response to obtain a flat gain over the band. of the The load-pull measurement data at 2.1 GHz for a similar sized device was used to design the output matching network. The series equivalent optimum load impedances under the 48V bias condition was about $(31+j45)\Omega$. Since the impedance is not far from 50Ω a simple two element low pass matching section consisting of a series inductance and shunt capacitance was sufficient to transform from the 50Ω system (fig. 1).

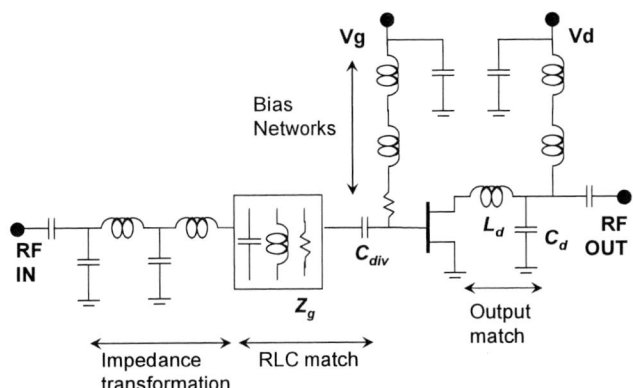

Fig. 1.   Schematic of the PA including the bias circuits

The input match is designed for a positive gain slope over the band, to compensate for the frequency response of the output network. As a starting point a small signal model of the device based on the measured s-parameters is used in the design equations of the all-pass network [7]. The device input capacitance is absorbed into the shunt element of the all-pass network, and an input termination of $Zg = 12.5\Omega$ is used. A 4-element 4:1 impedance transformer as shown in fig. 1 is used to transform from 12.5Ω to 50Ω. This type of match obtains wider bandwidth as the impedance transformation is between real impedances, compared to directly matching the device capacitance to a 50Ω system using a reactive matching technique. Further more the input return loss is excellent over the band, and the amplifier can be designed for unconditional stability with this lossy match approach. Final simulations were run using an EE-HEMT non linear model for the device.

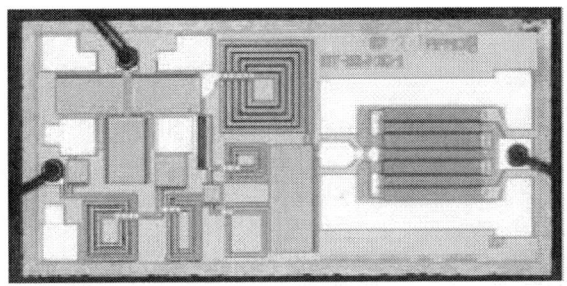

Fig. 2.   Micrograph of the 10 W GaN HEMT MMIC.
Die size : 2 mm x 1 mm ; GaN HEMT periphery : 2.2 mm

The circuit was layed-out and fabricated in the RFMD's GaN MMIC process. Fig. 2 shows the micrograph of the 2 x 1 mm^2 die. The dies were screened and packaged. To obtain low thermal resistance Au-Sn eutectic die attach is used to bond the GaN die to an AlN SO8 package (fig. 3). To obtain the best efficiency, a high-Q high current air-core inductor in series and a low ESR ceramic capacitor in shunt are used for the output match on the PCB.

978-1-4244-5190-6/09 $25.00 © 2009 IEEE   53

Fig. 3. Photograph of the MMIC die in a ceramic SO8 package. Package size : 5 mm x 6 mm.

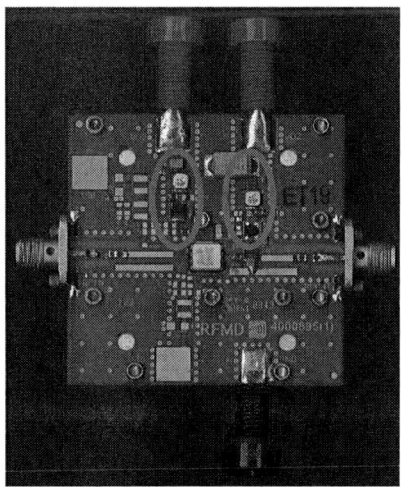

Fig. 4. Photograph of the evaluation board that includes the broadband bias chokes at the input and output (circled).

Additionally DC blocking capacitors and high current broadband DC bias chokes and implemented on the evaluation board (fig. 4). The ceramic package base is soldered on to the evaluation board over an array of filled vias that contacts the ground plane on the backside of the board. The board is mounted on an aluminum heat sink with fins for a good thermal path.

V. RF TEST RESULTS

All tests are performed on the evaluation board and the measured data includes the board losses, and effects of the broadband bias chokes. First, the small signal response of the amplifier was tested using a network analyzer. Measured S-parameters of the device at a bias of 48V and 44mA shows (fig. 5) about 15 dB gain and 0.5–3.0 GHz bandwidth, with input return loss better than -10 dB across the band.

Next the large signal performance of the PA was measured over a frequency range of 0.5–2.5 GHz using CW signal source, and driving the device to saturation.

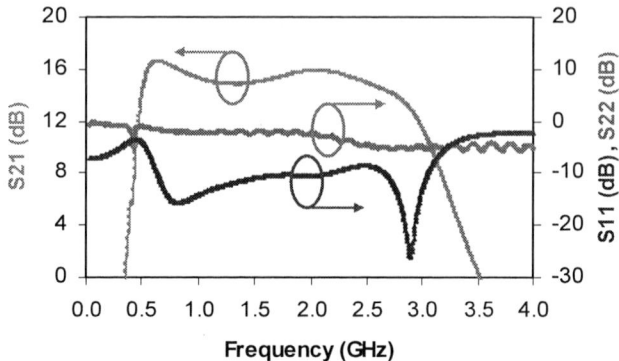

Fig. 5. Small signal S-parameters of the MMIC PA

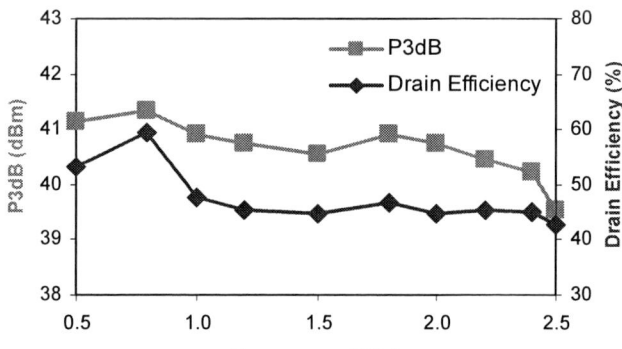

Fig. 6. Output power at 3dB compression and drain efficiency measured over 0.5 – 2.5 GHz.

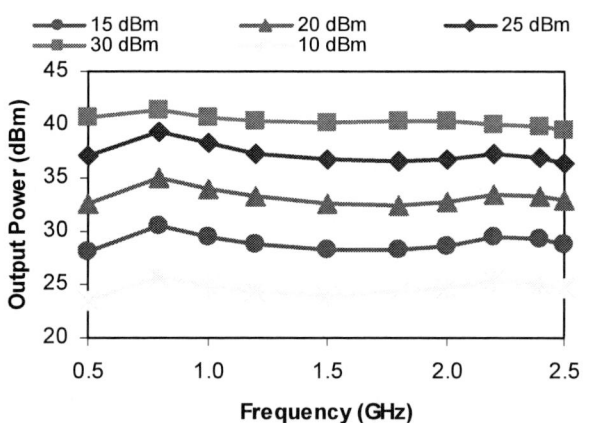

Fig. 7. CW output power, for various input power

The 3 dB compression point for the output power is between 9W and 13.6W over the band (fig. 6) with drain efficiency of 44.9% to 63.6%. Fig. 7 shows the frequency response of the PA for a power drive up from 10 dBm to 30 dBm in 5 dB steps.

Fig.8 shows the swept power performance of the device at 2.0 GHz. A 3dB compressed power (P3dB) of 10.8W is measured with 48% peak drain efficiency, and 15.4dB linear

978-1-4244-5190-6/09 $25.00 © 2009 IEEE   54

gain. Table II summarizes the RF performance of the PA over frequency.

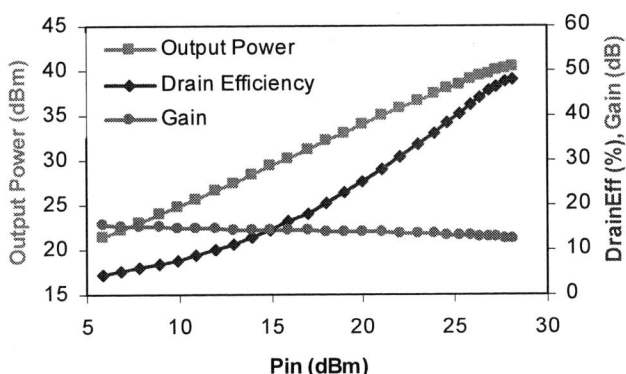

Fig. 8.   CW output power, drain efficiency and gain measured at 2.0 GHz.

VI. CONCLUSION

Using a 0.5 μm GaN HEMT MMIC technology, we have successfully demonstrated a compact 10W power amplifier in a ceramic SO8 package, operating at 48V. Using a lossy all-pass matching network for the input a wide bandwidth of 0.5–2.5 GHz was obtained with 15 dB gain. Using a simplified low-pass matching on the PCB at the output, consisting of a high-Q high current air-core inductor and a low ESR capacitor, a drain efficiency of 44.9–63.6% was obtained over the band. These GaN HEMT amplifiers are targeted for use in next generation digital cellular transmitters and in wideband communication systems.

ACKNOWLEDGEMENT

RFMD gratefully acknowledges the support for GaN technology development from ONR (Paul Maki and Harry Dietrich) and AFRL (John Blevins and Chris Bozada).

TABLE II
SUMMARY OF RF PERFORMANCE

Frequency (GHz)	SS Gain (dB)	P3dB (dBm)	P3dB (W)	Drain Eff (%)	PAE max (%)
0.5	15.0	41.1	13.0	56.4	50.5
0.8	17.0	41.3	13.6	63.6	56.6
1.0	15.1	40.9	12.3	49.7	45.9
1.5	14.8	40.6	11.4	47.2	42.0
2.0	15.4	40.7	11.8	47.0	42.2
2.5	14.9	39.5	9.0	44.9	39.5

REFERENCES

[1] R. J. Trew *et al.,* "Wide bandgap semiconductor electronic devices for high frequency applications", *1996 IEEE Gallium Arsenide Integrated Circuit Symp.Dig.*, pp. 6-9, Nov. 1996.

[2] B.M. Green *et al.,* "High-power broad-band AlGaN/GaN HEMT MMICs on SiC substrates", *IEEE Trans. on MTT*, vol. 49, no. 12, pp. 2486-2493, Dec. 2001.

[3] C. Xie *et al.,* "A high efficiency broadband monolithic gallium nitride distributed power amplifier," *2008 IEEE MTT-S IMS. Dig.*, pp. 307-310, 15-20 June 2008.

[4] A.K. Ezzeddine *et al.,* "Ultra-Broadband GaAs HIFET MMIC PA", *2006 IEEE MTT-S IMS Dig.*, pp. 1320-1323, June 2006.

[5] K. Krishnamurthy *et al.,* "RLC Matched GaN HEMT Power Amplifier with 2 GHz Bandwidth", *2008 IEEE CSICS. Dig.*, pp. 1-4xx, 12-15 Oct. 2008.

[6] R. Vetury *et al.,* "High power, high efficiency, AlGaN/GaN HEMT technology for wireless base station applications," *2005 IEEE MTT-S IMS. Dig.*, pp. 487-490, June 2005.

[7] P. Ikalainen, "An RLC matching network and application in 1-20 GHz monolithic amplifier," *1989 IEEE MTT-S IMS.*, pp. 1115-1118, 1989.

[8] T. Arell *et al.,* "A unique MMIC broadband power amplifier approach", *IEEE Journal of Solid-State Circuits*, vol. 28, no. 10, pp. 1005-1010, Oct. 1993.

978-1-4244-5190-6/09 $25.00 © 2009 IEEE

High Efficiency Digital GaN MMIC Power Amplifiers for Future Switch-Mode Based Mobile Communication Systems

S. Maroldt, C. Haupt, R. Kiefer, W. Bronner, S. Mueller, W. Benz, R. Quay and O. Ambacher

Fraunhofer Institute for Applied Solid State Physics (IAF)
Tullastrasse 72, D-79108 Freiburg, Germany
Email: Stephan.Maroldt@iaf.fraunhofer.de

Abstract — A high efficiency digital MMIC amplifier for mobile communication switch-mode concepts was designed by utilizing a 0.25 μm GaN HEMT technology with f_T of 32 GHz. A comparative investigation of two different driver concepts for a 1.2 mm GaN HEMT PA is shown. The MMICs were on-wafer evaluated for class-D and class-S operation. A drain efficiency of 70% for an output power of 4.4 W for a band pass delta-sigma (BPDS) class-S input signal at a bit rate of 3.6 Gbps equivalent to a 0.9 GHz fundamental was obtained. For the first time the operating mode up to 8 Gbps (2 GHz) is shown with an efficiency of 62%, demonstrating the prospect of future use of GaN HEMTs for switch mode amplifier concepts.

Index Terms — *GaN, high electron mobility transistor (HEMT), monolithic microwave integrated circuit (MMIC), switch-mode amplifier, class-D, class-S, mobile communication*

I. INTRODUCTION

Energy efficiency is a central demand in the field of next generation power amplifiers (PAs) for mobile communication. Commonly used class-A/B power amplifiers suffer from low efficiency especially in combination with state-of-the-art (UMTS) and future (LTE) modulation techniques. Increasing thermal challenges, energy consumption and energy costs in addition to increasing impact on global environmental problems motivates new amplifier concepts. Based on the combination of excellent switching speed and high breakdown voltage, GaN HEMTs allow the realization of high efficiency switch-mode PAs for mobile communication with a theoretical efficiency up to 100% for ideal switching devices. In recent years class-D, class-E and class-F switch-mode PAs were shown at mobile communication frequencies [1-5].

Another proposal for advanced amplifier concepts is the class-S amplifier (Figure 1) [6-7]. A band pass delta-sigma modulator generates a digital bit stream from an analog modulated input signal. The broad band digital data stream is amplified by a high efficiency GaN MMIC power amplifier. The amplified analog signal is reconstructed by a band pass filter. Without the need for compatibility to a today's analog input signal a digital data stream could be generated and directly fed into the GaN MMIC PA without any analog conversion in between. This would tremendously increase the flexibility of future switch-mode-based mobile communication systems, while decreasing the complexity of the whole system architecture.

II. GAN HEMT AND MMIC TECHNOLOGY

A. GaN MMIC Technology

The wafer structure used in this work was grown by metal-organic chemical wafer deposition (MOCVD) on a 3-inch semi-insulating SiC substrate. The layers consist of a highly-resistive c-plane GaN buffer, followed by a 22 nm $Al_{0.22}Ga_{0.78}N$ barrier and finally a thin GaN cap layer. Room temperature Hall measurements on the two-dimensional electron gas (2DEG) formed at the buffer to barrier interface resulted in a sheet resistance, a sheet carrier concentration, and a mobility of 500 Ω/sq, 8×10^{12} cm^{-2}, and 1600 cm^2/(Vs), respectively. After epitaxial growth, ohmic contacts were formed, showing a low contact resistance of 0.2 Ω·mm. The nitride assisted T-gate with a gate length of 0.25 μm was defined by e-beam lithography and trench etching into a SiN_x passivation. Furthermore the coplanar MMIC process includes NiCr based 50 Ω/sq thin film resistors, metal-insulator-metal (MIM) capacitors as well as a thick plated Au based air bridge technology.

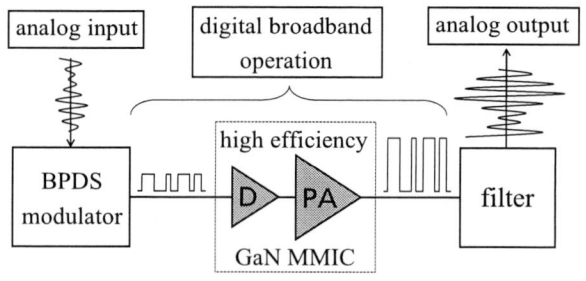

Figure 1.   Simplified diagram of the class-S amplifier concept.

This work was funded by the German Federal Ministry of Education and Research (BMBF) in the framework of the mobileGaN program.

978-1-4244-5190-6/09 $25.00 © 2009 IEEE

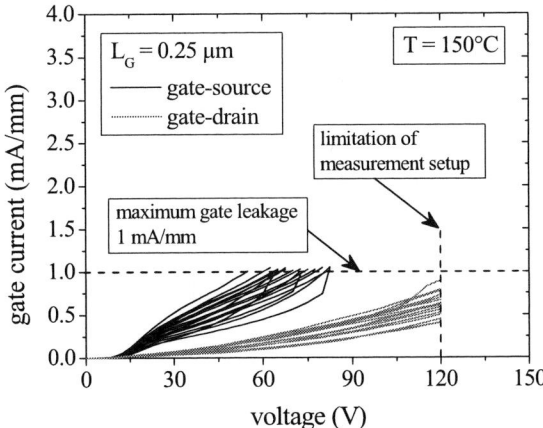

Figure 2. Measured gate-source and gate-drain breakdown characteristics of GaN HEMT devices with a gate length of L_G = 0.25 μm.

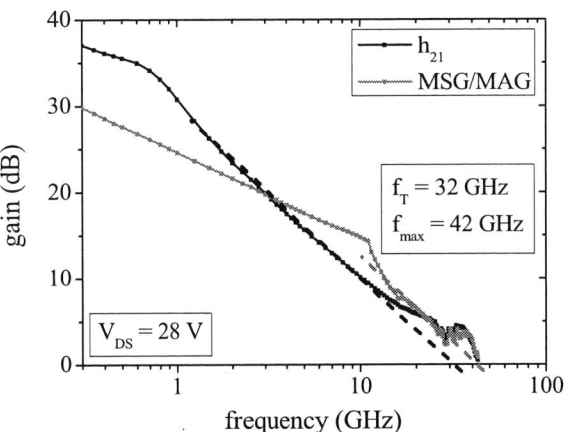

Figure 3. Measured current gain (h_{21}) and small signal power gain MSG/MAG of a 8 x 150 μm HEMT device (L_G = 0.25 μm) and extrapolated f_T/f_{max} at a drain bias voltage of 28 V.

B. GaN HEMT electrical properties

The GaN HEMT devices prepared by the described process are showing excellent DC and RF properties in addition to a very good on-wafer and wafer-to-wafer homogeneity [8]. In Figure 2 the breakdown behavior of the reverse biased gate-drain and gate-source diode at an elevated temperature of 150°C (wafer chuck temperature) is shown for 20 transistors uniformly distributed across the wafer. All of the devices yield a breakdown voltage for the gate-source diode (BV_{GS}) of about 60 V and more than 120 V for the gate-drain diode (BV_{GD}). Additionally to the outstanding breakdown voltage, devices with a total gate width of 1.2 mm (8 x 150 μm) and a gate length of 0.25 μm show high cut-off frequencies. Extrapolated from measured small signal current gain (h_{21}) and maximum available gain (MAG) reveal a transit frequency (f_T) of 32 GHz and a maximum frequency of oscillation (f_{max}) of 42 GHz at 28 V drain bias (Figure 3). Table I summarizes the most important DC and RF parameters of the 0.25 μm GaN HEMTs. Supplemental to the described properties our transistors have high maximum drain current ($I_{D, max}$) of 1050 mA/mm, a high maximum transconductance ($g_{m, max}$) of 330 mS/mm combined with a low on-resistance of 3 Ω·mm. The GaN HEMTs are depletion mode devices with a threshold voltage of -2.7 V.

III. CIRCUIT DESIGN

The circuit simulation and design was carried out using Agilent's ADS simulation environment including our in house developed GaN HEMT large signal model. The major application of the MMIC PA is either a current mode class-S operation or a current mode class-D operation. For class-S operation a band pass delta-sigma (BPDS) modulated signal requests a very high signal bandwidth. The frequency spectrum of the BPDS signal is defined from nearly DC to at least 4th harmonic due to a four times signal over sampling in generating the BPDS signal [7, 9, 10]. For a desired application in mobile communication (S-band), e.g., a 2 GHz fundamental will involve at least an over sampling frequency at 8 GHz with even higher spectral components for obtaining maximum amplifier efficiency. The class-D operation will always be possible, when the requirements for the class-S mode are

obtained. Because of the very high signal bandwidth of the BPDS modulated input signal, no matching circuit can be applied to the input of the MMIC. This has to be taken into account when a direct comparison to frequency matched class-D amplifiers is done. On the other hand this leads to a very high flexibility in using the MMIC PA, since the final operation frequency for the amplifier module can be selected independently from the MMIC.

The direct driving of the depletion mode GaN HEMTs requires about 3.5 V peak-to-peak voltage swing of the input square wave signal, which is basically defined by the given device technology. A reduction of the required input signal amplitude by a factor of two can be achieved by using an enhancement-mode GaN HEMT technology described in [11].

Specifications defining the MMIC PA design were mainly given by the 50 Ω output impedance of the square wave broad band input signal source. The intended output power of several watts leads to a GaN HEMT device of about 1 mm total gate width. The switching speed of this device is limited by the total device input capacitance which is charged by the 50 Ω input signal source. A direct driving of the PA transistor is to slow for a desired fundamental frequency in S-band. Consequently a two stage MMIC amplifier design was chosen for achieving good switching behavior and sufficient output power from a 50 Ω input source. Two different driver stages were designed based on a digital inverter circuitry with depletion mode pull down transistors (Figure 4).

TABLE I.   DC AND RF PARAMETERS OF GAN HEMTS WITH A GATELENGTH OF 250 NM.

$I_{D, max}$	1050 mA/mm
$g_{m, max}$	330 mS/mm
R_S	0.7 Ω·mm
R_{on}	3 Ω·mm
BV_{GS} (T = 150°C)	> 60 V
BV_{GD} (T = 150°C)	> 120 V
f_T	32 GHz
f_{max}	42 GHz

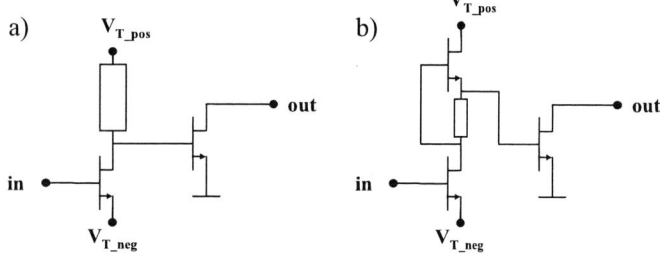

Figure 4. Schematic diagram of a 2-stage digital PA for switch-mode operation with a DTR driver stage (a) and a DTDR driver stage (b).

The absence of enhancement mode or complementary mode transistors is a major disadvantage of the GaN technology limiting the degree of freedom in driver design and efficiency. On the one hand a resistive load inverter is used in the driver stage (DTR: depletion transistor, resistive load). Major advantage of the DTR driver is the ruggedness against technology variation and device modeling imperfectness. On the other hand a modified depletion mode load inverter with an interconnected depletion mode transistor and resistive load is used (DTDR: depletion transistor, depletion and resistive load). The DTDR driving stage enables decreased static losses in the driver itself as well as an increased switching speed.

The output stage PA transistor for both driver concepts is using a total gate width of 1.2 mm (8 x 150 μm). Given that a 50 Ω broad band RF measurement system is used, which is required to evaluate the MMIC PA on-wafer performance, those PAs are best output matched for an operating voltage of 15 V equal to 30 V maximum drain source voltage of the GaN HEMT. Since the employment of the MMIC PAs in a class-S or class-D amplifier requires a differential pair of the shown circuit in Figure 4, a symmetrical layout was chosen. The chip images and coplanar layouts of both types of MMICs are shown in Figure 5 (half image of a symmetrical layout).

IV. RF POWER PERFORMANCE

Measurements were performed on a high bandwidth (50 kHz to 20 GHz) measurement setup. The input signal source consists of an Anritsu MP1758A pattern generator with a series-connected preamplifier. A high power attenuator was used as broad band 50 Ω output load. The input and output bias network has to be chosen very carefully to preserve the minimum and maximum available frequency of the measurement setup. A 50 GHz Agilent sampling scope (86100 with 83484A) combined with a software based spectral S-parameter correction was utilized for broad band measuring of the waveforms. All given data refers to measured square wave signals without any filtering which, however, has to be applied for proper class-D/S operation in the final amplifier module. The measurements were taken out on one amplifier of the differential amplifier pair. For a full differential amplifier including a filter, one can expect a doubling in output power with nearly no decrease in efficiency, which was evaluated by circuit simulations.

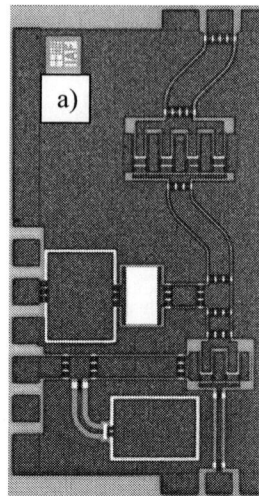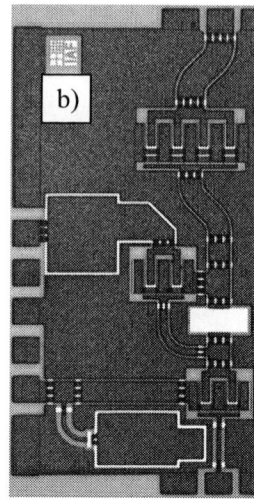

Figure 5. Chip photograph (half area of a symmetrical chip) of a 2-stage digital PA in 0.25 μm gate length technology with a DTR (a) and DTDR (b) driver stage and an output power transistor of 1.2 mm total gate width. The chip size of the full differential design is 2 x 2 mm².

A. Digital class-D operation

The class-D operation was induced by a periodic square wave input signal. Figure 6 shows the measured RF output power (P_{out}) and drain efficiency (DE) for bit rates from 0.45 Gbps to 4 Gbps. For periodic square wave signals the fundamental frequency equals to half of the given bit rate. For a low bit rate the efficiency is limited by the static losses due to the on resistance of the devices. For 0.9 Gbps equal to 450 MHz fundamental and 15 V (22.5 V) drain bias an excellent DE of 82% (78%) and a P_{out} of 3.2 W (6.3 W) was measured (DTDR driver). An increase in switching frequency leads to increased switching losses and a deteriorating in DE and P_{out}, e.g. at 4 Gbps (2 GHz) a DE and P_{out} of 55% and 2.1 W was obtained for the DTDR version, respectively. Additionally there is an improvement in DE and P_{out} visible, when comparing the DTDR to the DTR MMIC version. There is nearly no difference at low bit rates noticeable, because the static losses in the equal PA output transistors are comparable for both driver models. However, the lower switching loss for the faster switching DTDR MMIC leads to a higher DE at high bit rates compared to the DTR MMIC.

B. Digital class-S operation

For class-S operation a numerical generated 128k BPDS bit stream related to a 1-tone carrier at the specific fundamental frequency was used for the input signal. Figure 7 shows the measured RF performance of the two described MMIC versions. Because of the lower mean bit rate of the BPDS signal (class-S) compared to the periodic square wave (class-D) there is a less decrease of the DE for increasing bit rate. These are the first published results showing a class-S switching operation at bit rates related to standard mobile communication frequencies. At 8 Gbps (2 GHz fundamental) and a drain bias of 15 V a DE and P_{out} of 62% and 2.1 W were achieved, respectively.

978-1-4244-5190-6/09 $25.00 © 2009 IEEE

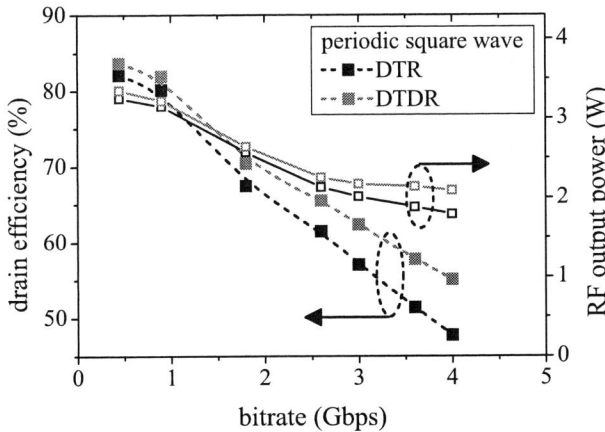

Figure 6. Measured drain efficiency and RF output power over signal bitrate of a periodic square wave input signal for a DTR and DTDR switch-mode PA with a gate length of 0.25 μm and a output gate width of 1.2 mm.

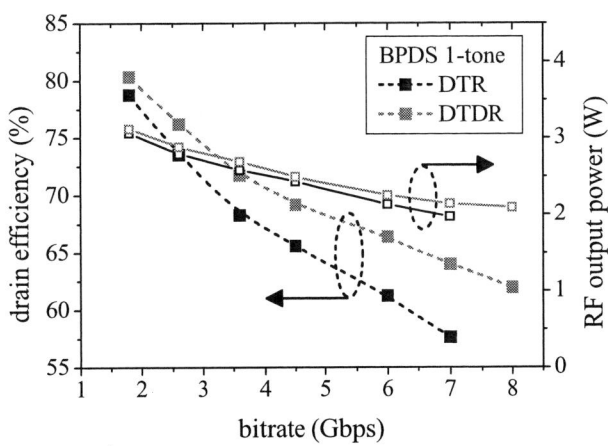

Figure 7. Measured drain efficiency and RF output power over signal bitrate of a 1-tone BPDS modulated signal for a DTR and DTDR switch-mode PA with a gate length of 0.25 μm and a output gate width of 1.2 mm.

At 3.6 Gbps (0.9 GHz) and a drain bias of 15 V (20 V) a DE and Pout of 72% (70%) and 2.7 W (4.4 W) were achieved, respectively. The eye diagram related to the 3.6 Gbps BPDS signal operation is shown in Figure 8 (without S-parameter correction). Distinct rectangular waveforms are visible with remarkably clearly opened eyes. The maximum drain voltage swing for a drain bias of 15 V was about 28 V. As predicted by circuit simulations the DTDR driver shows a shorter rise/fall time (65/55 ps) compared to the DTR counterpart (75/85 ps). Further improvements of the results can be expected by optimizing the input signal waveform which is deteriorated by combination of finite signal path length and the absence of impedance matching.

V. CONCLUSION

The basic GaN HEMT characteristics, the design and the measurement results for high efficiency digital switch-mode MMIC PAs are reported for bit rates related to mobile communication radio frequencies of up to 8 Gbps. The two stage MMIC PAs were compared for two different driver concepts. The DTDR driver based MMIC achieved a drain efficiency of 70% at an output power of 4.4 W for a 3.6 Gbps BPDS signal which is used in class-S amplifier concepts. The GaN HEMTs applied in these circuits are showing a transit frequency of 32 GHz and an on resistance of 3 Ω mm. Due to the flexible broad band features of the MMICs, they can be applied to various switch-mode amplifier modules independent to the application frequency.

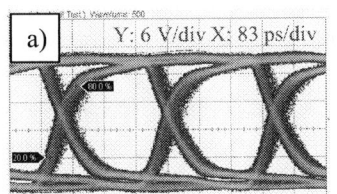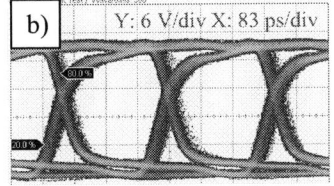

Figure 8. Measured output waveform eye diagrams of a switch-mode PA with a gate length of 0.25 μm and a total output gate width of 1.2 mm. Eye diagram of a switch-mode PA operating at 3.6 Gbps BPDS modulated signal using a DTR driver stage (a) and a DTDR driver stage (b).

ACKNOWLEDGMENT

The authors thank all colleagues from the IAF, the FBH in Berlin and Alcatel-Lucent in Stuttgart for their contribution.

REFERENCES

[1] H. Kobayashi, J. M. Hinrichs, and P. M. Asbeck, "Current-mode class-D power amplifiers for high-efficiency RF applications," IEEE Trans. Microwave Theory and Tech., vol. 49, pp. 2480-2485, 2001.

[2] J. Y. Kim, D. H. Han, J. H. Kim, and S. P. Stapleton, "A 50 W LDMOS current mode 1800 MHz class-D power amplifier," MTT-S International Microwave Symposium, 2005, pp. 1295-1298.

[3] H. M. Nemati, C. Fager, and H. Zirath, "High Efficiency LDMOS Current Mode Class-D Power amplifier at 1 GHz," 36th European Microwave Conference, 2006, pp. 176-179.

[4] U. Gustavsson, T. Lejon, C. Fager, and H. A. Z. H. Zirath, "Design of highly efficient, high output power, L-band class D-1 RF power amplifiers using GaN MESFET devices," European Microwave integrated circuit conference, 2007, pp. 291-294.

[5] Y. Abe, R. Ishikawa, and K. Honjo, "Inverse Class-F AlGaN/GaN HEMT Microwave Amplifier Based on Lumped Element Circuit Synthesis Method," IEEE Trans. Microwave Theory and Tech., vol. 56, pp. 2748-2753, 2008.

[6] A. Jayaraman, P. F. Chen, G. Hanington, L. Larson, and P. Asbeck, "Linear high-efficiency microwave power amplifiers using bandpass delta-sigma modulators" IEEE Microwave and Guided Wave Letters, vol. 8, pp. 121-123, 1998.

[7] C. Meliani, J. Flucke, A. Wentzel, J. Wurfl, W. Heinrich, and G. Trankle, "Switch-mode amplifier ICs with over 90% efficiency for Class-S PAs using GaAs-HBTs and GaN-HEMTs," IEEE MTT-S International Microwave Symposium, 2008, pp. 751-754.

[8] P. Waltereit, W. Bronner, R. Quay, M. Dammann, S. Muller, R. Kiefer et al., "High-efficiency GaN HEMTs on 3-inch semi-insulating SiC substrates," Physica Status Solidi a, vol. 205, pp. 1078-1080, May 2008.

[9] T.-P. Hung, J. Rode, L. E. Larson, and P. M. Asbeck, "Design of H-Bridge Class-D Power Amplifiers for Digital Pulse Modulation Transmitters," IEEE Trans. Microwave Theory and Tech., vol. 55, pp. 2845-2855, 2007.

[10] A. Samulak, G. Fischer, and R. Weigel, "Basic nonlinear analysis of class-S power amplifiers based on GaN switching transistors," German Microwave Conference, 2009, pp. 4 pp.

[11] S. Maroldt, C. Haupt, W. Pletschen, S. Müller, R. Quay, O. Ambacher, C. Schippel and F. Schwierz, "Gate-Recessed AlGaN/GaN Based Enhancement-Mode High Electron Mobility Transistors for High Frequency Operation", Jpn. J. Appl. Phys., Vol. 48, pp. 04C083, 2009.

978-1-4244-5190-6/09 $25.00 © 2009 IEEE

A Pulsed Load Modulation (PLM) Power Amplifier with 0.35μm pHEMT Technology

Shuhsien Liao and Yuanxun Ethan Wang, Member, IEEE

Department of Electrical Engineering, University of California Los Angeles, Angeles, United State

e-mail:shliao@ucla.edu

Abstract—A new modulation scheme, namely Pulsed Load Modulation (PLM), for high efficiency power amplification has been introduced. In PLM scheme, the concept of switched resonator is utilized to modulate the load impedance of a power amplifier in a digital fashion so as to maintain maximum efficiency and linearity. In this paper, a 1.87GHz high efficiency PLM power amplifier is presented. The power amplifier module is implemented with two 0.35μm GaAs pHEMT devices from Triquint Semiconductor and a bandpass filter centered at 1.87GHz. The measurement is done at 1.87GHz carrier frequency with 10V supply voltage and 384Mbps, 1.1V gate switching voltage. The testing results show a maximum of 72.5% drain efficiency and maximum 30.21dBm output power. A maximum of 60.9 % Drain efficiency is achieved at selected power level with diplexer included. Comparing with an ideal class B mode power amplifier, the improvement of power efficiency under power back off level of 6.4dB and 8.1dB are 40% and 50% respectively. An Envelope Delta Sigma Modulation (EDSM) version of WCDMA signal is used for the system linearity test.

Keywords- High efficiency power amplifier, pulsed load modulation, pHEMT

I. INTRODUCTION

Modern communications require transmitters with availability of output power, high linearity and large bandwidth. In addition, high power efficiency should be maintained under various signal modulation schemes. However, the power efficiency of a linear amplifier drops proportionally with the output power back off level. For complex modulation schemes, the average power efficiency will be seriously degraded due to the high peak to average power ratio. Efficiency enhancement techniques based on load impedance modulation have shown potential ability to restore power efficiency under the high peak to average power ratio circumstance. The two most seen examples are the Doherty [3] and Chireix's outphasing amplifiers [4, 5]. Load modulation is realized through different biasing controls on the two transistors in a Doherty amplifier and phase modulations in a outphasing amplifier. Because the load modulation is done in an analogue fashion, this creates a significant amount of nonlinear distortion in the amplification process. Consequently, pre-distortion techniques are usually required in order to maintain high linearity, and thus increase the complexity and limit the bandwidth of the system. Besides, the potential efficiency enhancement from load modulations in either case can only be optimized over a small range of power levels.

Due to the fact that the analogue control of load modulation introduces strong nonlinearity, a novel power amplifier technique, called "Pulsed Load Modulation (PLM)" has been demonstrated for high power efficiency operation [7]. Through this technique, load modulation is realized in a digital fashion and the optimum efficiency can be maintained over a wide range of output power for high frequency power amplification. In addition to maintain high efficiency, linearity of the system is less subject to the nonlinearity of the device. The envelope of the signal is modulated with delta sigma modulation and then used as a control biasing voltage to the gates of the transistors. The RF signal is then recovered by a high Q band pass filter. The power transistors work as a switch therefore the system linearity is less sensitive to nonlinearity of the devices. In this paper, we present a design of power amplifier module on 0.35um GaAs pHEMT technology and then tested with the Pulsed Load Modulation (PLM) scheme. Measurement results of this work exhibit the improvement of efficiency over conventional power amplifiers and the ability of the scheme to preserve linearity of the system for linear modulation scheme.

II. PRINCIPLE OF OPERATION

The PLM technique is based on impedance modulation technique. This technique can achieve optimal efficiency over a wide range of output power level because the load impedance can be adjusted inversely proportional to the output power levels. The concept of the PLM technique is based on the characteristics of the switched resonator [6] termination shown in Fig.1.

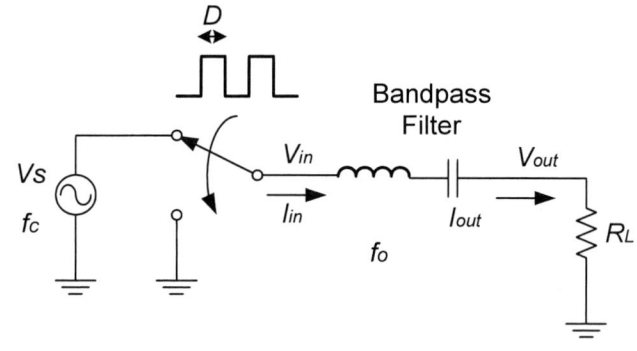

Figure 1.   Switched resonator.

978-1-4244-5190-6/09 $25.00 © 2009 IEEE

The switched resonator is comprised of a pulsed voltage source and a high-Q band pass-filter. The band-pass filter is driven by a pulsed voltage source at one of its ends. When the pulse is on, the band-pass filter sees a constant input voltage. On the other hand, the input is forced to zero by short-circuiting to the ground when the pulse is off. The other end of the filter is connected to a constant load resistance given by $R_{opt}/2$. R_{opt} is the optimum load impedance of the amplifier. The high Q filter will prevent any sudden change to the RF current flowing through it and restore the pulsed signal to the original analogue signal. The resonator exhibits different equivalent impedance levels during the process of energy charging and discharging. Therefore, by keeping a certain percentage of energy charged in the resonator through the switching process, one can realize any particular equivalence resistance value. During the turning on period, the effective load impedance R_{eff} observed from the input of the filter can be related to the duty cycle D of the pulse signal. Assuming the switching speed of the envelope modulator is much greater than the bandwidth of the high Q band-pass filter, the output current of the PA is then approximately proportional to the duty cycle of envelope modulator and stays constant during the whole cycle. As the input current of the high Q filter equals its output current, the relationship in (2) should hold.

$$\begin{cases} V_{out} = V_{\max} D \\ I_{out} = \dfrac{V_{out}}{R_{opt}/2} \end{cases} \quad (1)$$

The effective load impedance at the "pulse on" state is thus yielded as

$$R_{eff} = \frac{V_{\max}}{I_{out}} = \frac{1}{D} \cdot R_{opt}/2 \quad (2)$$

The load impedance could increase from $R_{opt}/2$ to infinity when the duty cycle drops from 100% to 0. The output power is then given by

$$P_{out} = \frac{1}{2} I_{out} V_{out} = 2 V_{\max}^2 D^2 / R_{opt} \quad (3)$$

Comparing (2) with (3), it is obvious that the optimal load modulation condition required by a power amplifier can be obtained as long as the above assumptions are satisfied. In the PLM architecture, the amplifier is controlled by pulses with either "on" or "off" state. At "on" state, the amplifier is fully turned on and the switch is turned off to form the current path to switched resonator. At "off" state, the amplifier is turned off. The output of amplifier appears to be high impedance, thus no DC power consumption is expected. Meanwhile, the switch is turned on to form a short circuit path at the input of the switched resonator so that current can flow. The load impedance increases from R_{opt} to infinity when the duty cycle drops from 100% to 0. The power amplifier, once turned on, always experiences high impedance level, therefore operates in the voltage saturation mode. Because of operating in saturation

mode, optimal power efficiency is then restored when the output power is backed off to any level.

The proposed configuration in this work is shown in Fig. 2. The PLM amplifier consists of a pair of FET devices which are connected through a quarter-wave impedance transformer. The switching pulses are generated by the envelope modulator. This pulsed signal controls the gates of both devices to turn them on and off simultaneously. For two identical devices, it can be shown that the load modulation maintains the optimal efficiency throughout 50% to 100% duty cycle as both devices are operating under voltage saturation conditions. Beyond that, the main power amplifier, which originally works like a switch, can no longer stay in saturation. The efficiency performance thus degrades like a standard power amplifier [1]. The theoretical efficiency performance of the proposed architecture predicts a flat region for up to 6dB power back off, as shown in Fig. 3.

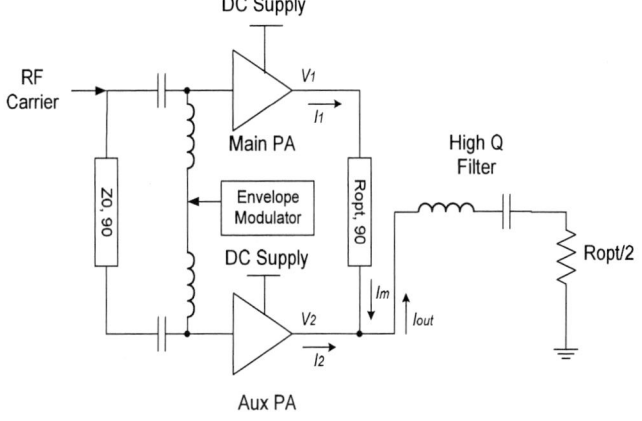

Figure 2.   PLM configuration.

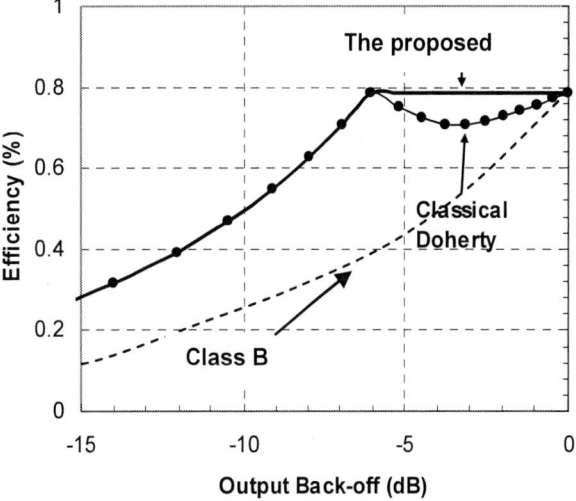

Figure 3.   Efficiency curve of PLM scheme vs. ideal class B and Doherty amplifier.

The PLM scheme is advantageous comparing to the Doherty amplifier where a dip exits in the middle of the efficiency curve. Other than its higher efficiency potential, the linearity of the power amplifier has less dependency on the device non-linearity, as it works at only two discrete voltage levels. It is proposed to develop high efficiency power amplifier module that operate at 1.87GHz and can support PLM architecture for efficiency improvement. The chosen technology for this paper is 0.35μm GaAs pHEMT from Triquint Semiconductor.

III. AMPLIFIER DESIGN

The power amplifier module is designed in a balanced fashion which includes a 90 degree hybrid for impedance matching and two amplifiers for the PLM scheme. In this design, 0.35μm GaAs pHEMT device is chosen. Two 0.6mm x 0.35μm pHEMT devices are used for the power amplification stage. The gate control of the driver amplifier is also driven by the same signal which turns on and off the power amplification stage. The layout of the PLM amplifier is shown in Fig. 4and the size of the whole module is around 82mm x 87mm.

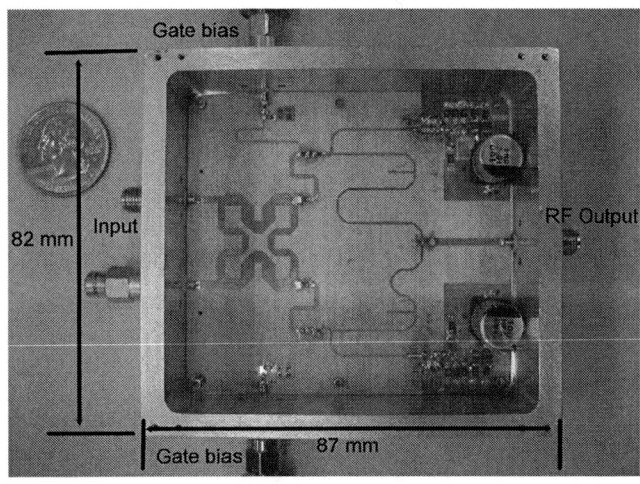

Figure 4. PLM module.

IV. MEASUREMENT RESULT

The drain supplies of the driver amplifier and the power amplifier stage are separate for maximum efficiency tuning. In order to generate pulse modulation signal for gate control, an arbitrary waveform generator (Tektronic AWG520) with 1GS/s clock is used, and 10% to 100% duty cycle pulses with 384MHz repetition rate are set up for efficiency measurement. The diplexer filter is measured to have a pass band centered at 1.87GHz with 30 MHz bandwidth and insertion loss of 0.85dB. In the test of PLM operations, the biasing is dynamically changing from the conventional Class-B condition to a deep pinch-off biasing point, e.g., Vg= -1.3V to -2.4V. The maximum power gain of the module under test is about 12.06 dB with maximum output power approximately 30.21dBm. The measured power efficiency versus power back off characteristics of the PLM amplifier is compared to that of the ideal conventional class-B approaches shown in Fig. 5. It is observed that the PLM amplifier has in general higher

efficiency than conventional Class-B amplifier when the output power starts to back off. The maximum drain efficiency of 60.9% is measured at selected power level for the whole module including the diplexer and coaxial adaptors. The efficiency enhancement over the conventional Class B amplifier at the power back off levels of 6.4dB and 8.1dB are 40% and 50% respectively. It is evident that the pulsed load modulation effect contributes to efficiency improvement.

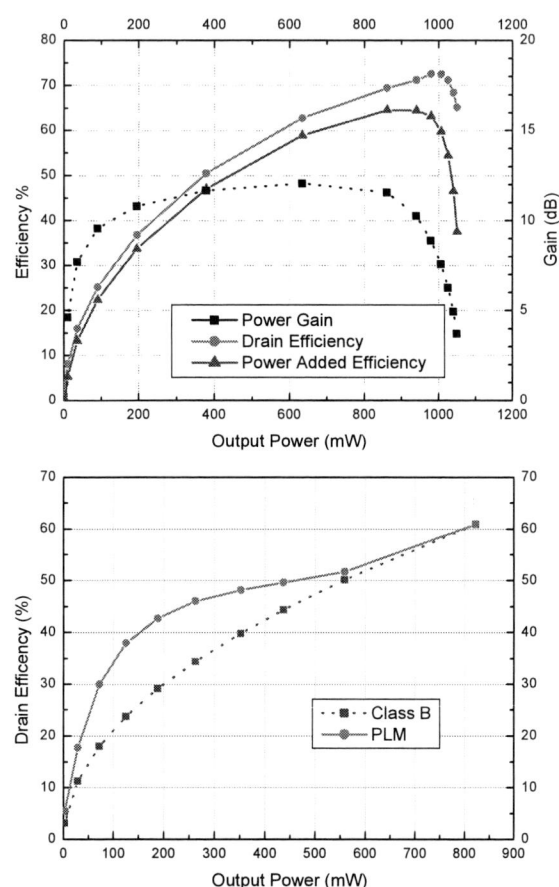

Figure 5. Efficiency versus output power back off curve for the PLM Class-B scheme and the conventional Class-B/A scheme.

In order to test the system linearity, an EDSM version of WCDMA signal modulating on a carrier of frequency 1.87GHz with peak to average ratio (PAR) of 8dB is used. In this set up, the Envelope Delta Sigma modulation and base band I/Q of the phase modulation of the WCDMA signal are generated by AWG. The baseband signals are sent to Agilent - E4438C ESG Vector Signal Generator and the output is used to drive the amplifier. The envelope signal is used as gate bias to turn on/off of the PLM module. In Figure 6, the testing set up is presented. The output shows a 25.68dBm output power and 45.65% drain efficiency. And ACLR of 40dBc and 45dBc at 5MHz and 10MHz offset from carrier frequency without pre-distortion applied. The spectrum of the output signal and ACLR can be observed in Figure 7.

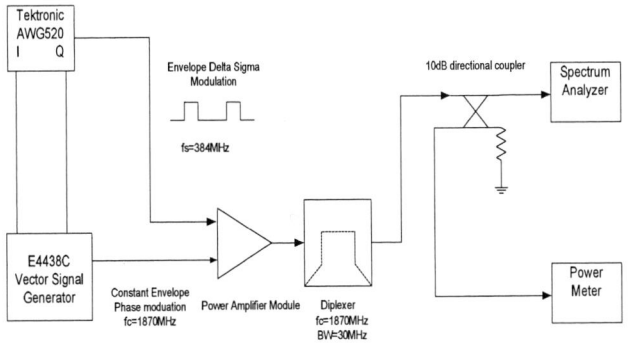

Figure 6. Measured set up and theoretical prediction of the output spectrum in each stage.

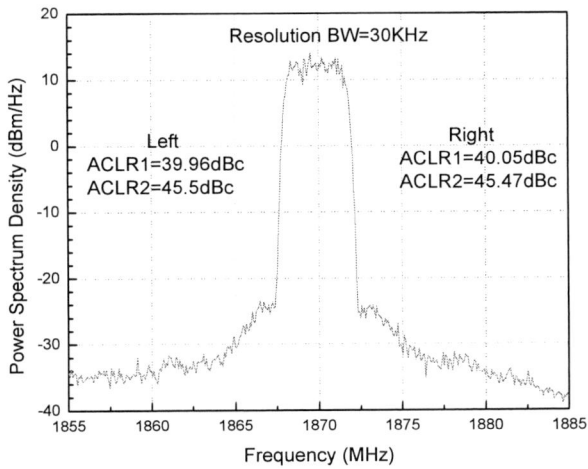

Figure 7. Measured output spectrum

We have demonstrated that the PLM technique is among the few efficiency enhancement approaches that can be utilized for linear amplification.

REFERENCES

[1] F. H. Raab, P. Asbeck, S. Cripps, P. B. Kennington, Z. B. Popovic, N. Phthercary, J. F. Sevic and N. O. Sokal, "Power Amplifiers and Transmitters for RF and Microwave," IEEE Trans. Microwave Theory & Tech., vol. 50, pp. 814-826, March 2002.

[2] S. C. Cripps, RF Power Amplifiers for Wireless Communications, 2nd Edition, Norwood, MA: Artech House, 2006.

[3] F. H. Raab, "Efficiency of Doherty RF power-amplifier systems," IEEE Trans. Broadcast., vol. BC-33, no. 3, pp. 77–83, Sep. 1987.

[4] A. Birafane and A. B. Kouki, "On the linearity and efficiency of outphasing microwave amplifiers," IEEE Trans. on Microwave Theory & Tech., vol.52, no. 7, July 2004.

[5] F. H. Raab, "Efficiency of outphasing RF power amplifier systems," IEEE Transactions on Communications, vol. COM-33, no.10, Oct. 1985.

[6] S. Kim and Y. E. Wang, "Theory of switched RF resonators," IEEE Trans. On Circuits and Systems I: Regular Papers, vol.53, Issue 12, pp. 2521 – 2528, Dec. 2006.

[7] J. Jeong and Y. E. Wang, "Pulsed Load Modulation (PLM) Technique for Efficient Power Amplification," IEEE Trans. On Circuits and Systems II: Express Briefs, vol.55, Issue 10, pp. 1011 –1015, Oct. 2008.

V. CONCLUSION

In this work, a high efficiency RF power amplifier based on the pulsed load modulation scheme using the 0.35μm GaAs pHEMT technology is presented. The circuit is designed to operate with 10V supply voltage. The amplifier module has a maximum of 12.06dB power gain and a maximum of 30.21dBm output power under the class B configuration. In addition, the measured drain efficiency is 72.5% or 60.9% when the diplexer is included. Under the PLM operating condition, a 384Mbps 1.1V switching voltage is applied to the gates of the pHEMT devices. The drain efficiency under power back offs of 6.4dB and 8.1dB is improved by 40% and 50% respectively when comparing with an ideal class B mode power amplifier. An EDSM version of WCDMA testing signal is used to verify the system linearity. Observed results show 25.68dBm output power and 45.65% drain efficiency. The ACLR at 5MHz and 10MHz offset from carrier frequency are 40dBc and 45dBc respectively without pre-distortion applied.

10-Gbit/s Wireless Transmission Systems Using 120-GHz-Band Photodiode and MMIC Technologies

*[1]Naoya Kukutsu, *[1]Akihiko Hirata, *[2]Toshihiko Kosugi, *[1]Hiroyuki Takahashi, *[1,3]Tadao Nagatsuma, *[1]Yuichi Kado, *[4]Hiroshi Nishikawa, *[4]Akihiko Irino, *[4]Toshihiro Nakayama, and *[4]Naohiro Sudo

[1]NTT Microsystem Integration Laboratories, NTT Corporation,
[2]NTT Photonics Laboratories, NTT Corporation
[3]Osaka University
[4]Fuji Television Network, Inc.
E-mail: kukutsu.naoya@lab.ntt.co.jp

Abstract

This paper describes 10-Gbit/s wireless transmission systems using the 120-GHz band. System configurations, key devices/modules, transmission characteristics, and experiments on transmitting uncompressed high-definition (HD) video signals are discussed.

INTRODUCTION

The data capacity required for the exchange of data, distribution of HD video, and other services being developed on the Internet continues to rise each year, and the communications services that handle this traffic continue to expand accordingly. The approval of 10-Gbit/s Ethernet PON (10G-EPON), providing 10 Gbit/s as a PON standard, and the standardization of 100-Gbit/s Ethernet (100GbE) is expected by the year 2010. However, current wireless technologies are limited to transmission speeds of a few gigabits per second [1,2]. Thus, there is a need for ultrafast wireless technology that can support network standards like 10GbE and 10G EPON, on which new services will be developed. Moreover, t he switch from analog to digital broadcasting will be completed by 2011 in Japan, and the need for wireless technology to transmit HD video is increasing. However, uncompressed HD video signals require up to 1.5 Gbit/s, and the lack of wireless technology able to transmit real-time multiple HD video signals from a live broadcast site is becoming a problem.

Approach to 10 Gbit/s wireless link

As the data rate depends on the carrier frequency, we believe that the most promising way of building a 10-Gbit/s wireless link would be to exploit undeveloped frequencies above 100 GHz. Figure 1 presents the transition of our 10-Gbit/s wireless data transmission systems using a 120-GHz band wireless link. First, we used photonic technologies to generate a 120-GHz-band millimeter-wave (MMW). Introducing photonic technologies like a uni-traveling carrier photodiode (UTC-PD) and optical high-speed modulator into wireless systems enables us to generate desired broadband modulated signals over 100 GHz with certain output power because the photonic components have a larger bandwidth than electronic ones and practical parts are steadily manufactured for optical network links. Many MMW and terahertz wave (THzW) systems have recently been developed using photonic technologies [3]. However, all-electronic systems are also attractive because they can be compact and have low power consumption, especially when the transceiver functions are implemented with monolithic microwave integrated circuits (MMICs).

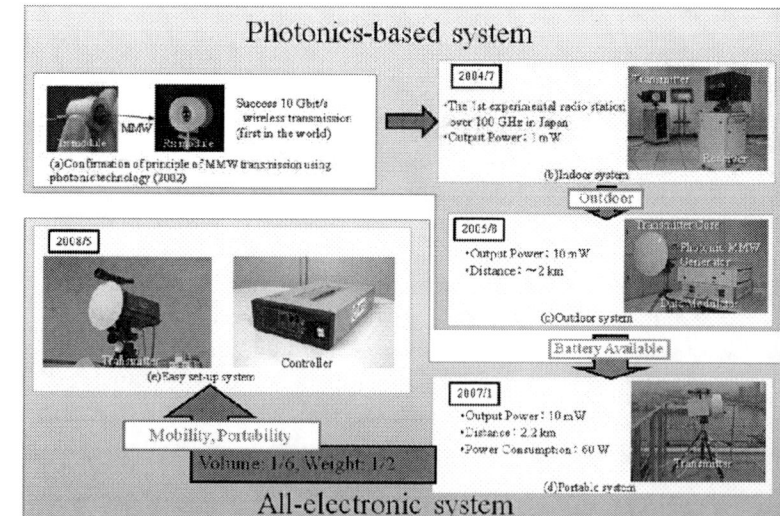

Fig. 1 Transition of our 10 Gbit/s wireless system

120-GHz-band wireless system using photonic technologies

Figure 2 is a schematic of a wireless system using photonic technologies. We implemented photonic components in three units whose photograph is shown in Fig.1 (c). The first unit was a photonic MMW generator (PMG) [4], which consisted of a planar lightwave circuit (PLC) that integrates an AWG and a 3-dB combiner [5]. The second was a data modulator (DM) that modulated an optical subcarrier signal with data signals. The third was the core of the transmitter (Core) that contained a uni-traveling carrier photodiode (UTC-PD) module, which was composed of the UTC-PD and an InP high-electron-mobility-transistor (HEMT) amplifier [6]. The generated electrical MMW signal was transmitted from a high-gain antenna. In the receiver, the RF block consisted of a MMIC receiver module [6], which included a low-noise amplifier (LNA) and an ASK demodulator. The receiver MMIC was integrated in a waveguide module package. The demodulated signal went from a limiting amplifier to a CDR and then to an E/O converter, and then it was output to an external optical fiber.

typically have a unity current gain frequency, f_t, of 170 GHz and a maximum oscillation frequency, f_{max}, of 350 GHz.

The maximum output power of the transmitter MMIC is over 3 dBm. Figure 3 shows a transmitter module and a power amplifier (PA) module, which were integrated in packages with an interface of MMW waveguide. We used a PA module in which a PA MMIC had been integrated to increase the output power of the transmitter up to 10 dBm. Since all the functions necessary for a wireless transmitter have been integrated in a set of very small MMW waveguide modules , the transmitter of the all-electronic wireless transmission system becomes very compact and has low power consumption as shown in Fig. 1 (d), compared to the photonic technologies-based system. In the receiver, the automatic gain control (AGC) circuit monitored the demodulated signal and controlled the gate voltage of the LNA so that extra power did not enter the final amplifier and overload it.

Moreover, we improved the all-electronic system for applying a practical use of broadcasting. For live TV broadcasts, there is a strong demand to minimize the time required from material arriving from an on-site location until it is ready for broadcasting. It must be (1) simple in structure, (2) quick to assemble, and (3) easy to operate.

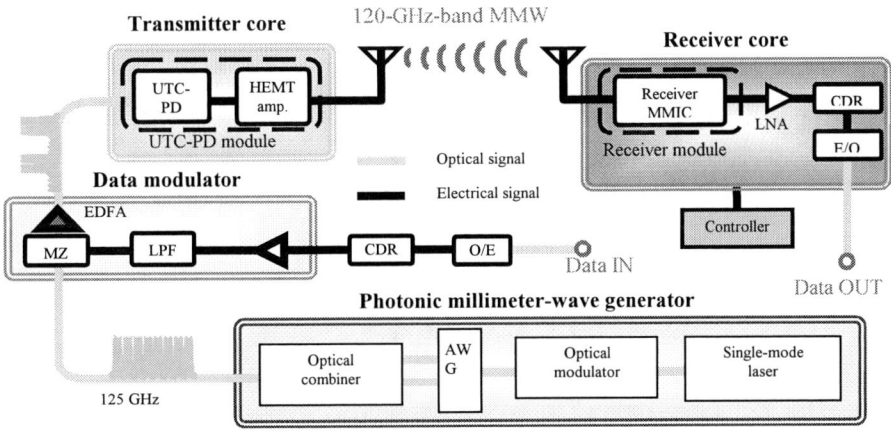

Fig. 2 Schematic of 120-GHz-band wireless system using photonic technologies.

120-GHz-band wireless system using electronic technologies

Improvements in electron beam gate lithography are increasing the speed of HEMTs as gate length decreases, and the use of these transistors enables us to make electronic devices that operate at frequencies over 100 GHz. We developed flexible CPW-MMIC chipsets for 120-GHz-band wireless systems, which included an amplifier, modulator, demodulator, and frequency doubler [6]. The devices

Therefore, we developed the new 120-GHz-band wireless system for use in the Beijing Olympics trials, which featured usability, configuration, and a simple setup similar to the FPU equipment currently in wide use by broadcasters [7]. The new equipment is shown in Fig. 1 (e).

Fig. 3 Photographs of the transmitter module and power amplifier module

Experimental Results

We conducted a transmission experiment using the all-electronic system. The error-free condition, i.e., BER < 10^{-12}, was obtained at a minimum received power of -38 dBm for a 10.3125-Gbit/s data rate as shown in Figure 4. We also succeeded in outdoor error-free transmission at a distance of 800 m between the roofs of two buildings over a few hours without alignment of the direction of either the transmitter or receiver antenna [8]. The maximum transmission distance was estimated to be about 2 km in fair conditions.

We also carried out the transmission trial in the Beijing Olympic Games 2008 to figure out the problems we need to solve and review the feasibility of this system for the broadcasting area. To get fruitful results, we installed the 120 GHz wireless transmission system parallel with the actual broadcaster's network to emulate real conditions [9]. Figure 5 shows the overall wireless network configuration. At that time, we installed the 120 GHz receiver on the rooftop of the International Broadcast Center (IBC) and put the transmitter at the Broadcasting Media Center (BMC), which is on the opposite side of the Beijing Olympic Park. Due to this configuration, we could send HD-SDI signals from almost everywhere in the Beijing Olympic Park without any blind spots. We confirmed that the received

power fluctuation was below 1 dB for 20 hours in the location. These results indicate that the output power fluctuation and the divergence of the antenna axis were small. Not one error was observed in the 120 GHz-band wireless link, and HD image transmission was very stable.

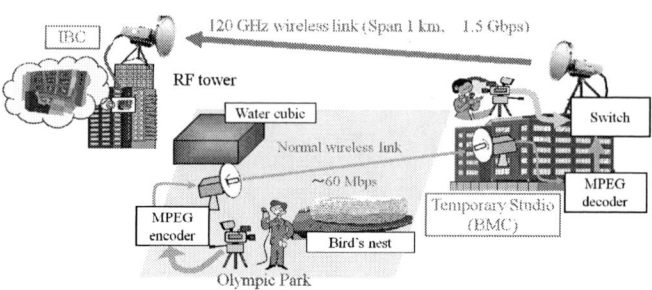

Fig. 5 Overall wireless network configuration in the Beijing Olympic Games transmission trial

Conclusion

We discussed the 10-Gbit/s data transmission systems using an over 120-GHz-band that incorporates both photonic and all-electronic technologies. The maximum transmission distances of the systems were estimated to be 2 km. In the trial at the Beijing Olympics, we have shown that our system, using the 120-GHz-band millimeter-wave signals not used previously in the industry, has attained a technical level compatible with practical use.

Acknowledgements

This work was supported by "The research and development project for the expansion of radio spectrum resources" of the Ministry of Internal Affairs and Communications, Japan.

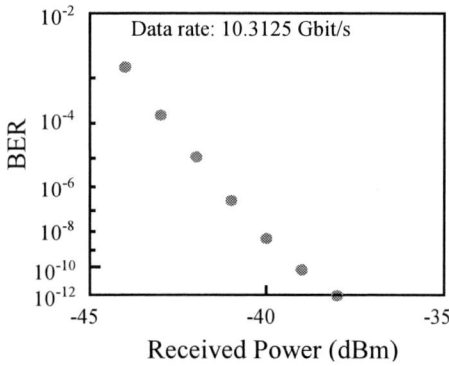

Fig. 4 BER of the system

References

[1]   Loea Corporation   http://www.loeacom.com/

[2]   Gigabeam Corporation   http://www.gigabeam.com/

[3]   S. Takano et al., "The first radio astronomical observation with photonic local oscillator," Publ. Astron. Soc. Japan, Vol. 55, pp. L53–L56, 2003.

[4]   A. Hirata et al., "Low-phase noise photonic millimeter-wave generator using an AWG integrated with a 3-dB combiner," IEICE Trans. Electron., E88-C, No. 7, pp. 1458–1464, 2005.

[5]   H. Ito et al., "Over-10-dBm output uni-traveling-carrier photodiode module integrating a power amplifier for wireless transmissions in the 125-GHz band," IEICE Electronics Express, Vol. 2, pp. 446–450, 2005.

[6] T. Kosugi et al., "120-GHz Tx/Rx chipset for 10-Gbit/s wireless applications using 0.1μm-gate InP HEMTs," 2004 IEEE CSIC Symp. Dig., pp. 171–174, 2004.

[7] N. Kukutsu et al., "Compact, Low-power, 120-GHz-band Wireless Systems for 10-Gbit/s Data Transmission," GSMM2009 , 2009.

[8] R. Yamaguchi et al., "10-Gbit/s MMIC Wireless Link Exceeding 800 Meters," IEEE RWS2008 Symp. , 2008.

[9] H. Nishikawa et al., "Transmission trial using 120 GHz wireless system in Beijing Olympic Games 2008," GSMM2009., 2009.

Low-cost CMOS-based receive modules for 60 GHz wireless communication

Piet Wambacq[1,3], Kuba Raczkowski[1,2], Valéry Ramon[1], Alexander Vasylchenko[1,2], Amin Enayati[1,2], Michael Libois[1], Jonathan Borremans[1], Karen Scheir[1,3], Stephane Bronckers[1,3], André Bourdoux[1], Bertrand Parvais[1], Bob Verbruggen[1,3], Steven Brebels[1], Wim Van Thillo[1,2], Christophe Pavageau[1], Bart Nauwelaers[2], Guy Vandenbosch[2], Walter De Raedt[1], Charlotte Soens[1]

[1]IMEC, Belgium
[2]ESAT, Katholieke Universiteit Leuven, Belgium
[3] Vrije Universiteit Brussel, Belgium

Abstract— **To enable mass-market applications based on short-range wireless communication around 60 GHz at datarates above 1 Gbps, cheap implementation technologies are needed. The radios for these applications often use antenna arrays with beamforming to improve the link budget. This work discusses the state-of-the-art on CMOS beamforming architectures and describes the different parts of a wireless link that is built around a four-antenna array on a PCB that also contains 45nm digital CMOS receiver ICs with beamforming capabilities.**

Keywords-CMOS, 60 GHz, mm-wave, phased arrays.

I. INTRODUCTION

The availability of an unlicensed band of 7GHz around 60GHz together with the ability of modern downscaled silicon-based technologies (ultra deep submicron CMOS, SOI, BiCMOS) to handle mm-wave signals has fueled a lot of research in the silicon IC design community to come up with 60GHz building blocks [1-4] and radios [5-9] that often use beamforming to improve the wireless link budget.

CMOS is obviously the most viable candidate to be used in future mass market consumer products, since it has proven to be fast enough and it allows for very complex flexible transceivers. Besides the IC technology, the technology for the antenna and the antenna interface should be cheap as well. In this work we demonstrate a beamforming receiver module based on a mm-wave receiver IC in 45nm digital CMOS. The functionality of the module, which contains two such ICs in a low-IF receiver, as well as four antennas on a PCB laminate and a waveguide to PCB transition, is demonstrated experimentally in a setup that emulates a wireless communication around 60 GHz. Before discussing the receiver module and the setup in Section III, Section II gives an overview of beamforming architectures realized in CMOS.

II. BEAMFORMING ARCHITECTURES

Beamforming in the receiver requires a programmable delay or phase shift followed by signal combination. In the transmitter the signal is split and then a phase shift or delay is applied to the different antenna paths. To minimize the duplication of functional blocks over the antenna paths, one could perform these operations as close as possible to the antenna, right after the low-noise amplifiers (LNAs) in the receiver, and just before the power amplifiers (PAs) in the transmitter. A programmable phase shift at RF can be made for example by first generating a quadrature version [7][10] of the RF signal after which the in-phase version (I) and the quadrature version (Q) are combined. Indeed, with a signal in I-Q format, one can make new in-phase and quadrature components I' and Q' that are phase shifted over an angle φ using

$$I' = \cos(\varphi)I - \sin(\varphi)Q$$
$$Q' = \sin(\varphi)I + \cos(\varphi)Q \qquad (1)$$

The two-antenna receiver demonstrated in [7] using a 65nm technology is based on this idea. With two LNAs, two RF phase shifters and a signal combiner this receiver consumes 78 mW per antenna paths with a power gain of 10 dB and a noise figure of 7.2 dB. The 45nm CMOS two-antenna receiver described in Section III.A is also based on RF phase shifting. It consumes less power at the expense of a larger noise figure. However, phase shifting and signal combination or split operations at mm-wave frequencies are hard to implement in bulk CMOS without a significant performance loss (insertion loss, increase of noise figure, …). Moreover, operations on signals at these frequencies are sensitive to small parasitics from on-chip interconnect and transistors, and any modeling error on these parasitics already yields a considerable change in the circuit performance at mm-wave frequencies.

Another phase shift approach is realized by mixing the signal in the different antenna paths with a phase shifted replica of the local oscillator (LO) signal. This has already been demonstrated in 90nm CMOS with a two-antenna receiver [9]. In this way the lossy phase shift operation is removed from the signal path. This LO phase shifting is less susceptible to amplitude variations and gives a limited phase noise degradation in the LO signal as long as the LO swing is kept high enough. An extension of the two-antenna receiver of [9] to four antennas in 45nm digital CMOS has been simulated. It is a direct downconversion architecture (see Figure 1. uses the LNA-mixer section described in [1] and the 56-67GHz PLL from [2]. The phase selectors are similar to the ones used in [9]. The four outputs from the quadrature VCO of the PLL are distributed to the four antenna paths via buffers (see Figure 1.) that can drive long transmission lines and two identical subsequent buffers that can again drive similar transmission lines and a load (next buffering stage or the phase shifters). The signal combination is performed at baseband by summing the output currents from the mixers. Simulations on this receiver (including layout parasitics) in 45nm digital CMOS give a noise figure at the output of the channel selection filter of 8dB per antenna path (including downconversion and signal combination) with a 26dB conversion gain and a total power consumption of 307mW (including frequency synthesis and LO distribution).

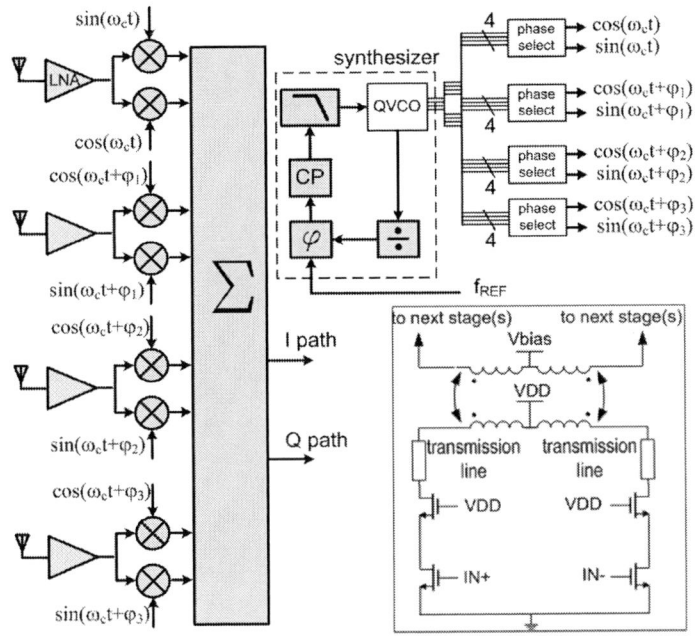

Figure 1.   Receive architecture with phase shifting in the LO path. The inset shows the buffer used in the distribution of the LO (both the I and the Q path).

While in the LO phase shifting architecture signal combination has been moved to baseband, one could go one step further and also perform phase shifting at baseband using equation (1). This idea has been demonstrated in the direct upconversion transmitter from [8]. The advantage of baseband phase shifting is its low area consumption as no inductors are needed and a lower sensitivity to small parasitics of a few femtofarads or picohenries.

III.   WIRELESS LINK WITH FOUR ANTENNAS

A wireless link (see Figure 2.) has been made with a four-antenna receiver that uses two 45nm CMOS receiver ICs in which beamforming at RF is implemented. These two ICs are mounted on a printed circuit board (PCB) made on a Nelco 4000-12 substrate. The PCB also contains four 60 GHz patch antennas and two commercial GaAs MMIC LO buffers. The LO signal is made with a frequency multiplier that is fed to the LO buffers via a transition from waveguide to microstrip and a power divider. The transmitter is built around an Arbitrary Waveform Generator (AWG) and a commercial 60GHz quadrature mixer. After upconversion by the mixer, the signal is amplified and sent through a coupler to a horn antenna. A picture of the link and a zoom on the receiver module are shown in Figure 3. Figure 4.

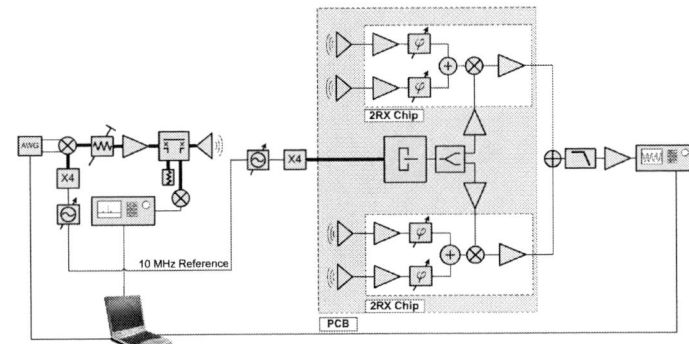

Figure 2.   Setup of a 60 GHz link containing a module on a PCB laminate.

A.   Two-antenna receiver IC in 45nm CMOS

The architecture of the two-antenna receiver IC designed in a digital low-power 45 nm CMOS technology is shown in Figure 7. . The LNA and mixer are described in [1]. The RF phase shifting is realized in two steps before downconversion. First, a quadrature all-pass filter (QAF), similar to the one of [10], generates two differential RF signals in quadrature. Secondly, 360° phase interpolation is achieved based on equation (1). The two paths' signals are then combined at the outputs of the phase selectors, before the mixer. This has two advantages: a combiner circuit is no longer necessary and only one mixer is required. The performance of the RF phase shifting is measured on the standalone IC, using off-chip, manually adjustable waveguide phase shifters. A change in the phase shift of the on-chip phase shifters influences the amplitude of the combined signal, which reaches a minimum when the signals at the output of the two phase shifters have an opposite phase (see Figure 8.).

Figure 3.   Picture of the wireless link from Figure 2.

Figure 4.   Detail of the wireless link: receiver.

The measured conversion gain of the receiver IC is around 15dB with a noise figure between 14 and 20dB from 55 to 63GHz with an IF frequency up to 1GHz and a power consumption of 60mW. The high noise figure of the receiver IC is due to the loss in the phase shifter and to the mismatch between the resonance frequencies of the different LC tanks used in the RF part of the chip. The many signal operations performed at mm-wave frequencies, each time requiring resonant circuits to boost the voltage swing, make the architecture sensitive to small parasitics and hence subject to large process tolerances. Therefore, architectures that perform a minimal amount of operations at mm-wave are preferred.

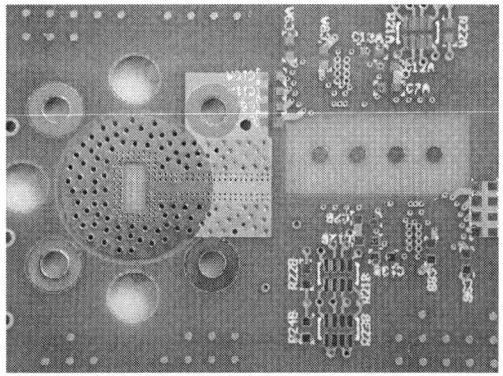

Figure 5.   Detail of the receiver PCB : antennas and backside of the waveguide to microstrip transition.

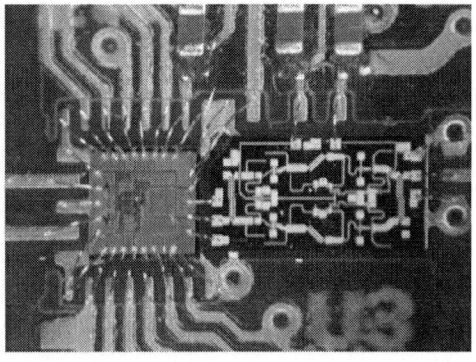

Figure 6.   Detail of the receiver PCB: two-antenna receiver chip and MMIC LO buffer.

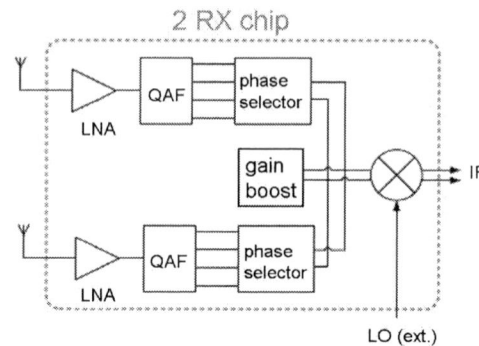

Figure 7.   Architecture of the two-antenna receiver IC in 45nm CMOS.

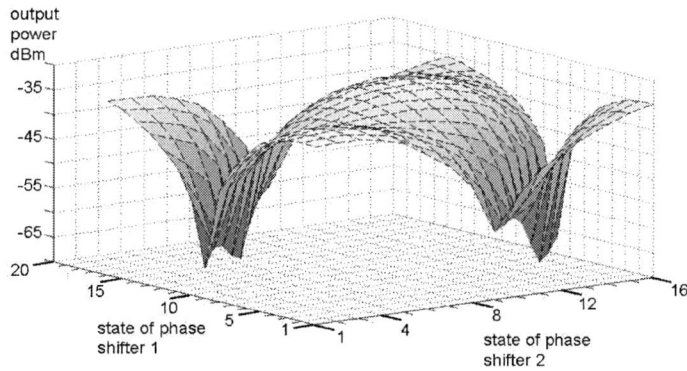

Figure 8.   Measured power of the combined signal after downconversion, as a function of the settings of the two variable phase shifters.

B.   Antenna array

An array of four patch antennas has been realized on the PCB stack that is composed of several Nelco N-4000-12 core layers and prepreg layers. The antenna is a stacked patch configuration with a rectangular patch on antenna layer 1 and a circular patch on top of the antenna layer 2 (see Figure 9.). A size of the matching stub of the microstrip line has a length of 530 µm. It stretches from the center point of the ground plane aperture until the end of the microstrip line. To increase the bandwidth of the slot antenna, the slot width is increased symmetrically from the slot's center to either end, to provide of dog-bone slot configuration. The dog-bone slot has a main width of 90 µm. The measured antenna gain of one element is shown in Figure 10. . It reaches 6 dB at 61 GHz. The gain of the antenna array is 7 dB, which is constrained by losses in the feeding network. Moreover, the array reaches maximum gain between 50 and 55GHz also due to array's feeding network better matching at these frequencies.

C.   Measurements on the link

The wireless 60GHz link uses Matlab both to transmit the digital data streams (compliant with the IEEE 802.15.3c standard) and to postprocess the received data (via in-house baseband algorithms). Some parameters of the link are given in Figure 11. The link can sustain a bit rate of at least 1.25Gbits/s (QPSK modulation can be used) with an acceptable bit error

978-1-4244-5190-6/09 $25.00 © 2009 IEEE

rate (between 10^{-5} and 10^{-4}, uncoded) up to a range (distance between transmitter and receiver) of at least 1 meter. Further, the effect of beam steering can be clearly seen when the settings of the on-chip phase shifters are adapted to the angle of incidence of the signal that is transmitted via the horn antenna (see Figure 11.).

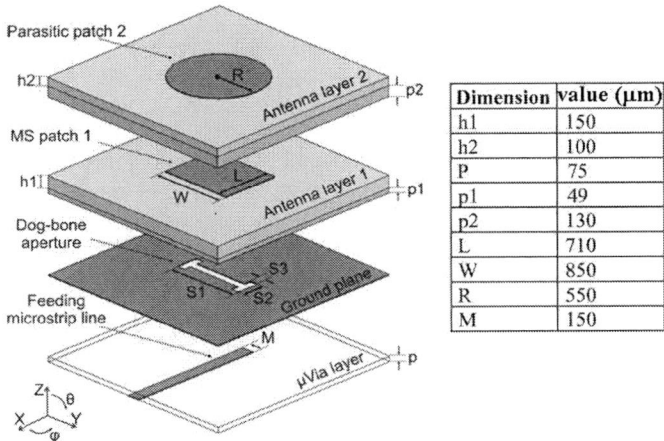

Dimension	value (μm)
h1	150
h2	100
P	75
p1	49
p2	130
L	710
W	850
R	550
M	150

Figure 9.   Buildup of one antenna element.

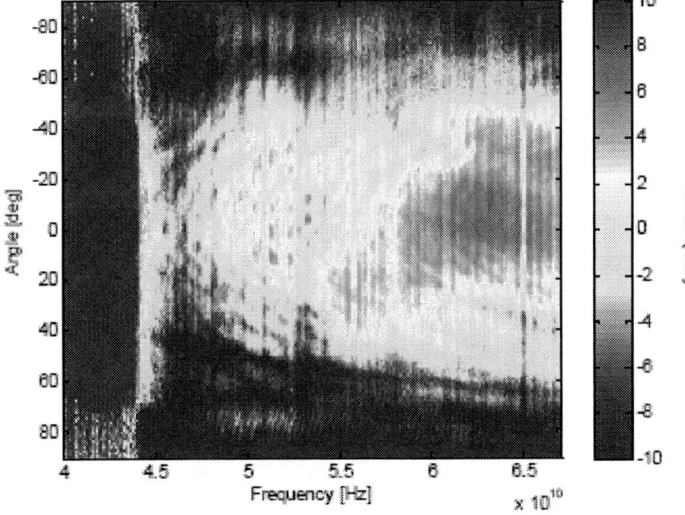

Figure 10. Measured antenna gain in the H-plane.

IV.   CONCLUSIONS

In a few years, consumer products for high-datarate communication at mm-wave frequencies are expected to conquer the market. These will most probably be based on CMOS as this allows for a highly complex transceiver at an affordable price. Beamforming architectures are emerging to control antenna arrays which are used to improve the link budget. The coming years' research in this domain will have to focus on lowering the power consumption of the demonstrated prototype circuits and on the baseband circuitry that needs to have a high bandwidth to deal with the high datarates required

for wireless communication at datarates above 1 Gbit per second.

Parameters	Values
Carrier frequency	57 GHz
IF (low-IF receiver)	500 MHz
RF bandwidth	1.0417 GHz
Symbol rate	833.3 Msymbols/s
Receiver sampling rate	2.5 GHz
Channel coding	no
Tx antenna gain	23 dB
Power of transmit signal	around 8/9 dBm
Preamble duration	3.072 μs
Payload duration	15.437 μs

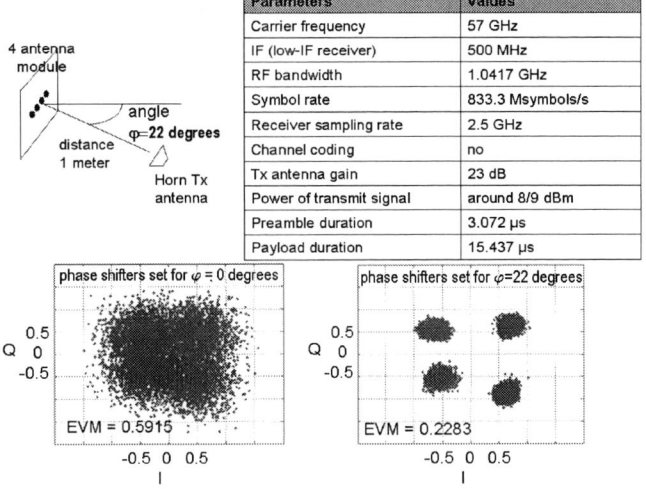

Figure 11. Measured constellation diagrams for QPSK data with the settings of the link given in the above table. The error vector magnitude (EVM) decreases significantly when the state of the phase shifters corresponds to the angle of incidence φ which is 22 degrees in this case.

REFERENCES

[1] J. Borremans, K. Raczkowski and P. Wambacq, "A digitally-controlled compact 57-66GHz receiver front-end for phased-arrays in 45nm digital CMOS", International Solid-State Circuits Conference (ISSCC), pp. 492-493, February 2009.

[2] K. Scheir, G. Vandersteen, P. Wambacq and Y. Rolain, "A 57-66 GHz PLL in 45nm digital CMOS", International Solid-State Circuits Conference (ISSCC), pp. 494-495, February 2009.

[3] T. Yao, M.Q. Gordon, K.K.W. Tang, K.H.K. Yau, M-T Yang, P. Schvan, and S. P. Voinigescu, "Algorithmic Design of CMOS LNAs and PAs for 60-GHz Radio," IEEE Journal of Solid State Circuits. Vol.42, No.5, pp.1044-1057, May. 2007.

[4] A. Cathelin et al., "Design for Millimeter-wave Applications in Silicon Technologies", Proceedings European Solid-State Circuits Conference, pp. 464-471, September 2007.

[5] S. Reynolds et al., "A Silicon 60-GHz Receiver and Transmitter Chipset for Broadband Communications", IEEE J. Solid-State Circuits, Vol. 41, No. 12, pp. 2820-2831, December 2006.

[6] G. Glisic et al., "A Fully Integrated 60 GHz Transmitter Front-End with a PLL, an Image-rejection Filter and a PA in SiGe", Proceedings European Solid-State Circuits Conference, pp. 242-245, September 2008.

[7] Y.Yu et al., "A 60GHz Digitally-Controlled RF-beamforming receiver Front-end in 65nm CMOS", Proceedings RFIC, pp. 211-214, June 2009.

[8] S. Kishimoto, N. Orihashi, Y. Hamada, M. Ito and K. Maruhashi, "A 60GHz Band CMOS Phased Array Transmitter Utilizing Compact Baseband PhaseShifters", Proceedings RFIC, pp. 215-219, May 2009.

[9] K. Scheir, S. Bronckers, J. Borremans, P. Wambacq and Y. Rolain, "A 52GHz Phased-Array Receiver Front-End in 90nm Digital CMOS", IEEE Journal of Solid-State Circuits, Vol. 43, No. 12, pp. 2651-2659, December 2008.

[10] K.-J. Koh, and G. M. Rebeiz, "0.13-μm CMOS Phase Shifters for X-, Ku-, and K-Band Phased Arrays", IEEE Journal of Solid-State Circuits, vol. 42, No. 11, November 2007.

A Passive W-Band Imager in 65nm Bulk CMOS

A. Tomkins[1], P. Garcia[2], and S. P. Voinigescu[1]

1) EDWARD S. ROGERS SR. DEPT. OF ECE, UNIVERSITY OF TORONTO, TORONTO, ON M5S 3G4, CANADA

2) STMICROELECTRONICS, 850 RUE JEAN MONNET, F-38926 CROLLES, FRANCE

Abstract - **A passive imager operating in the W-band around 90GHz has been realized in a digital 65nm CMOS process. The imager, occupying only 0.41mm², integrates an SPDT switch with 4.2dB loss and 25dB isolation, a 5-stage telescopic cascode LNA with 27dB gain at 90GHz, and a W-band square-law detector, all consuming less than 33mA from 1.2V. The imager, when measured without the input SPDT, has a peak responsivity of over 200kV/W, and a minimum NEP of less than 0.1pW/√Hz. With the input switch, it achieves a 90kV/W responsivity and an NEP of 0.2pW/√Hz. This receiver represents the highest frequency imager to be implemented in standard CMOS with this level of integration.**

I. INTRODUCTION

W-band imagers are attractive due to their potential for high resolution operation and their ability to penetrate various materials. Typical imagers at these frequencies have been realized using III/V technologies, with InP HEMT LNAs in conjunction with planar doped barrier or Schottky detectors [1]. Recent publications have investigated the use of zero-bias backward tunnel diodes [1], and even SiGe HBTs [2].

Nanoscale CMOS, as in this paper, represents an opportunity to leverage the economies of scale, and the potential for large-scale wafer-level integration. A modern CMOS process utilizes 300mm wafers and can potentially permit the integration of thousands of imaging elements on a single wafer.

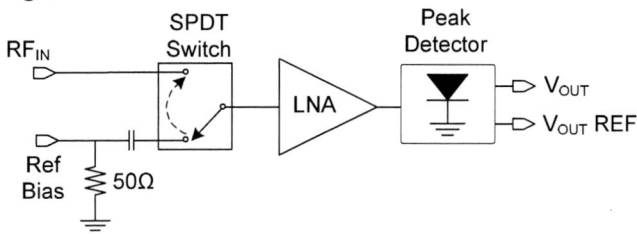

Figure 1: System schematic

II. SYSTEM DESCRIPTION

The proposed receiver, shown in Figure 1, integrates an LNA centered at 90GHz with greater than 27dB gain and less than 7dB noise figure, a square-law detector that utilizes a differential topology to suppress system and power-supply noise, an on-chip SPDT switch with 4.2dB insertion loss centered at 90GHz and a reference 50-Ohm resistor for calibration.

The low-noise amplifier consists of 5 cascode stages that use lumped inductors for tuning and series caps for inter-stage matching, and is shown in Figure 2. The input stage is simultaneously noise and impedance matched to 50-Ω using the transformer feed-back technique described in [3]. Modest linearity requirements for a passive imager mean that the amplifier was designed for high-gain without excessive power-consumption. From 1.2V, the LNA consumes only 30mA and has over 27dB gain with a 3-dB bandwidth exceeding 10 GHz.

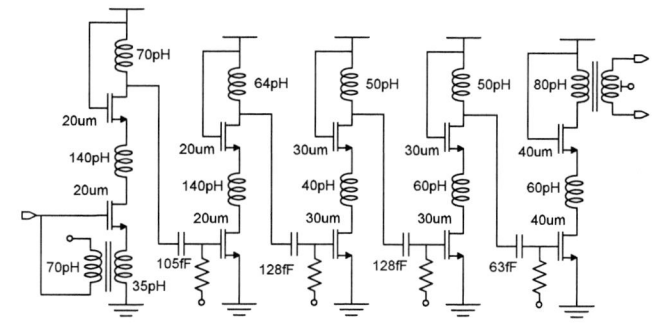

Figure 2: Schematic of the LNA.

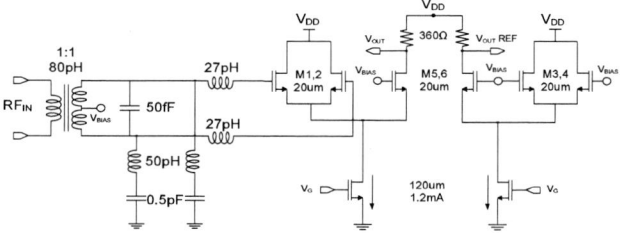

Figure 3: Schematic of the square-law detector.

The square-law detector utilizes a differential topology similar to [4] and is shown in Figure 3. The input transformer, which is also the output load of the LNA, performs a single-ended to differential conversion. The input of the detector is matched to 50-Ω using a shunt-series matching network in addition to a tuning capacitor across the secondary coil of the transformer. The common-mode signal at the source of the input transistors, which is proportional to the signal amplitude, is amplified by the common-gate amplifier formed by transistor M5 and the output load resister. A duplicate input pair and CG amplifier are also included on-chip, in perfect layout symmetry, in order to ensure good power supply rejection which can seriously impact the sensitivity of single-ended detectors.

The SPDT switch, shown in Figure 4, uses lumped elements for matching, and implements shunt-shunt switches as the SPST unit-cell [5]. The inclusion of an on-chip SPDT switch permits the calibration of the detector using "Dicke" switching, in which the input to the LNA is switched between the antenna port and an on-chip 50Ω reference resistor.

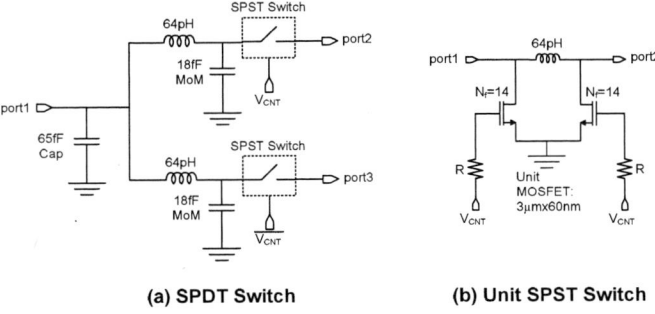

(a) SPDT Switch   **(b) Unit SPST Switch**

Figure 4: Schematic of the 90GHz SPDT switch.

III. FABRICATION AND MEASUREMENT RESULTS

The full receiver, Figure 5, and a stand-alone detector breakout, Figure 6, were fabricated and occupy 865x470um^2 and 420x495um^2 respectively. Additionally, a version without the input switch was fabricated (which is not shown), but was measured in order to characterize the additional loss produced by the SPDT switch. The circuits were fabricated in the GP (general-purpose) variant of a 65nm bulk CMOS process with a standard digital back-end and MiM capacitors. The n-MOSFETs with a minimum gate-length of 45nm have an f_T and f_{MAX} of 180 GHz and 250 GHz, respectively, the latter depending strongly on layout geometry.

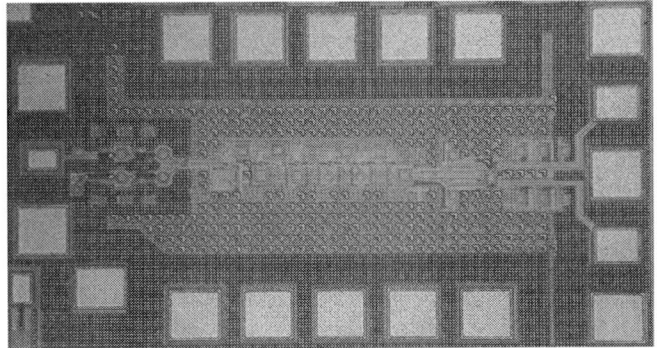

Figure 5: Die photo of the full receiver. The circuit occupies 865x470um^2 including all pads.

The LNA was measured using a Wiltron 360B network analyzer up to 94GHz, and using an Agilent NFA with a single side-band down-converter between 75-88.5GHz. Shown in Figure 7, the measurements of the gain produced by the VNA are in agreement with those from the noise-figure measurement. A peak gain of over 27dB is measured at 88GHz, coinciding with the minimum noise figure of 6.8dB. The return loss of the LNA is better than 10dB from below 70GHz. The return loss of the full receiver and the

stand-alone detector are shown in Figure 8, with both circuits well matched around 90GHz. The measured insertion loss and isolation of the SPDT switch is shown in Figure 9. The insertion loss is better than -5dB over the whole W-band, and the minimum measured insertion loss is -4.2dB at 94GHz. Isolation is flat at approximately -25dB.

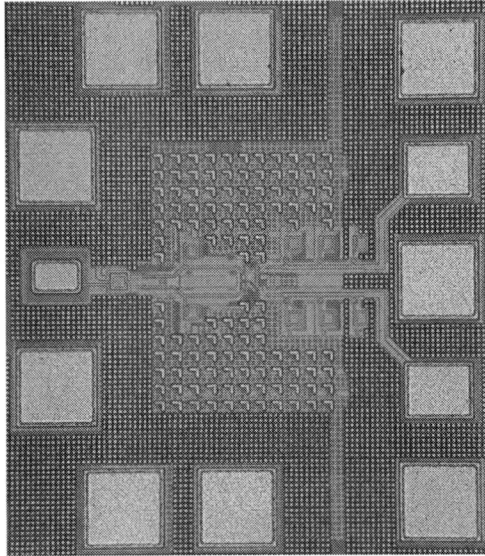

Figure 6: Die photo of the stand-alone detector. The circuit occupies 420x495um^2 including all pads.

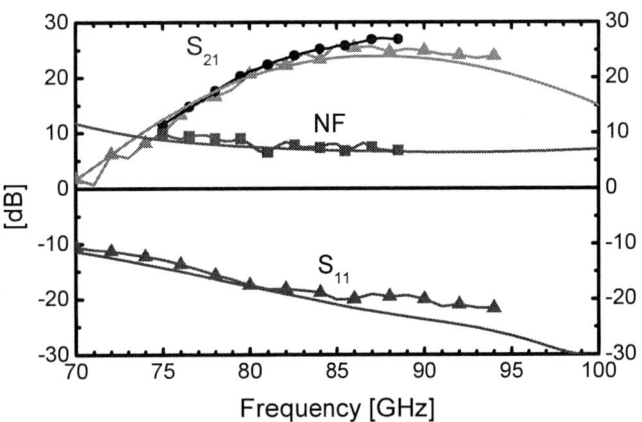

Figure 7: Measured (symbols) and simulated (line) LNA S-parameters and NF.

The measurement setup for testing the receiver is shown in Figure 10. An external low-noise pre-amplifier was used to boost low signal levels, but was factored out of all results. The amplifier had limited bandwidth and restricted the maximum measurement frequency. The amplifier was also used to drive the 50Ω test ports on the spectrum analyzer and oscilloscope.

An examination of the low-frequency noise at the output of the circuit was carried out by sweeping the detector reference current and measuring the noise voltage at various frequencies. The results, shown in Figure 11, demonstrate that the output noise voltage is still decreasing with frequency, and is thus still within the 1/f noise range of the

978-1-4244-5190-6/09 $25.00 © 2009 IEEE   73

n-MOSFET. Unfortunately, the low sensitivity of the measurement equipment and the limited bandwidth of the low-noise pre-amplifier restrict measurements significantly beyond 400 kHz.

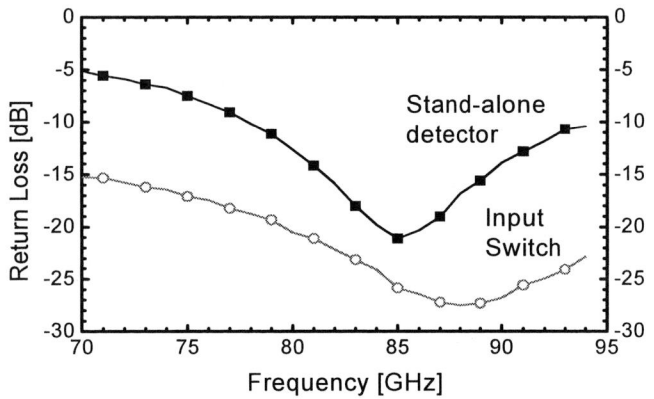

Figure 8: Measured return loss for input switch and stand-alone detector

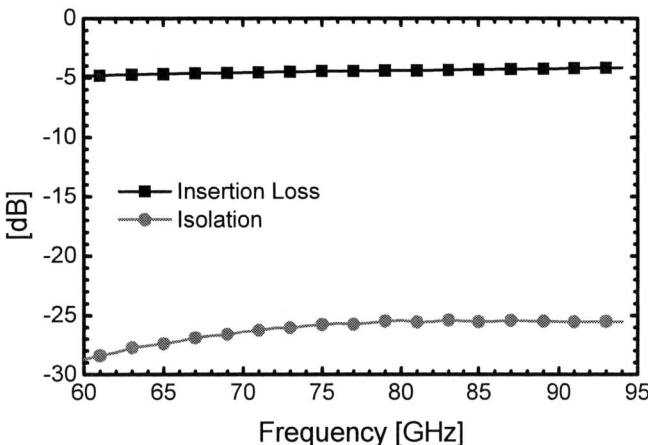

Figure 9: Measured performance of the SPDT switch.

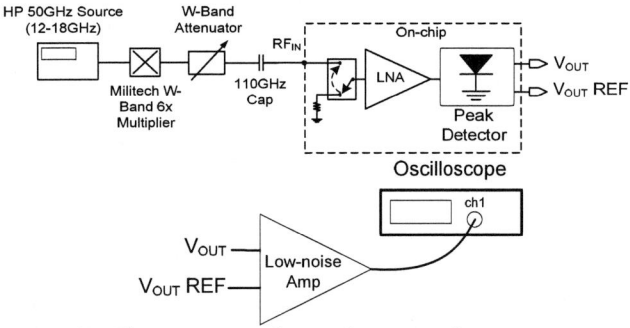

Figure 10: The experimental setup for testing the receiver system.

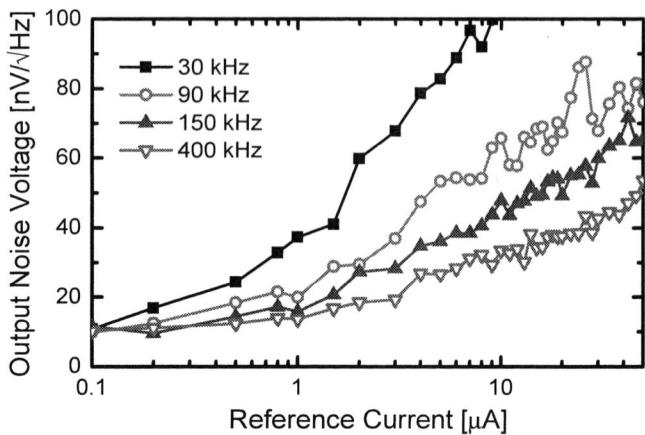

Figure 11: Measurement results of the output noise voltage as a function of the detector reference current for several frequency points.

The results from Figure 11 also indicate that biasing the detector at a minimal current density will reduce the output noise voltage, but this has the effect of degrading the system responsivity (the output voltage divided by the input RF power). A plot of the system noise-equivalent power (NEP), which is the output noise voltage divided by the responsivity, can permit the optimal biasing, is shown in Figure 12, and allows the balancing of the output noise and system responsivity. The measurement was carried out on the version of the receiver without the input SPDT switch to minimize losses, and was performed at 88GHz. The results indicate that biasing at 1-3μA produces the best system NEP.

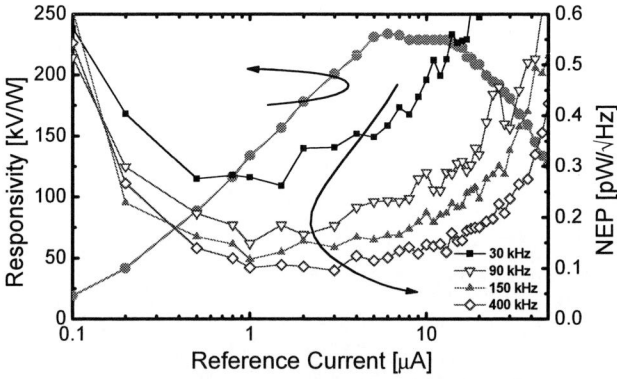

Figure 12: The measured system (LNA+detector) responsivity (at 88GHz) and NEP over detector bias currents, and output frequencies.

Using a bias current of 3μA, and making all noise measurements at 400 kHz, the linearity and frequency response of the receiver are presented in Figure 13 and Figure 14. The linearity measurements, performed at 88GHz, indicate that the receiver requires signal levels below -40dBm for optimal operation. Peak results for responsivity and NEP are 210kV/W and 0.09pW/√Hz. Over frequency, the 3dB bandwidth can be seen to extend from 81-93GHz.

978-1-4244-5190-6/09 $25.00 © 2009 IEEE

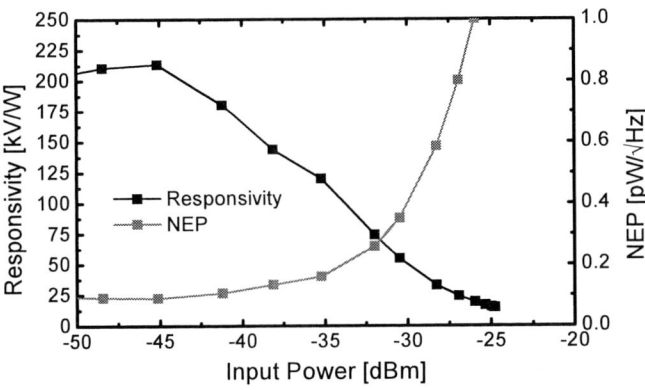

Figure 13: Measured responsivity and NEP at 88GHz as a function of input power.

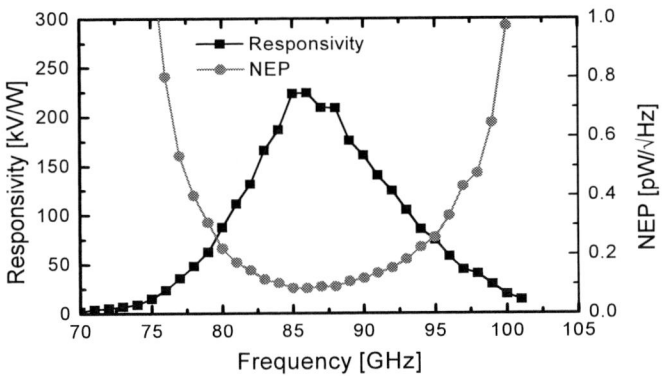

Figure 14: Measured responsivity and NEP over frequency. Measured without the SPDT switch.

Measurements carried out on the receiver with SPDT switch (Figure 15) show a peak responsivity of approximately 90kV/W, and an NEP of 0.2pW/√Hz. The 3dB bandwidth is unchanged at 81-93GHz. The receiver degradation due to the inclusion of the SPDT switch is in-line with the measured losses of the stand-alone SPDT switch (from Figure 9). The responsivity is approximately 40% that of the receiver without SPDT, which corresponds to an approximate 4dB drop. The receiver degradation due to the SPDT switch versus frequency is compared with the measured SPDT losses in Figure 16.

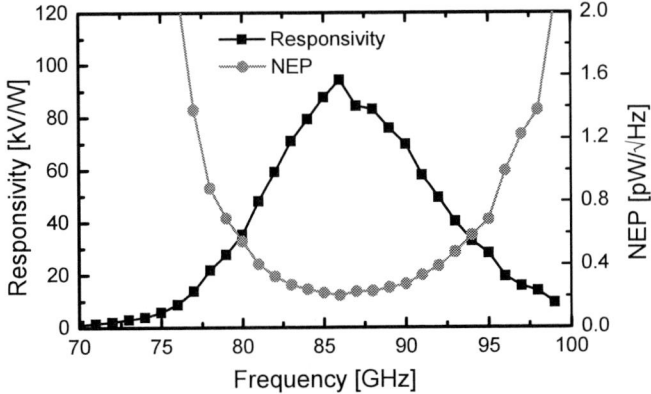

Figure 15: Measured responsivity and NEP over frequency of the receiver with SPDT switch.

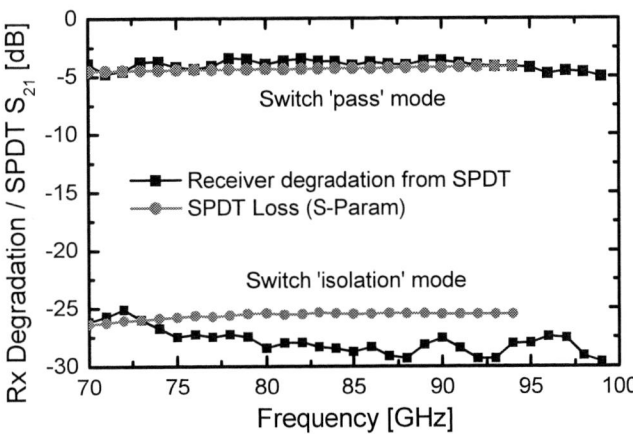

Figure 16: Measured receiver degradation due to the inclusion of the SPDT switch compared to the measured SPDT switch losses from S-parameters.

Conclusion

A fully integrated W-band passive receiver has been implemented in a bulk 65nm CMOS process. The imager has a responsivity of >90kV/W and an NEP of 0.2pW/√Hz with an integrated SPDT switch, and >200kV/W and <0.1pW/√Hz without. The system integrates the highest reported gain CMOS LNA in the W-band, with over 27dB gain at 90GHz. Occupying only 0.41mm² and consuming less than 33mA from 1.2V, this chip is the highest frequency imager reported in CMOS, and represents the first step towards the unique CMOS prospect of having large wafer-scale integration of imagers, potentially containing thousands of imaging elements.

Acknowledgements

This work was funded by NSERC. We would like to thank Bernard Sautreuil and STMicroelectronics for fabrication. We would also like to thank Jaro Pristupa for CAD support, and ECTI for test equipment.

References

[1] J. Lynch and et al., "Passive Millimeter-Wave Imaging Module With Preamplified Zero-Bias Detection," *IEEE MTT*, vol. 56, pp. 1592-1600, July 2008.

[2] J. W. May and G. M. Rebeiz, "High-Performance W-Band SiGe RFICs for Passive Millimeter-Wave Imaging," in *RFIC 2009*, Boston, 2009.

[3] M. Khanpout and et al., "A Wideband W-Band Receiver Front-End in 65-nm CMOS," *IEEE JSSC*, vol. 43, no. 8, pp. 1717-1730, August 2008.

[4] H. Tran and et al., "6-kOhm, 43-Gb/s Differential Transimpedance-Limiting Amplifier with Auto-Zero Feedback and High Dynamic Range," *IEEE JSSC*, vol. 39, pp. 1680-1689, November 2004.

[5] A. Tomkins and et al., "A 94GHz SPST Switch in 65nm Bulk CMOS," in *IEEE CSICS 2008*, Montery, 2008.

A Compact Cascode Power Amplifier in 45-nm CMOS for 60-GHz Wireless Systems

Torgil Kjellberg[1,2], Morteza Abbasi[1], Mattias Ferndahl[1], Anton de Graauw[3], Edwin v.d. Heijden[3], and Herbert Zirath[1]

[1]Microwave Electronics Laboratory, Chalmers University of Technology, 41296 Göteborg, Sweden
[2]Chalmers Industrial Technologies, Göteborg, Sweden
[3]NXP Semiconductors, Research, Eindhoven, the Netherlands

Abstract — **This paper presents a power amplifier in 45nm CMOS technology operating in the 50 to 60-GHz frequency range. It uses a single-ended topology with two cascode stages. A gain of 19.6 dB is obtained at 54 GHz where the gain peaks, and 17 dB at 60 GHz. The measured output-referred 1-dB compression point is 8.7 dBm at 60 GHz and a supply voltage of 2.1 V, and the saturated output power is 13.5 dBm. The bias conditions are shown to be reliable through lifetime measurements. The active chip-area measures 160 μm x 110 μm.**

Index Terms — **Power amplifiers, CMOSFET amplifiers, 45-nm CMOS technology, 60 GHz, reliability.**

I. INTRODUCTION

The world-wide unlicensed band around 60 GHz is highly attractive for short-range high-bandwidth applications such as wireless HDMI-cable replacement and ultra-fast file transfer. The development of new deep-submicron technology nodes with f_{max} in the order of 200 GHz have made the use of standard digital CMOS technologies viable for designing transceivers at mm-wave frequencies. For mass-market deployment, system-on-chip integration is required where the digital parts usually constitute the major part of the die area and thus dictate the technology to be used. This drives the need for using the smallest available technology nodes, such as 45 nm, in order to reduce die size and cost in large volume. The power amplifier (PA) is to a large extent limiting the transceiver performance in terms of transmission range, linearity and required supply power. Therefore, PAs with high output power, linearity and efficiency are key elements to produce successful integrated solutions. A number of 60-GHz power amplifiers (PAs) manufactured in 90-nm and 65-nm CMOS technologies have been presented [2-7], but so far very little in 45 nm [1].

In this paper, we present a PA fabricated in a standard 45-nm low-power 7-metal-layer CMOS technology. Our goal has been to explore the use of 45-nm CMOS technology for the design of a compact and reliable 60-GHz PA suitable for system-on-chip integration. The design is based on lumped elements and it is the most compact 60-GHz PA presented to date. We also present lifetime measurements to demonstrate reliable operation at the bias conditions chosen for the PA.

II. CIRCUIT DESIGN

The PA consists of two cascode stages as seen in the simplified schematic in Fig. 1. The cascode topology has the advantage over a common-source (CS) one of achieving both higher gain and reverse isolation. However, the requirement on voltage headroom is higher for a cascode stage. The common-gate (CG) transistors are placed in isolated p-wells which are connected to the sources in order to allow higher supply voltages without obtaining excessive gate-bulk voltages potentially causing gate-oxide breakdown. Although the isolated p-well increases the parasitic capacitance at the drain node of the gain transistors, no substantial deterioration of the gain is noticed.

The gates of the CG transistors are biased to the supply voltage for simplicity and are strongly decoupled for maximum gain. All transistors have a total gate width of 100 μm and are designed for biasing at a current density of approximately 0.3 mA/μm where f_{max} peaks, which is in line with previous studies [7]. Both stages of the amplifier use the same device size since maximum linearity was prioritized over efficiency. This assures that gain compression mainly is caused by the last stage so that full output power capability of the last stage is obtained.

Compact, lumped-element LC networks are realized with spiral inductors and inter-digitated fringing-field capacitors for input, interstage and output impedance matching. While the input of the PA is conjugate matched to the source impedance, the matching network on the output is optimized to achieve

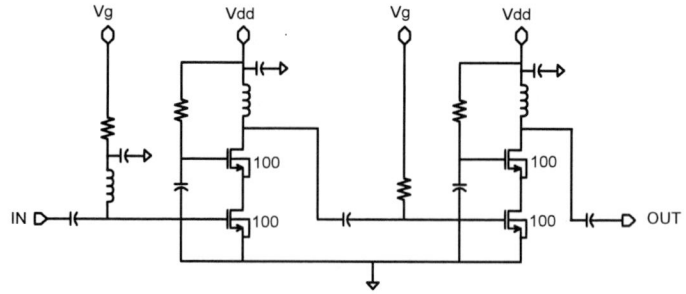

Fig. 1.   Simplified schematic of PA.

978-1-4244-5190-6/09 $25.00 © 2009 IEEE

maximum output power. The inter-stage matching network is optimized for maximum gain.

Scalable models are used for both active and passive elements throughout the design. Although only lumped elements are used in the design, all interconnects in the RF signal path are modeled as transmission lines.

A photo of the chip is shown in Fig. 2. The die size is 700x540 μm out of which the active area constitutes approximately 160x110 μm including decoupling capacitors.

III. MEASURED AND SIMULATED RESULTS

A. Small-signal measurements

All measurements were done using wafer probing. The S-parameters were measured using a 67-GHz network analyzer. Fig. 3 shows the measured and simulated S-parameters for the PA at a supply voltage of 2.1 V and 30 mA of drain current per stage. A peak gain of 19.6 dB was obtained at 54 GHz. Return loss is below -10 dB for both input and output terminals between 47 and 60 GHz. The 3dB bandwidth is 12.7 GHz (from 48.2 to 60.9 GHz).

The PA was designed for maximum gain at 60 GHz but was shifted down in frequency due to model discrepancies. After measurements, the circuit model was refined and a good agreement with simulations was obtained. In particular, more accurate modeling of interconnects was done and previous models of some passive components were updated.

B. Large-signal measurements

Large-signal measurements were performed using a 67 GHz signal generator and a power meter with a WR15 waveguide interface. The available source power was measured and the losses from cables, probes and transitions were carefully taken into account to calibrate the input and output power at the probe-chip interfaces.

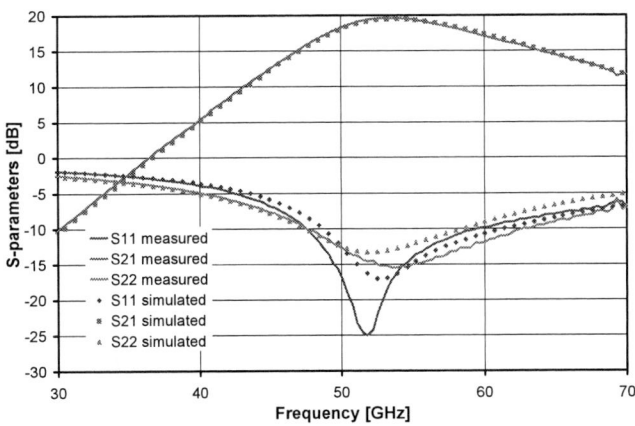

Fig. 3. Measured (solid lines) and simulated (symbols) S-parameters at V_{dd}=2.1 V and I_{dstot}=60 mA.

The output power was measured vs. input power, and is plotted in Fig. 4 together with the large-signal transducer gain and power added efficiency (PAE) at 60 GHz for supply voltages of 1.8 V and 2.1 V. The drain currents were biased for maximum gain. The measured values of P_{1dB}, P_{sat} and PAE are 8.7 dBm, 13.5 dBm, and 13.4 %, respectively at 2.1 V supply. At 1.8 V, the obtained values are 7.6 dBm, 12.0 dBm and 12.3 %. In Fig. 5, P_{1dB} and P_{sat} are shown for frequencies between 50 and 60 GHz.

C. Reliability and lifetime

Accelerated lifetime measurements at room temperature were made by stressing the PA with excessive supply voltage at an RF input power causing a gain compression of approximately 1dB. The transistor failure is assumed to be caused by channel hot electron degradation and the lifetime is estimated with a simple but commonly used reliability model [8]. In our case, lifetime is defined as the time when the drain current of the output stage of the PA, which is subject to a higher RF voltage stress than the driver stage, has dropped by 10%. The lifetime, τ, is assumed to follow the exponential relationship:

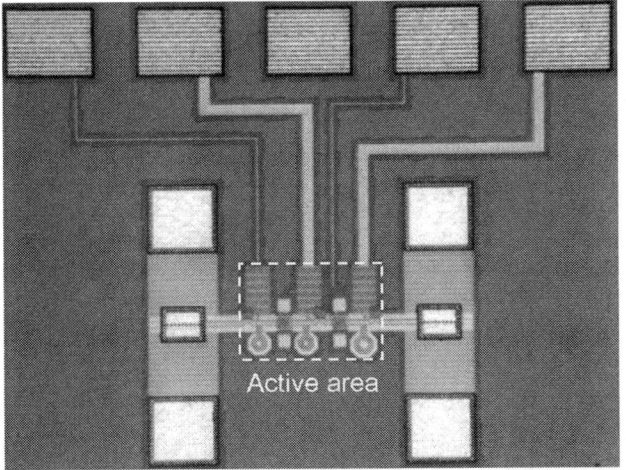

Fig. 2. Micrograph of PA. Die size is 700x540μm whereof 160x110μm is active area.

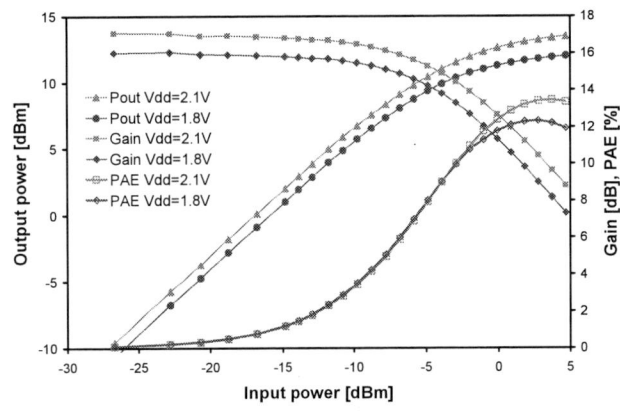

Fig. 4. Output power, transducer gain and PAE at 60 GHz for V_{dd}S=2.1V (I_{dstot}=64 mA) and V_{dd}=1.8V (I_{dstot}=57mA).

978-1-4244-5190-6/09 $25.00 © 2009 IEEE

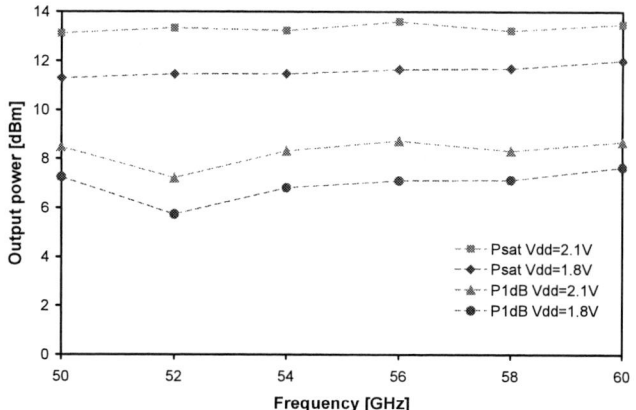

Fig. 5. Measured P_{1dB} and P_{sat} vs. frequency at V_{dd}=2.1V (I_{dstot}=64 mA) and 1.8V (I_{dstot}=57 mA).

$$\tau = c_1 \cdot e^{\left(\frac{c_2}{V_{dspeak}}\right)} \qquad \text{(Eq. 1)}$$

where c_1 and c_2 are fitting parameters. $V_{ds,peak}$ is the peak drain-source voltages across the CS transistor in the output stage. The degradation of this transistor is the main cause for the drain current reduction. At a given supply voltage and RF input power, $V_{ds,peak}$ is obtained from simulations. The simulated waveforms of V_{ds} across the CS and CG devices at constant input power are shown in Fig. 6 for three different supply voltages. For V_{dd}>2.1 V, $V_{ds,peak}$ is larger for the CS than for the CG device. At V_{dd}=2.1 V, the DC-voltage is split between the devices such that none of them exceed the rating of 1.1 V.

Compression occurs first for the CG device, which is seen as a tendency of clipping at the voltage minimum when the transistor is driven into the linear region. This is a result of the biasing method where the gate of the CG transistor is biased at V_{dd}, and is the reason for the soft gain compression with a comparatively large difference between P_{1dB} and P_{sat}. A more abrupt compression of the output signal occurs if both devices compress simultaneously.

The DC voltage across the cascode stage is estimated to be approximately 0.2 V less than the supply voltage at 60 mA total bias current due to DC voltage drop on the chip.

The drain current degradation with time is shown for four samples at different supply voltages in Fig. 7. Since the gate voltage for the CS device was kept constant and V_{ds} across the device increases with the supply voltage, the initial drain current also increases with supply voltage.

Fig. 8 shows the measured lifetimes and the predicted lifetime for two different bias cases. At V_{dd}=2.1 V, the predicted lifetime is about 5000 hours of continuous operation, which is adequate for many consumer applications. At V_{dd}=1.8 V, the predicted lifetime exceeds 10 years even if uncertainties in the extrapolation are taken into account. For those time scales, other effects not studied here could, of course, be of more importance.

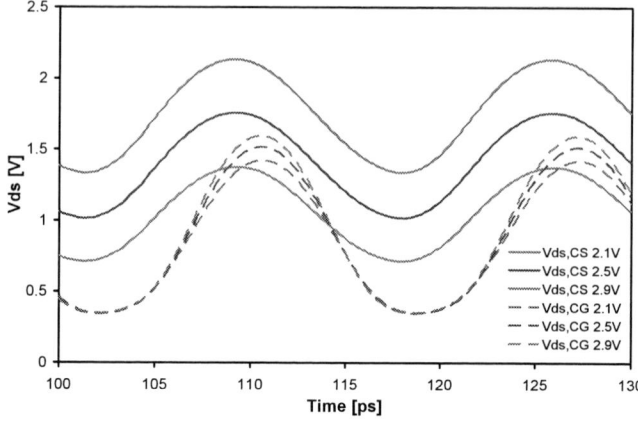

Fig. 6. Simulated V_{ds} waveforms for CS and CG transistors in the output stage ar V_{dd}=2.1V, 2.5V, and 2.9V. The gate voltage of the CS transistors is kept constant and P_{in}=-9 dBm, which is close to the 1dB compression point.

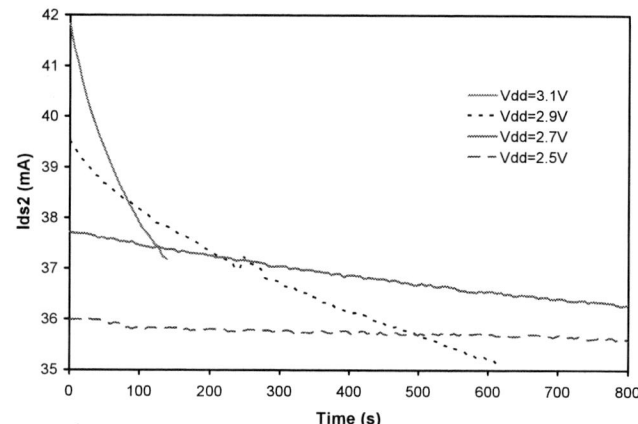

Fig. 7. Drain current degradation of the second stage vs. time for four samples at different supply voltages. Measurements were performed at room temperature.

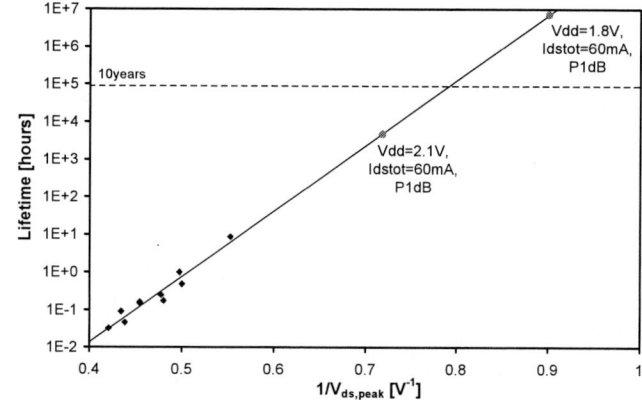

Fig. 8. Accelerated lifetime measurements at room temperature for 10 samples biased at different supply voltages from 2.5 to 3.1V and predicted lifetime for 1.8V and 2.1V supply.

978-1-4244-5190-6/09 $25.00 © 2009 IEEE

TABLE I.   COMPARISON BETWEEN RECENTLY PUBLISHED CMOS PAs FOR 60 GHz .

Ref.	Topology	Gain [dB]	P_{1dB} [dBm]	P_{sat} [dBm]	PAE_{peak} [%]	Die size (active area) [mm^2]	Technology	Supply voltage [V]	Estimated lifetime [hours]
[1]	2 stage Push-pull, CS-FETs	6	11	13.8	7	0.056	45 nm	1.1	n/a
[2]	3 stage differential, CS-FETs	15.4	2.5	11.5	9	0.053	65 nm	1.1	n/a
[3]	3 stage Cascode FETs	30	10.3	13.6	12.6	0.33	90 nm	1.8	n/a
[4]	3 stage differential, CS-FETs	15	10.2	12.2	19.3	0.15	90 nm	1.2	n/a
[5]	2 stage differential, CS-FETs	5.5	9	12.3	8.8	0.25	90 nm	1	n/a
[6]	3 stage cascode, distr. combining	26	10.5	14.5	10.5	0.64	90 nm	1.8	n/a
This work	2 stage Cascode-FETs	17	8.7	13.5	13.4	0.018	45 nm	2.1	5000
This work	2 stage Cascode-FETs	16	7.6	12	12.3	0.018	45 nm	1.8	>10 years

IV. CONCLUSIONS

A two-stage cascode PA fabricated in 45 nm CMOS technology and operating in the 50-60 GHz range has been presented. The active area is the smallest for 60-GHz PAs reported to date, making it suitable for system-on-chip integration where a minimum die size is desirable. Accelerated lifetime measurements show reliable operation at bias conditions where state-of-the-art performance is obtained. At 60 GHz, a linear gain of 17 dB, a P_{1dB} of 8.7 dBm, a P_{sat} of 13.5 dBm and a PAE of 13.4 % were obtained together with a predicted lifetime of 5000 hours.

ACKNOWLEDGMENT

The authors would like to thank Dr. Raf Roovers from the IRFS group at the Research Department of NXP Semiconductors for supporting this work.

REFERENCES

[1] K. Raczkowski, S. Thijs, W. de Raedt, B. Nauwelaers, and P. Wambacq, "50-to-67GHz ESD-Protected Power Amplifiers in Digital 45nm LP CMOS," in IEEE Int. Solid-State Circuits Conference Dig. Tech. Papers., Feb. 2009, pp. 382-383.

[2] W. L. Chan, J. R. Long, M. Spirito, and J. J. Pekarik, "A 60GHz-Band 1V 11.5dBm Power Amplifier with 11% PAE in 65nm CMOS," in IEEE Int. Solid-State Circuits Conference Dig. Tech. Papers., Feb. 2009, pp. 380-381.

[3] J.-L. Kuo, Z.-M. Tsai, K.-Y. Lin, and H. Wang, "A 50 to 70 GHz Power Amplifier Using 90 nm CMOS Technology," IEEE Microwave and Wireless Components Letters, vol. 19, pp. 45-47, 2009.

[4] T. LaRocca and M. C. F. Chang, "60GHz CMOS differential and transformer-coupled power amplifier for compact design," in IEEE RFIC Radio Freq. Integr. Circuits Symp. Dig., 2008, pp. 65-68.

[5] D. Chowdhury, P. Reynaert, and A. M. Niknejad, "A 60GHz 1V + 12.3dBm Transformer-Coupled Wideband PA in 90nm CMOS," in IEEE Int. Solid-State Circuits Conf. Dig. Tech. Papers, Feb. 2008, pp. 560-635.

[6] Y.-N. Jen, J.-H. Tsai, T.-W. Huang, and H. Wang, "Design and analysis of a 55-71-GHz compact and broadband distributed active transformer power amplifier in 90-nm CMOS process," IEEE Trans. Microw. Theory Tech., vol. 57, no.7, pp. 1637-1646, July 2009.

[7] T. Yao, M. Q. Gordon, K. K. W. Tang, K. H. K. Yau, M.-T. Yang, P. Schvan, and S. P. Voinigescu, "Algorithmic design of CMOS LNAs and PAs for 60_GHz radio," IEEE J. Solid-State Circuits, vol. 42, pp. 1044-1057, May 2007.

[8] F. J. Guarin, G. La Rosa, Z. J. Yang, and S. E. Rauch III, "A practical approach for the accurate liftime estimation of device degradation i deep sub-micron CMOS technologies," in IEEE Int. Conference on Devices, Circuits and Systems, April, 2002.

978-1-4244-5190-6/09 $25.00 © 2009 IEEE

A 2.5-V Low-Reference-Voltage, 2.8-V Low-Collector-Voltage Operation, HBT Power Amplifier for 0.8-0.9-GHz Broadband CDMA Applications

Kazuya Yamamoto, Atsushi Okamura*, Takayuki Matsuzuka, Yutaka Yoshii, Nobuyuki Ogawa,

Masatoshi Nakayama, Teruyuki Shimura, and Naohito Yoshida

High-Frequency Optical Device Works, Mitsubishi Electric Corporation, 4-1 Mizuhara, Itami, Hyogo 664-8641, Japan
*) Sun-A Microelectronics Corporation, Japan
Tel: +81-72-780-3781, Fax: +81-72-780-2690, **E-mail: Yamamoto.Kazuya@bk.MitsubishiElectric.co.jp**
High-Frequency and Optical Device Works, Mitsubishi Electric Corporation,
4-1 Mizuhara, Itami, Hyogo 664-8641, Japan   Tel: +81-72-780-3781, Fax: +81-72-780-2690

Abstract—**This paper describes circuit design and measurement results of a pure HBT MMIC power amplifier module (PA) operating with a 2.5-V low reference (V_{ref}) and 2.8-V low collector supply voltages (V_{cc}). While covering 824-925-MHz broadband CDMA operation at 2.8-V V_{cc}, the PA allows a 1.1-V low V_{cc} and 18-dBm Pout operation. This is realized by an on-chip step quiescent current selector monitoring collector voltage. Measurement results under the 2.8/1.1-V V_{cc} and 2.5-V V_{ref} bias conditions show that the PA meets J-/W-CDMA power and distortion specifications sufficiently over a wide temperature range from −20 to 85°C while operating over a wideband from 824 to 925 MHz. For J-CDMA (IS-95B) modulation, the PA can deliver a 28-dBm P_{out}, a 36% PAE, and a −50-dBc ACPR, while a 29-dBm P_{out}, a 38% PAE, and a −40-dBc ACLR are achieved for W-CDMA (R99) modulation. In addition, the PA is capable of delivering a 18-dBm P_{out} and more than 23% PAE under 824-925-MHz and 1.1-V J-CDMA modulation test conditions. To the best of author's knowledge, this is the first report on a broadband CDMA PA operating with low V_{ref} and low V_{cc}.**

Index Terms—**GaAs, power amplifier, heterojunction bipolar transistors (HBTs), MMIC.**

I. INTRODUCTION

GaAs-based HBT power amplifiers (PAs) are widely used for single- and multi-band operation CDMA handset terminals, because the HBTs possess high power density with single voltage operation and excellent reproducibility leading to low cost and high yield [1-4]. Since CDMA PAs usually consume quiescent current continuously, lower supply voltage operation as well as high efficiency operation is one of the most effective ways to make battery life of the handsets longer. Generally, however, it is not easy to reduce the supply voltage of the HBT PA, because there is a reference voltage (V_{ref}) limitation which is known as a 2-V_{be} problem. Figure 1 illustrates the simplified block diagrams for the HBT PA and its peripheral circuits. The use of the on-chip emitter-follower based bias circuits constrains the V_{ref} reduction. Therefore, the PA and its peripheral circuits usually require a final battery voltage of 3.1 V or higher. On the other hand, as shown in Fig. 1, a reduced supply voltage operation, i.e. a 2.5-V V_{ref} and 2.8-V V_{cc} operation, allows a final battery voltage of 2.8 V, thus making the battery life longer. However, this reduction needs a special V_{ref} bias scheme in the case of pure HBT PAs, while commercially available BiFET processes allow the reduction but involve additional wafer cost [5-7]. With regard to increased multi-band handsets, a wideband operation of the PA is one of the key issues in realizing low-cost and small-sized handsets. To our knowledge, however, there are few reports on such low V_{ref}/V_{cc} and wideband operation PAs, although there are some BiFET PAs with an on-chip V_{ref} generator [5].

In this paper, we present an HBT MMIC power amplifier module newly developed for use in CDMA applications. The PA features (i) a 2.5-V low-V_{ref} and 2.8-V low-V_{cc} operation over a

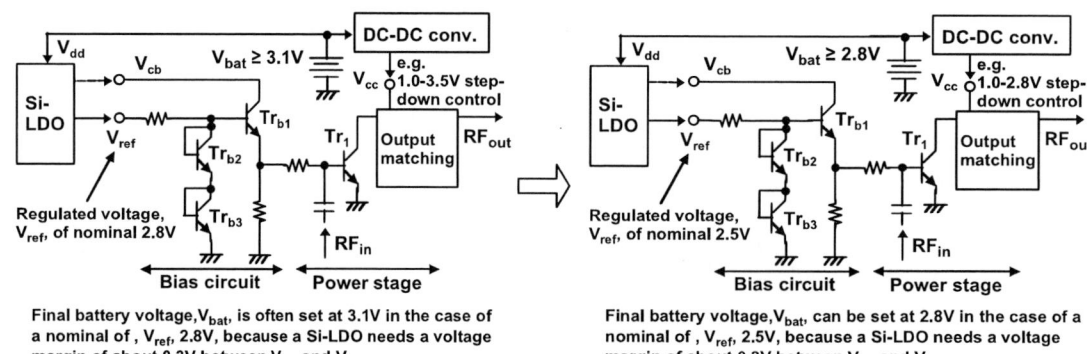

Fig. 1. Simplified block diagrams based on conventional V_{ref}/V_{cc} and reduced V_{ref}/V_{cc} supplies.

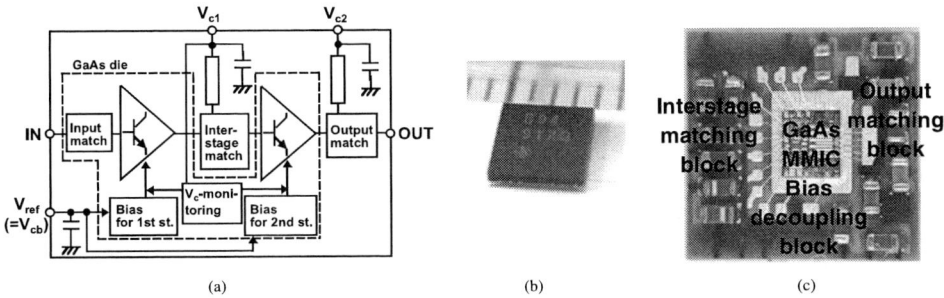

Fig. 2. (a) Simplified block diagram of the PA, (b) package photo, and (c) module photo.

wide temperature range from –20 to 85°C, (ii) an on-chip V_{cc}-dependent step quiescent current (I_{cq}) selector which makes I_{cq} step down at V_{cc} of less than 1.4 V, and (iii) a 824-925-MHz wideband operation. This paper gives a detailed description of the circuit design and measurement results.

II. DESIGN CONSIDERATION OF DIODE-BASED DETECTOR

(A) PA ARCHITECTURE AND DEVICE DESCRIPTION

The simplified block diagram, package photo, and module photo for the PA are depicted in Fig. 2. The PA is a two-stage amplifier with 50-Ω input and output matching, and is assembled on a 4 mm x 4 mm glass epoxy substrate. The GaAs die accommodates the first- and second-power stages and their related bias circuits together with an input matching block.

Regarding fabrication processes, in-house advanced InGaP HBT processes were used for the PA to obtain high output power density with single voltage operation [8]. Typical DC characteristics are as follows — a DC current gain of 100, a collector-to-emitter breakdown voltage of 15 V, and a collector-to-base breakdown voltage of 22 V. The HBT processes possess thin film resistors, MIM capacitors, spiral inductors, and substrate via-holes, as a passive element.

A 2.8-V low V_{cc} operation needs more emitter fingers than a 3.5-V operation does for the second power stage. To achieve our target Pout of more than 28 dBm under a 2.8-V V_{cc} condition, therefore, we have implemented a simple estimation in terms of emitter fingers shown in Fig. 3. The estimation is based on our previous work. In the work, 45 fingers were used for exhibiting 28-dBm Pout at V_{cc} of 3.5 V [9]. As shown in the figure, taking an appropriate power margin into account, we have determined the final stage fingers to be 80 for 2.8-V V_{cc} operation. All the input, interstage, and output matching circuit parameters were determined based on load-pull simulation and measurement so that sufficient Pout and gain could be obtained over a wideband from 824 to 925 MHz (BC0 to BC3, Band 5 to Band 8).

(B) BIAS CIRCUIT AND STEP I_{CQ} SELECTOR

To enable the 2.5-V low-V_{ref} operation, we have applied our unique bias scheme [9,10] to the PA, as drawn in Fig. 4. The features of the bias scheme are (i) two different kinds of bias feeding (current feed, and voltage-/current-feed) to AC-coupled power stages, and (ii) successful implementation of a diode linearizer built in the power stage. For example, in Fig. 4(b), the emitter follower (Tr_{b21}) does not turn on fully, because V_{ref} of 2.5 V is too low. Instead, additional current injection paths (R_{cd21}, R_{cd22}) placed between the power stages (Tr_{21} and Tr_{22}) and V_{ref}

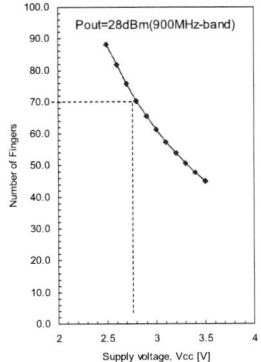

Fig. 3. Estimation of the number of fingers based on 45 fingers used for 3.5-V operation.

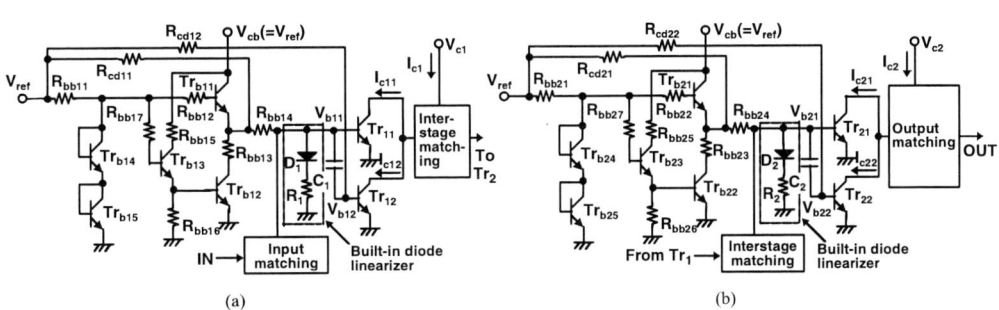

Fig. 4. Circuit schematics for (a) the first stage and (b) the second stage.

978-1-4244-5190-6/09 $25.00 © 2009 IEEE

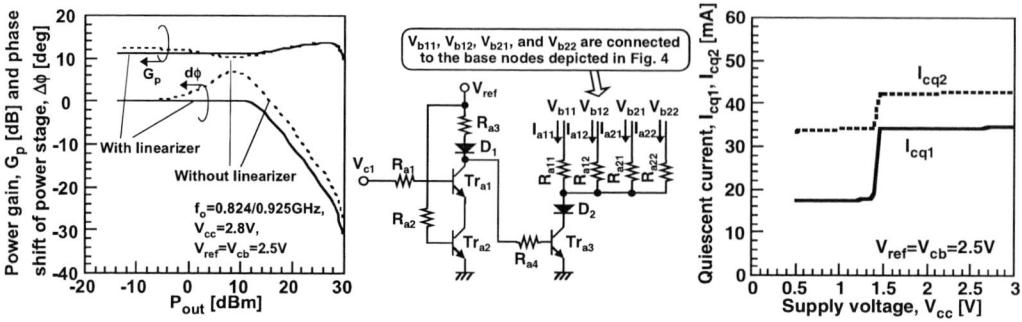

Fig. 5. Simulated output characteristics for the second stage.   Fig. 6. Circuit schematic for step I_{cq} selector.   Fig. 7. Simulation of Vcc vs. Icq.

terminal help feed quiescent current required for the power stage. Appropriate combination of the current injection and voltage drive by the emitter follower gives a smooth and monotonic gain and phase characteristics versus output power as shown in Fig. 5.

The built-in diode linearizer (D_2) with a degeneration resistor (R_2) yields slight gain expansion even at low temperatures of 0°C or less, thereby suppressing a gain dip and its related rapid phase shift. The same bias scheme is applied for the first stage, as depicted in Fig. 4(a).

Proper bias control is very important for obtaining relatively high P_{out} (e.g. 17-18 dBm) with higher efficiency even under a lower V_{cc} (e.g. 1.0-1.1 V) condition, where V_{cc} means both V_{c1} and V_{c2}. In contrast, V_{cb} and V_{ref} are kept constant at 2.5 V, because the two are fed through a Si-LDO. As shown in Fig. 6, the on-chip V_{cc}-dependent step I_{cq} selector newly incorporated into the PA can provide step control for I_{cq} according to V_{cc}. In the figure, when V_{cc} is less than 1.4 V, Tr_{a1} turns off and Tr_{a3} turns on and sinks pulling-down currents (I_{a11}-I_{a22}). The simulation of Fig. 7 indicates that I_{cq} is stepped down by the selector at low V_{cc} of 1.4 V or less.

III. MEASUREMENT RESULTS

Measurement was done using 824-MHz (BC0) and 925-MHz (BC3) J-CDMA (IS-95B compliant) and W-CDMA (R99 compliant) modulation test-sets. Supply voltage conditions were as follows: V_{cc} (V_{c1} and V_{c2}) was 2.8/1.1 V, and both V_{ref} and V_{cb} were fixed at 2.5 V. Figure 8 shows the measured frequency response of the PA at 2.8 V of V_{c1} and V_{c2}. S_{21} between 27.9 dB and 29.3 dB is kept in the band of interest. Good input return loss ranging from −10dB to −13 dB is also obtained in the band. Figure 9 shows the measured input-output characteristics under the 25°C, 824 MHz and 925 MHz J-CDMA modulation condition. The PA delivers a 28-dBm P_{out}, a 27.7-dB power gain, and a 36% PAE with ACPR(900-kHz offset) of −50.5 dBc at 824 MHz. At 925 MHz, the PA also delivers a 28-dBm P_{out}, a 27.7-dB power gain, and a 36% PAE while keeping ACPR (900-kHz offset) of less than −51.3 dBc. The PAE obtained is slightly lower than that of conventional PAs ever reported [4,5,9], which may result from low-V_{cc} and wideband operation of this work PA. Regarding NACPR (next adjacent channel power rejection ratio), NACPR (1.98-MHz offset) of less than −58dBc is kept for both the bands.

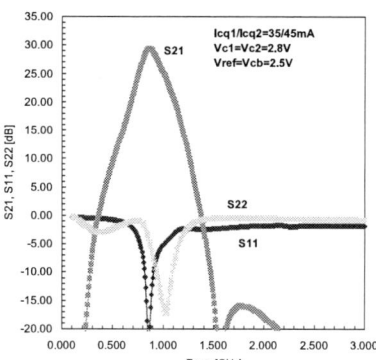

Fig. 8. Measured frequency response of PA.

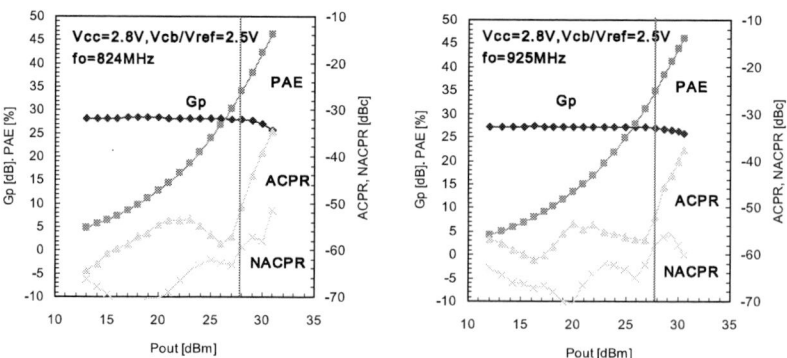

Fig. 9. Measured output power characteristics under 2.8-V V_{cc} and J-CDMA modulation test. (a) 824 MHz and (b) 925 MHz.

978-1-4244-5190-6/09 $25.00 © 2009 IEEE

Figure 10(a)-(d) shows the measured temperature dependence of the PA. The V_{cc} bias and frequency conditions are 2.8/1.1 V and 824/925 MHz. Over a wide temperature range from –20 to 85°C, the PA exhibits a smooth gain variation and sufficiently low, practical distortion characteristics (ACPR<–50 dBc, NACPR<–58 dBc) due to our unique bias scheme with the built-in linearizer, as expected. At 1.1 V of V_{cc}, 25°C, and 18 dBm of relatively high P_{out}, the PAE was higher than 23%. This good performance for low V_{cc} results from optimum bias current obtained by the I_{cq} selector, as previously mentioned. Figure 10(e) and (f) shows the measured output characteristics under a W-CDMA (R99 compliant) test-set. As shown in the figure, this PA can deliver a 29-dBm output power, a 27-dB power gain, and higher than 38% of PAE with ACLR (5-MHz offset) of less than –40 dBc. These results prove that the PA is also most suitable for W-CDMA applications.

IV. CONCLUSION

We have demonstrated the design and measurement results of the 2.5-V low-V_{ref} and 2.8/1.1-V low-V_{cc} operation, HBT MMIC power amplifier module for use in 824-925 MHz broad-band CDMA applications. The bias- and power-stage configuration suitable for low-V_{ref} and two-mode low-V_{cc} (2.8/1.1-V) operation has been proposed, and its effectiveness has been verified with measurement. The fabricated PA module has excellent performance which meets J- and W-CDMA modulation specifications.

REFERENCES

[1] Y. Yang, et al., "DC boosting effect of active bias circuits and its optimization for class-AB InGaP-GaAs HBT power amplifiers," *IEEE Trans. MTT*, vol. 52, no. 5, pp. 1455-1463, May 2004.

[2] G. Zhang, et al., "WCDMA PCS handset front end module," *in IEEE IMS Dig.*, pp. 304-307, 2006.

[3] J. H. Kim, et al. , "An InGaP-GaAs HBT MMIC smart power amplifier for W-CDMA mobile handsets," *IEEE J. SSC*, vol. 38., no. 6, pp. 905-910, June 2003.

[4] RF2162, "3V 900MHz linear power amplifier," *in RF Micro Devices data sheet.*

[5] AWT6331, "ZeroIC™ cellular CDMA 3.4 V/28 dBm linear power amplifier module," *in Anadigics data sheet.*

[6] O. Krutko, et al., "Structures and methods for fabricating integrated HBT/FET's at competitive cost," *US patent no. US 2005/0184310 A1*, Aug. 2005.

[7] A. G. Metzger, et al., "Drivers and applications for an InGaP/GaAs merged HBT-FET (BiFET) technology," *in IEEE Topical Workshops on Power Amplifiers for Wireless Communications Dig.*, Jan. 2006.

[8] K. Yamamoto et al., "A 0/20-dB step linearized attenuator with GaAs-HBT compatible, AC-coupled, stack type base-collector diode switches," *in IEEE IMS Dig.*, pp. 1693-1696, 2006.

[9] K. Yamamoto, et al., "A CDMA InGaP/GaAs-HBT MMIC Power Amplifier Module Operating With a Low Reference Voltage of 2.4 V," *IEEE J. SSC*, vol. 42., no. 6, pp. 1282-1290, June 2007.

[10] Kris, "Circuits at the nanoscale," *Chapter 18, CRC Press*, 2008.

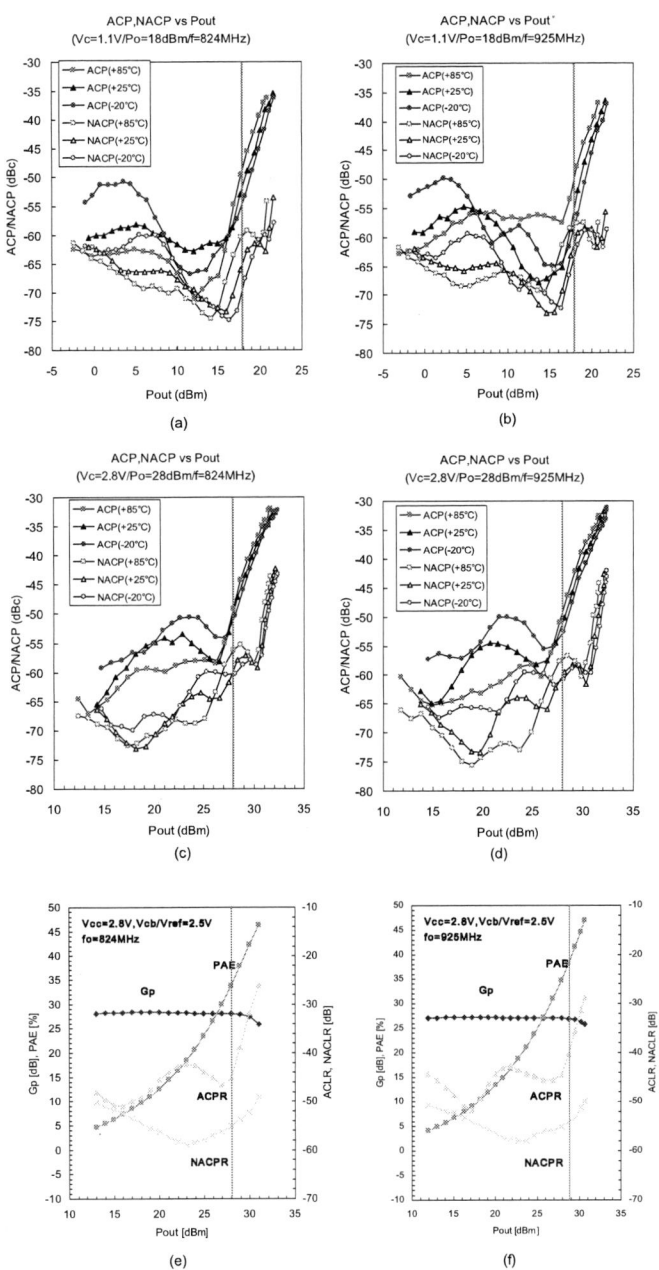

Fig. 10. (a)-(d) Measured temperature dependence for output power characteristics under 2.8-V/1.1-V V_{cc} and 824/925-MHz J-CDMA modulation test.
(e) and (f) measured output power performance under the 824/925-MHz W-CDMA modulation test.

978-1-4244-5190-6/09 $25.00 © 2009 IEEE

Dual Transformer Injection Locked Frequency Divider using GaAs E/D-Mode PHEMTs Process

Po-Yu Ke , Hsien-Chin Chiu, Jeffrey S. Fu

Chang Gung University,
No. 259 Wen-Hwa 1st Road, Kwei-Shan, Tao-Yuan, Taiwan, R.O.C:
hcchiu@mail.cgu.edu.tw

Abstract—This letter proposes a new divide-by-2 pHEMT injection locked frequency divider (ILFD) fabricated by 0.5-μm GaAs ED-Mode pHEMTs process and describes the operation principle of the dual-transformer ILFD. First transformer was applied to replace two inductors of cross-couple *LC*-tank oscillator circuit. The injection signal of ILFD transmits into a transistor through the second transformer which consisted of a band-pass filter achieving a high injection signal power and wide locking range. The measurement results show that the divider's free-running frequency were from 6.47 to 9.54 GHz (32.2%) with 3 V supply voltage. With an incident power of 0 dBm, the locking range is 3.04 GHz from the incident frequency 16.41 to 19.45 GHz (15.6%). The measured phase noise of free running VCO is -92.2 dBc/Hz at 1 MHz offset frequency at 9.45 GHz and this value of the locked ILFD is -128.4 dBc/Hz ,which is 36.2 dB lower than the free running VCO. The core power consumption was 42 mW.

Index Terms—Voltage-controlled oscillator , transformer , low phase-noise,. injection-locked frequency divider

I. INTRODUCTION

When communication system frequency is higher than X-band, frequency dividers play an important role in achieving the timing synchronization of system [1]. Digital–type frequency divider can not be applied at this frequency owing to their low devices cut-off frequency. In order to achieve high-frequency operation, an injection-locked frequency divider (ILFD), has been proposed for the divide-by-two function [2-3]. Compared to the regenerative divider, the ILFD approach provides a compact circuit topology and a low dc power dissipation which was popularly used in RFCMOS ICs. Recently, GaAs ILFD also attracted attention owing to its high output power and low substrate dissipation at millimeter wave frequency [4-5]. In this study, GaAs E-mode pHEMT was applied to design core VCO and D-mode pHEMT was adopted to inject locking signals. Therefore, negative voltage supply was avoided in this circuit which was a major drawback for traditional GaAs ICs. In additional, two lower-Q GaAs transformers were adopted to improve the Q-value of VCO's *LC*-tank and to suppress the injection locked signals loss due to gate capacitance of input D-mode pHEMT. Therefore, a wider VCO tuning range and locking range were obtained simultaneously. Under the injection-locking operation, the measured results demonstrated a low phase noise and a low dc power performance with an accurate dividing state.

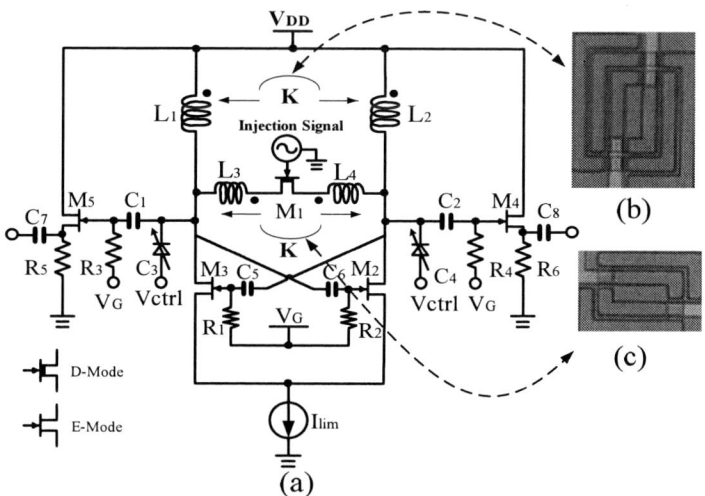

Fig. 1.(a) Proposed divider-by-2 ILFD.(b) Layout of the LC tank transformer and (c) with injection locked transformer

II. DESIGN OF TWO TRANSFORMERS IN VCO CIRCUIT

Fig.1(a) illustrates the circuit schematics of the two transformers in cross-coupled ILFD. This circuit is composed of three specific parts, the (M_1) D-mode for injection locked signals and the oscillator (M_2、 M_3) E-mode for active devices of VCO core with an output buffer (M_4、 M_5), which, together with the pair of varactor (C_3、C_4), and (C_5、C_6) comprise the LC tank. The primary coil with inductance (L_1、 L_2) are connected at the drains of cross coupled pair E-mode (M_2、 M_3) respectively is shown in Fig.1(b). The secondary coil with inductances (L_3、 L_4) are connected at the sources and drain of D-mode (M_1) shape band-pass filter structure. Two transformers (L_1、L_2、L_3、L_4) are magnetically coupled together with a coupling factor k, which, a wide set of devices with turn number(n) from 1 to 2,width (w)15μm, and inner diameter (Din) from 130 to 170μm. The simulated Q factor and inductance of the transformer, Fig.2 shows quality factor and inductance EM simulation results.

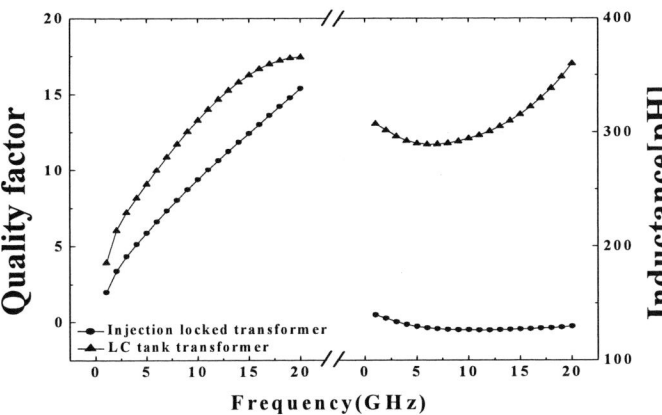

Fig. 2.Quality factor and Inductance EM simulation of transformers

Analyzed above simulated can be known, these two transformers Q factor not higher, This result is that we want the design of circuit. Tuning varactors and degrading the resonator Q factor are thus used to extend the locking range. There are some conditions we can discuss, which improve looking range effectively, the question which we must consider next is voltage injection (V_{inj}), and used inject signal into band-pass filter technology, shown in Fig.3(a) simulate band-pass filter S21 characteristic. The transient gate of the filter in the signal is very key structure, it can be got effectively to utilize the filter a perfect frequency signal , so can improve injection signal wider locking range improvement of accuracy shown in Fig.3(b) equivalent .

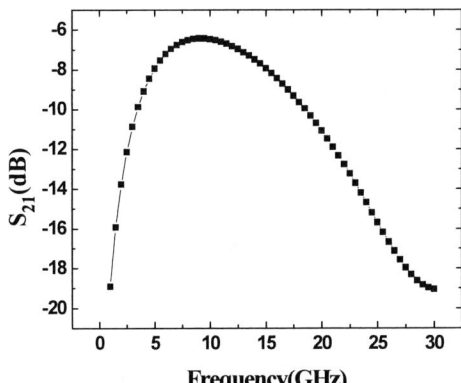

(a)

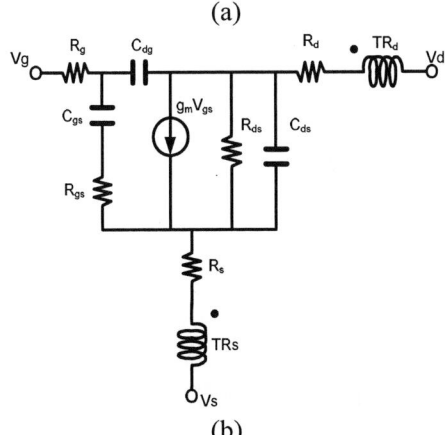

(b)

Fig.3(a) simulate band-pass filter S21, (b) equivalent small-signal circuit of the band-pass filter,

The filter a transfer function is derived from (1-5), shown in Fig.4 assumption block Z_g , Z_{gs}, Z_s, Z_{ds}, Z_d

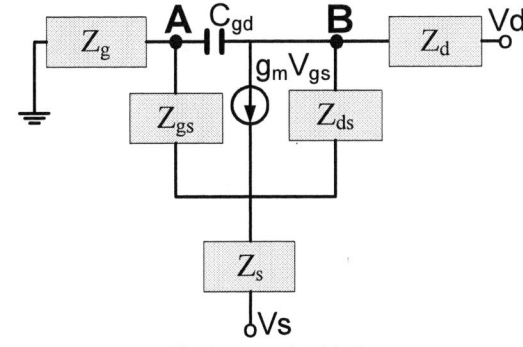

Fig.4 assumption block.

Define:

$$Z_g = R_s$$
$$Z_{gs} = \frac{1}{SC_{gs}} + R_{gs}$$
$$Z_s = R_s + SL_{trs} \quad (1)$$
$$Z_{ds} = \frac{R_{ds}}{SR_{ds}C_{ds} + 1}$$
$$Z_d = R_d + SL_{trd}$$

Utilize node analysis shown in Fig4 assumption node A、B

Known $V_{GS} = (V_G - V_S)$ $\because V_G = 0, \therefore V_{GS} = -V_S$

By KCL

Node A: $\dfrac{V_s}{Z_g} - \dfrac{V_s}{Z_{gs}Z_s} - (V_s + V_d)SC_{gd}$ (2)

Node B: $(-V_s - V_d)SC_{gd} = gm(-V_s) + \dfrac{V_d}{R_L}$ (3)

(1) into (4)

$$R_x = Z_d // (Z_{ds} + Z_s)$$
$$\frac{Z_{ds}Z_d + Z_sZ_d}{Z_d + Z_s + Z_{ds}} \quad (4)$$

(1)(2)(3)(4) find $\dfrac{V_o}{V_s}$ transfer function give(5)

$$\frac{V_d}{V_s} = \frac{R_xZ_{gs}(SC_{gd} - gm)}{S^2Z_gR_xZ_{gs}C_{gd} + S[R_x(Z_g + Z_{gs})C_{gd} + Z_{ds}Z_g(C_{gd} + Z_{gs}) + Z_gR_xZ_{gs}gmC_{gd}] + (Z_g + Z_{gs})} \quad (5)$$

Originally deriving the formula and simulation result accords with.

The block diagram of the proposed divider is detail analysis, shown in Fig.5, The input signal of frequency V_{in}, a signal with an angular frequency of 2ω is input via one of input port of the mixer, and the mixer injects a current with an angular frequency of ω into the LC tank load. To get wider locking range, design trade off of device parameter is required because the transistor size of injection M_1 and values of inductors and coupling coefficient can affect the locking range, in operation frequency, and the best supply voltage. The dc gate bias voltage of the injection M_1 is biased at 2.8 V to get a maximum locking range. When V_{inj} is gradually reduce to be smaller than 2.8 V the circuit start as a free-running oscillator, as V_{inj} decreases toward 0V, the injection power efficiency decreases, describe the locked-range analysis in which mixing

978-1-4244-5190-6/09 $25.00 © 2009 IEEE

between DC and ω components is not present. We denote the signal voltage with frequency ω signal across the LC tank load as V_{out}, the input signal with frequency 2ω as V_{inj}, and the mixer output current as I_{mix}. The mixer output current I_{mix} is proportional to the product of injection voltage V_{inj} and divider output voltage V_{out}, where θ is the overall phase shift and $f(x)$ is nonlinear function.

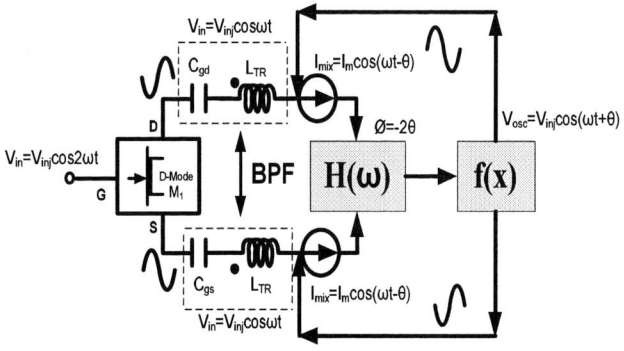

Fig,5 Show the proposed divider is M1 transistor detail analysis

Result learnt in the past experiments were performed to prove the hypotheses Section III is the experimental, our result agree with those obtained by reference journal. Show Fig.7 measure locking tuning range versus V_{inj}.

III. EXPERIMENTAL RESULT

The whole circuit was simulated and designed by using the 0.5 μm gate-length GaAs ED-mode pHEMT model. Enhancement-mode transistor a gate-width of 50 μm and two fingers in VCO core designed. Injection signal device of depletion-mode transistor a gate-width of 25 μm and two fingers. Fig.6 shows display the chip photograph of proposed divider-by-2 injection-locked frequency divider. The chip size of this circuit was 1.2×0.97 mm^2 including the RF and DC pads. The circuit is biased at VDD=3V, the circuit consumption 14 mA. Fig.7 shows measured injection locked oscillator divider consider on V_{inj} bias voltage versus locking frequency range Pin=0 dBm, Fig.8 shows measured tuning curve by varying the varactor tuning voltage at three V_{inj}. At V_{inj} of 0 V to 2.8 V as the tuning voltage sweeps from 0 to 3 V, the oscillation frequency varies from 6.47 to 9.54 GHz, indicating a tuning range of 3.07 GHz(32%). The tuning traces are slightly dependent on the injection bias .Fig.9 shows the operation frequency and locking ranges of the proposed divider-by-2 injection locked frequency divider, when an injection signal from -10.5 to 3.75 dBm, V_{tune} from 0 V to 2.8-V ,V_{inj}= 1.8 V and 2.8 V while varying V_{tune}. As the input power with injection voltage increases, the locking range also increases. An external injection signal power of 0 dBm injection voltage of 2.8V provides a locking range of 16.45 to 19.451 GHz (15.6%). The measured phase noises of the free-running and injection-locked. The phase noise of the free-running LC oscillator at 1-MHz offset is about -92.2-dBc/Hz at a 9.43 GHz output frequency shown in Fig.10. When the LC VCO signal is injection, the phase noise of injection locked is about -128.4 dBc/Hz. Show the measure output spectra of the divider before and after the locked respectively shown in Fig 11. The locked output spectra show a lower phase noise.

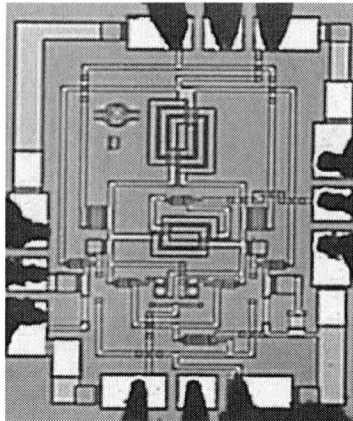

Fig,6 chip photo of the proposed injection-locked frequency divider

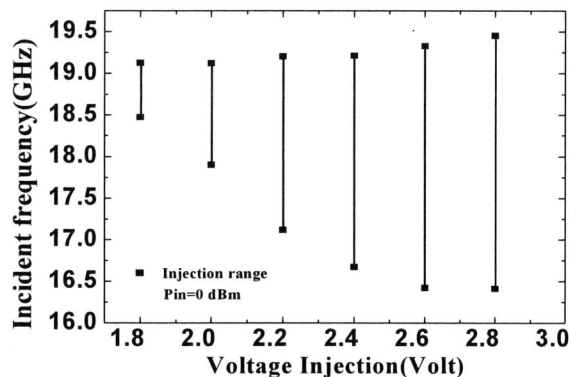

Fig,7 measured Vinj bias voltage versus locking frequency range

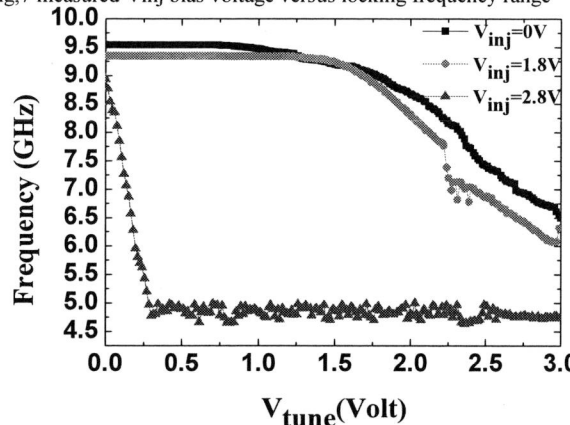

Fig,8 measured tuning curve by varying the varactor tuning voltage

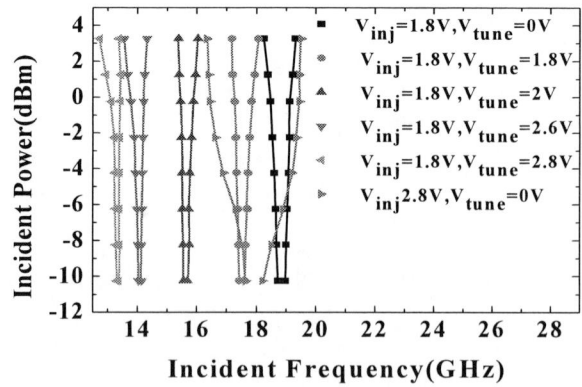

Fig,9 measured the operation frequency and locking ranges

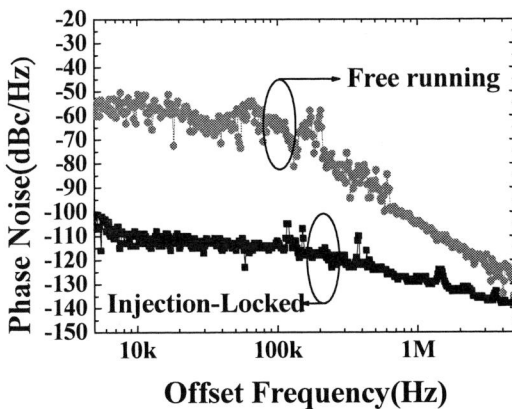

Fig,10   measured phase noises of the free-running and injection-locked.

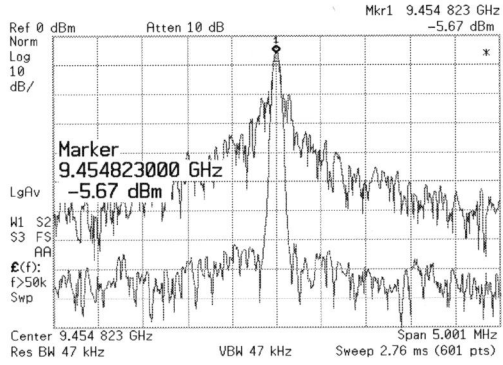

Fig,11 the measure output spectra of the divider before and after the locked respectively

IV. CONCLUSION

This letter reports the first ED-Mode device pHEMT divider-by-2 injection locked frequency divider circuit in open literatures. The operation principle of the circuit has been described in terms of the simulated   circuit behaviour and the function of divider-by-2 has been verified with the experimental result. When the incident signal power is 0 dBm, injection voltage is 2.8 V, the circuit can provide a locking range from 16.45 to 19.451 GHz and the phase noise phase of the of the locked at low frequency offset from the oscillation frequency is lower then the injection source by 36.2 dB.

TABLE I
COMPARISON OF ILFD PERFORMANCE

	Process	Vdd	Pin (dBm)	Locking Range	DC Power Consumption
[2]	0.5μmGaAs (E/D)-mode pHEMT	2.5V	10	200MHz	24.5 mW
This work	0.5μmGaAs (E/D)-mode pHEMT	3V	0	3.01GHz	42 mW

REFERENCES

[1]   Stefan Scheiblhofer, Stefan Schuster, Andreas Stelzer, "High-Speed FMCW Radar Frequency Synthesizer With DDS Based Linearization" IEEE Microw. Wireless Compon. Lett., vol.17,no.5, pp.397 – 399 May 2007.

[2]   F.H Huang, C.H Lin, H.Y Chang, Y.J Chan, "A Low-power Subharmonic Injection-Locked Oscillator using E/D-mode GaAs PHEMTs for *Ka*-Band Applications," *IEEE APMC Symposium*, pp. 1–4, Dec. 2008

[3]   F.H. Huang, C.K. Lin, Y.S. Wu, Y.C. Wang, and Y.J. Chan, "A *W*-band Injection-Locked Frequency Divider Using GaAs pHEMTs and Cascode Circuit Topology", *IEEE Microwave and Wireless Components Letters*, Vol. 17, No. 12, pp. 885-887, Dec. 2007.

[4]   M. Lang, P. Leber, Z. G.Wang, Z. Lao, M. Rieger-Motzer,W. Bronner, A. Hulsmann, G. Kaufel, and B. Raynor, "A completely integrated single-chip PLL with a 34 GHz VCO using 0.2μm E-/D-HEMT technology," in *Proc. IEEE Custom Integrated Circuits Conf. (CICC)*, 1997, pp. 529–532.

[5]   J. Jeong, Y. Kwon, "A Fully Integrated V-Band PLL MMIC Using 0.15 μ m GaAs pHEMT Technology" *IEEE J. Solid-State Circuits*, vol. 41, no. 5, pp. 1042–1050,. 2006.

Modeling of an InGaP/GaAs BiFET VVR device

William Clausen and Brian Moser
RFMD, Inc.
7628 Thorndike Rd., Greensboro, NC. 27409
wclausen@rfmd.com

Abstract— A recently introduced InGaP/GaAs BiFET process allows the integration of a JFET device with a standard HBT process. This JFET device, also called a voltage variable resistor (VVR), can be utilized in bias control circuits on the same die as HBT power amplifiers. An innovative modeling solution was developed to simulate temperature and surface state mechanisms related to device performance. For this modeling work, a parasitic FET model is used for surface state effects along with a dual diode model for voltage dependent ideality. Using these techniques, accurate simulations can be made to design near device limitations.

Keywords-component; BiFET; JFET; modeling;

I. INTRODUCTION

Recently, a BiFET process was introduced that presented a low cost solution in integrating a JFET into an HBT epi layer stack by the addition of a doped GaAs channel and an InGaP etchstop above the InGap emitter [1]. This concept is similar to previous approaches of integrating a FET into HBT layers [2-4]. A cross section of the BiFET epi stack is shown in Figure 1. Since the layers contribute a negligible change in base-emitter capacitance and resistance, the HBT performance is comparable to an HBT-only process. The VVR is controlled by the p-n heterojunction of the base and emitter ledge, and does not have a schottky contact between drain and source. Due to the wet etch used to isolate the drain and source of the VVR, the InGaP etch stop is prone to surface effects. Since the drain-source region is without a metal contact, these uncontrolled surface states affect the depletion region of the upper InGaP layer thereby affecting current conduction through the channel. To fully utilize this process, a modeling kit was created that accounts for the process variation and surface trap related effects.

Several key design areas for this process include bias control circuits that allow the use of low voltage controls from the transceiver to control the HBT PA. Since the additional cost of the VVR epi layers and process steps are small, the functionality of this InGaP process decreases the amount of additional support circuits in a GSM/WCDMA module thereby decreasing module overhead cost. Examples of these DC control circuits are voltage regulators, analog control, low leakage logic, and bias enable switches. Being able to predict the device leakage and the voltage range for safe operation in the simulator will contribute to the yield of functional circuits. The graph in Figure 2 shows three typical areas of operation for circuit applications. This is characterized by high voltage-low

current to low voltage-high current. The model should be able to predict acceptable areas of operation while varying threshold and temperature for this effort. Also, the model should give an indication to what might happen in the circuit if operational conditions exceed the upper bounds of the loadline in Figure2.

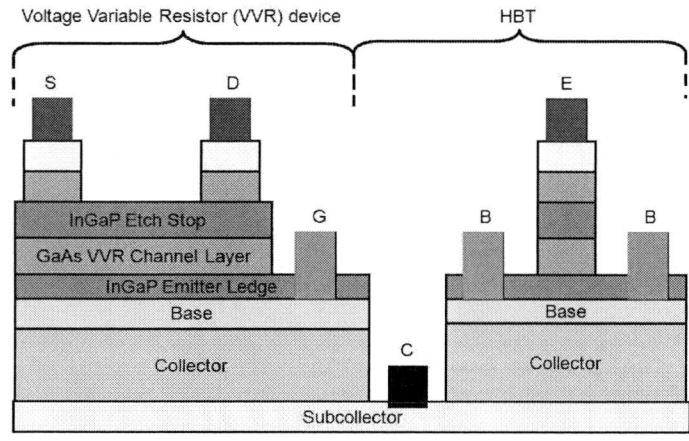

Figure 1. Device cross-sections for the HBT and VVR device.

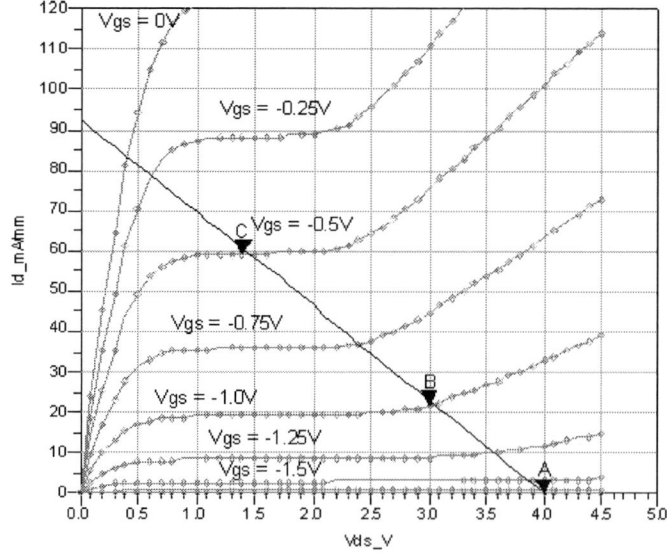

Figure 2. Loadline for circuit applications on a measured VVR device in mA/mm scale over drain voltage.

978-1-4244-5190-6/09 $25.00 © 2009 IEEE

II. DEVICE CHALLENGES

Several initial design challenges are the change in transconductance and output conductance at higher drain-to-source voltages, temperature variance on surface states, threshold changes, and leakage below pinch-off. The change in drain current at a specified drain voltage is due to surface states on the InGaP etchstop layer creating a depletion area in the GaAs channel. At higher voltages, these surface states release captured electrons and cause a significant increase in current [5]. This effect has been seen in pulsed IV measurements and DC hysteresis analysis. Examining the DC characteristics over temperature shows that the surface effects become more pronounced at cold temperatures.

The diode ideality becomes a challenge due to the high density of interface states which alters the effective barrier height at lower gate voltages resulting in a high ideality at lower voltages. At higher gate voltages the diode primarily transitions to the base-emitter diode. The forward gate diode thus behaves similar to two diodes in parallel with separate idealities. Each region is affected differently by temperature which makes scaling a little more cumbersome.

III. MODEL TOPOLOGY

Previously, a BiFET model was created for a 4 terminal device using a schottky contact between drain and source while using the base metal as a back gate control [6-7]. Similar in this work is the requirement for accurate DC/RF results yet the challenges are different. Figure 3 shows the circuit schematic for the model used for the VVR device. Available JFET models do not have the capabilities that were needed in shaping Gm, sub-threshold slope, threshold variation, and capacitance over voltage. Since the device is basically a three-terminal JFET, a standard FET model is used for the primary and parasitic elements. The Agilent EEHEMT model was chosen for the DC and capacitance functions and implemented in Verilog-A. For simplicity of extraction, the diode models were separate from the FET model instead of incorporating them into the FET verilog code.

The forward gate conduction is a product of both the gate-source and gate-drain diodes. Separate models were used for the gate-to-drain and gate-to-source diodes to account for the two separate diode characteristics. Figure 4 illustrates the 2 separate gate-source diode characteristics over temperature. Simplified temperature scaling was used based on the ADS diode model for the forward and reverse currents. The characteristics of the reverse leakage seen by the device are shown in the swept gate voltage measurement of Figure 5. Most available models for pHEMTs don't have adequate diode models as part of the code simply for the reason that it is not as important for amplifier simulations than for a device specifically for designing DC bias circuits where leakage is of prime importance. The reverse diode model is applied directly to the two gate-drain diodes using a high ideality factor. Although the graph shows the absolute value of the leakage,

from the drain node it is seen as negative current. The absolute value is shown here for simplicity and comparison.

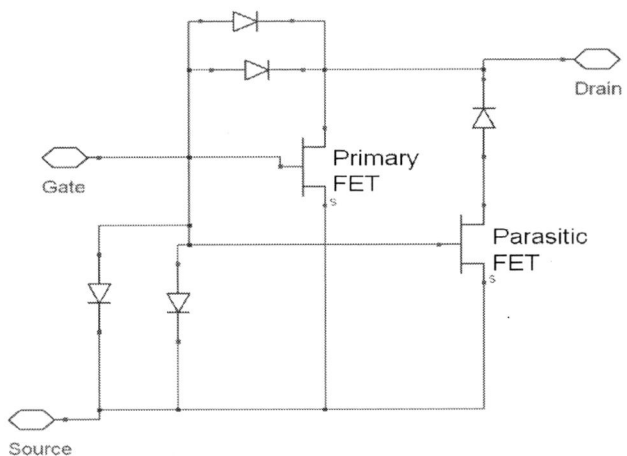

Figure 3.   BiFET model topology implemented in ADS using dual gate diodes and a parasitic FET element.

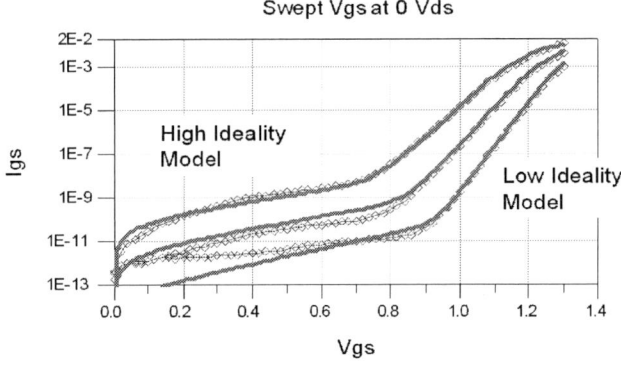

Figure 4.   Temperature variance of the two diode model at -30, 25, and 85C. Measurement (Blue) Model (Red)

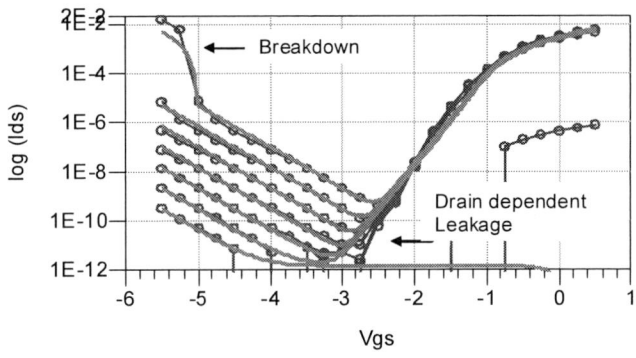

Figure 5.   Absolute value of drain current versus Vgs at 0 to 3V Vds in 0.5V steps. Measurement (Blue) Model (Red).

The parasitic FET model is an extension of the primary FET model with a change in parameters for gm and gds and a diode in series with the drain. In order to control the voltage in which the output conductance changes, a diode model with very low resistance is used in series with the drain node. The reverse breakdown of the diode can be set to model the drain voltage at which gm and gds change. Using the reverse breakdown ideality and current parameters of the diode model, the drain current kink can be fitted for abruptness and slope to the data. Since the diode current rises exponentially after breakdown, it is used only for determining the voltage noted by the drain current kink. The parasitic FET controls the current after the diode becomes a short once it breaks down. It behaves the same as the primary FET and is thus a parallel drain-to-source voltage controlled current source. Shown in Figure 6 are the fitting results using the model developed in this work.

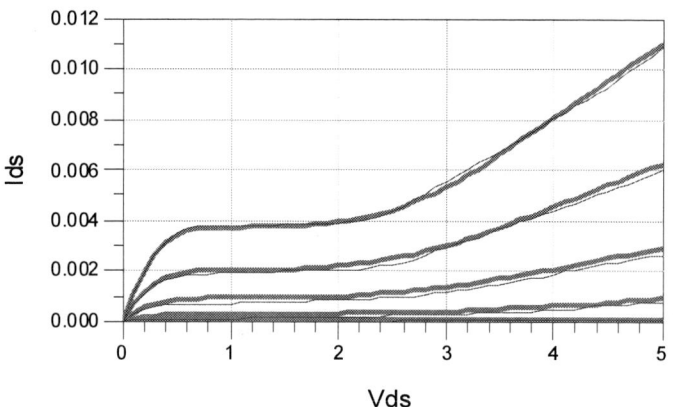

Figure 6. DCIV measurement and simulation of VVR device at 25C. Measurement (Blue) Model (Red

An addition to the DC model is the temperature variance in the threshold slope shown in Figure 7. This was implemented by adding the following function to the definition of Vts0 or the sub-threshold onset voltage

$$Vtso = Vta + VT + Tvt \qquad (1)$$

$$Tvt = (Tamb - Tnom) * b^2 \qquad (2)$$

where Vta is the extracted voltage for the threshold roll-off, VT is the parameter for threshold voltage change, and b is the coefficient for threshold slope change versus temperature. Since gm for the VVR does not peak or compress, the parameters were adjusted so that the GM peak would be above

the diffusion capacitance of the forward diode making threshold scaling simpler.

For temperature modeling, the scaling is opposite for the primary FET than for the parasitic FET. Assuming that the drain current in the low drain voltage range is limited by the captured electrons from the surface states, the higher temperatures will free these states resulting in increased current. For the parasitic FET, the temperature response is similar to that of any active device where the thermally excited carriers in the channel exhibit more collisions causing reduced drain current. Extraction of temperature coefficients are done by Id-Vg sweeps at high and low voltage to separate the contributions of each model.

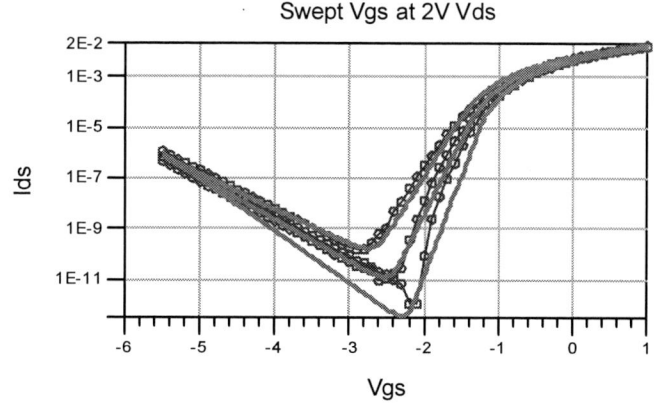

Figure 7. Temperature variance of the threshold slope and reverse leakage at -30, 25, and 85C. Measurement (Blue) Model (Red).

IV. CONCLUSION

A modeling innovation has been shown for a BiFET VVR device that gives accurate simulations over the complete voltage and temperature region of operation. By using the highlighted technique, one can implement a parasitic mechanism that can model troubled regions of device performance for more robust design analysis. This model will therefore help to employ a low cost addition to a single wafer process that can incorporate both HBT power amplifier and VVR bias control circuits.

ACKNOWLEDGMENT

The authors would like to thank their colleagues in the fab device development and corporate engineering modeling groups for assistance in this work. Most notably we acknowledge Joe Gering, Mike Fresina, Paul Partyka, and The Dude.

REFERENCES

[1] B. Moser, et al., "An InGaP/GaAs HBT/JFET BiFET technology for PA bias circuit applications", 2008 CS Mantech Technical Digest, pp. 273-276, April 2008.

[2] W.J. Ho, M.F. Chang, S.M. Beccue, P.J. Zampardi, J. Yu, A. Sailer, R.L. Pierson, and K.C. Wang, "A GaAs BiFET LSI Technology", 1994 GaAs IC Symposium Digest, pp. 47-50.

[3] D.C. Streit, D.K. Umemoto, K.W. Kobayashi, and A.K. Oki, "Monolithic HEMT-HBT Integration for Novel Microwave Circuit Applications", 1994 GaAs IC Symposium Digest, pp. 329-332.

[4] M. Sun, et al., "A High Yield Manufacturable BiFET Epitaxial Profile and Process for High Volume Production", 2006 CS Mantech Technical Digest, pp. 149-152..

[5] C.L. Liang, H. Wong, N.W. Cheung and R.N. Sato, "Parasitic Effects of Surface States on GaAs MESFET Characteristics at Liquid-Nitrogen Temperature," IEEE Trans. On Electron Devices, vol 36, p 1858-1860, Sept. 1989.

[6] C.J Wei, A. Metzger, Y. Zhu, C. Cismaru, A. Klimashov and Y.A. Tkachenko, "Four Terminal GaAs-InGaP BiFET DC model for wireless application", APMC 2005, Asia-Pacific Conference Proceedings Volume 2, Issue , 4-7 Dec. 2005.

[7] C.J.Wei, et al, "DC/RF and Statistic Modeling of Four Terminal InGaP/GaAs Bifet for wireless application", 2006 European Microwave Integrated Circuit Conference, pp. 300-303.

2x2 and 4x4 CMOS Switching Matrices for 0.01-12 GHz Applications

Donghyup Shin and Gabriel M. Rebeiz

Electrical and Computer Engineering

The University of California, San Diego

doshin@, rebeiz@ece.ucsd.edu

Abstract— **This paper presents 2x2 and 4x4 switching matrices implemented in 0.13um CMOS. The switches are based on a series-shunt design with inductive matching at the input and output ports. High substrate resistance together with deep trenches is used for low insertion loss. A 2x2 switch matrix (also called a transfer switch) results in an insertion loss and isolation of 1.3-2.3 and 48-40 dB, respectively, at 2-12 GHz. The 2x2 switch is used to build a single-chip 4x4 CMOS switch matrix with an insertion loss and isolation of 2.3-4.5 and 57-35 dB, respectively, at 2-12 GHz. The measured input P1dB and IP3 is 11-12 dBm and 29-30 dBm, respectively, both at 6 GHz and at 12 GHz. The chip area is 0.84x0.79 mm² (2x2) and 0.95x1.52 mm² (4x4), and the switch matrices does not consume any current from a 1.5 V supply.**

Index Terms— **CMOS switches, Switching matrix, DPDT Switch, 4x4 Switch, substrate resistance.**

I. INTRODUCTION

SWITHING matrices are an important circuit block in satellite communications, reconfigurable radios and base stations, and are commonly implemented using mechanical relays. The relays provide very low loss high linearity and power handling, and therefore, can be used after the RF power amplifiers. However, this results in a large system and with limited lifetime since the mechanical relays typically have a lifetime of 105-106 cycles.

This paper explores the use of CMOS switches as high performance switching matrices. In this application, the CMOS switches are placed before the power amplifiers (in the transmit mode) and after the low noise amplifiers (in the receive mode). In this case, the isolation and linearity of the CMOS switch matrix are of utmost importance, followed by a reasonable insertion loss and power handling capabilities. It is seen that complex switching matrices can be built using standard 0.13 μm CMOS for 0.01-12 GHz applications and with good performance.

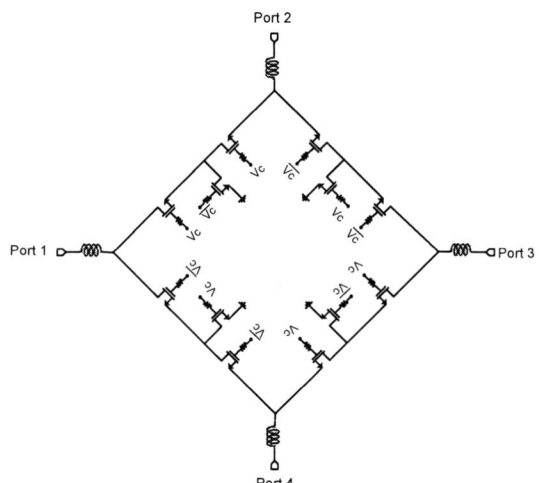

Fig. 1. Schematic of 2x2 Switching Matrix

II. 2X2 AND 4X4 SWITCHING MATRIX: DESIGN

Fig. 1 presents the layout of a 0.01-12 GHz 2x2 switching matrix, commonly called a transfer matrix or a double-pole-double-throw switch (DPDT). In this case, ports 1 and 3 are connected to ports 2 and 4 or ports 4 and 2, depending on the control voltage. In order to get wideband performance and high isolation, a series-shunt-series design is placed in every path. Also, inductors are used at the 4 RF ports in order to result in a wideband match to 12 GHz.

The transistors are based on the deep n-well technology which enables the minimization of the capacitance between the body and the substrate without any latch-up problems. The transistor body and n-well are tied to ground and V_{DD}, respectively, using R_w=20 kΩ to eliminate any RF leakage. A high substrate resistance of 1 kΩ is used to reduce the RF loss through the junction

978-1-4244-5190-6/09 $25.00 © 2009 IEEE

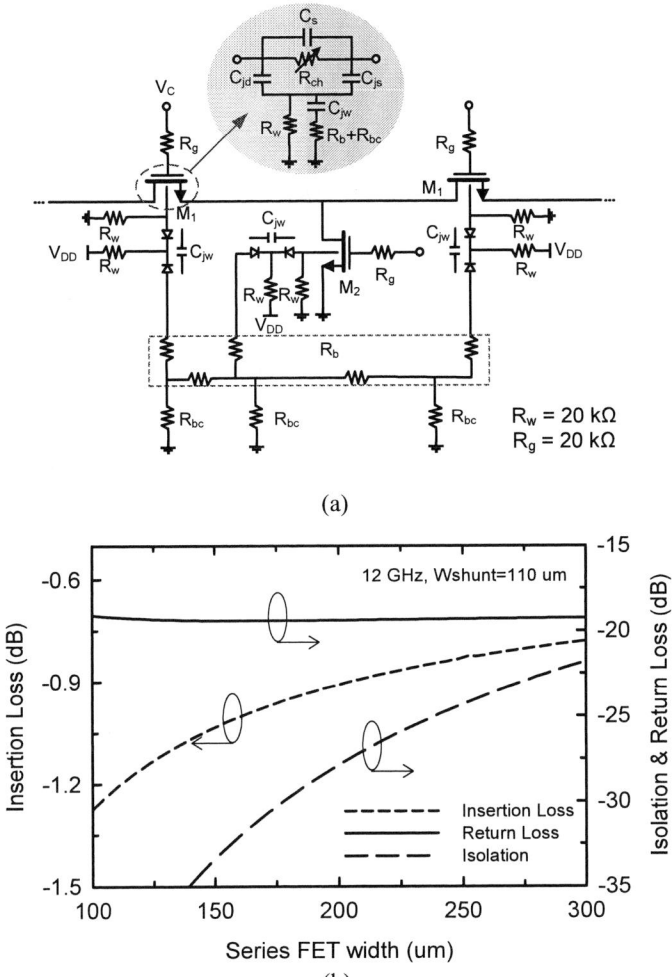

(a)

(b)

Fig. 2. (a) Circuit Model of series-shunt-series cell (b) Simulated insertion loss, isolation and reflection coefficient of a series-shunt-series cell at 12 GHz versus series FET width

capacitances, C_J. The high R_{SUB} is realized by adopting very small substrate contacts (0.8×0.8 μm^2) close to the nMOS transistor. The transistors are surrounded with deep trenches that isolate the transistor active area from the substrate. The deep trench has a depth of 6 μm, and increases the resistance between the junctions of two nMOS transistors, thus also improving the isolation between the ports.

An optimization was done in order to determine the best transistor sizes. Fig. 2a shows the full model of the series-shunt-series cell together with all the parasitic capacitances and resistances. Fig. 2b presents the insertion loss, isolation and reflection coefficient as a function of the series transistor width. This was done for several shunt transistors, and Fig. 2b is for a shunt transistor width of 110 μm. The simulation in Fig. 2 is done at 12 GHz. As expected, increasing the series transistor width results in a better insertion loss but at the

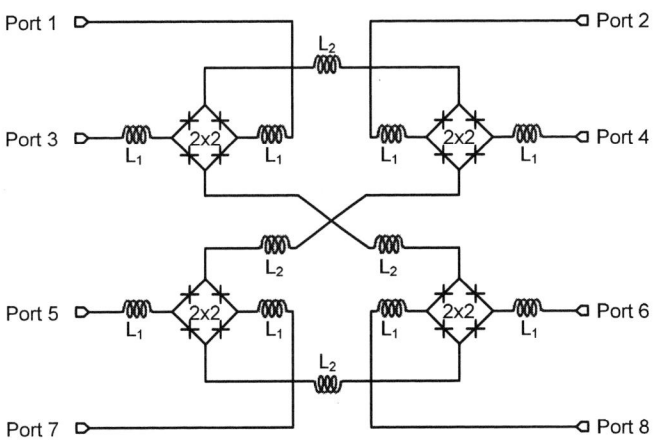

Fig. 3. Block Diagram of 4x4 Switching Matrix

expense of worse isolation and matching. Also, the substrate resistance is not well known and increasing the series transistor width results in additional losses due to power loss in the substrate (coupled to the substrate resistance through the device capacitance).

A final design consisting of series and shunt transistors of width 130 and 110 μm, respectively, was chosen. This, together with 270 pH matching inductors at the RF ports (Q=11 at 12 GHz) results in an insertion loss of < 2.5 dB at 12 GHz, an isolation > 30 dB at 12 GHz excellent match (< -13 dB at 12 GHz).

The 4x4 switching matrix is based on the 2x2 cell as shown in Fig. 3, and with several additional matching inductors. The series and shunt transistors are the same width as before (W_{series}=130 μm, W_{shunt}=110 μm), and the matching inductors are optimized to result in a wideband match to 12 GHz with L1=270 pH (Q=11 at 12 GHz) and L2=540 pH (Q=16 at 12 GHz). The simulated insertion loss, isolation and reflection coefficient of the 4x4 switching matrix are 4.5, 35, 15 dB, respectively, at 12 GHz. Again, part of the insertion loss is due to the transistor substrate resistance (see Fig. 2a).

III. IMPLEMENTATION

The 2x2 and 4x4 switching matrices chip are implemented in the IBM 8RF process with 7 metal layers. This process has 0.13μm CMOS transistors (f_T of 90-100 GHz). Grounded CPW transmission lines are used with dimensions of 11/12/11 μm for 50 Ω and with an estimated loss of 0.18 dB/mm at 10 GHz. Standard IBM transistor cells and resistor models (for gate biasing) are used, and full electromagnetic modeling is done on the transmission-lines using Sonnet [7]. The input and output CPW pads are designed to be 50 Ω and are included in the

978-1-4244-5190-6/09 $25.00 © 2009 IEEE

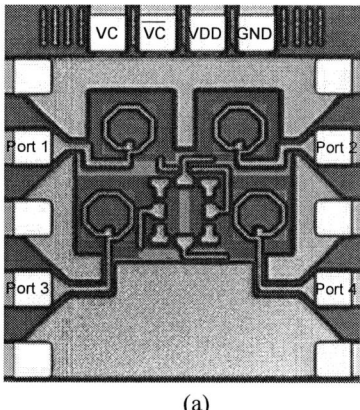

(a)

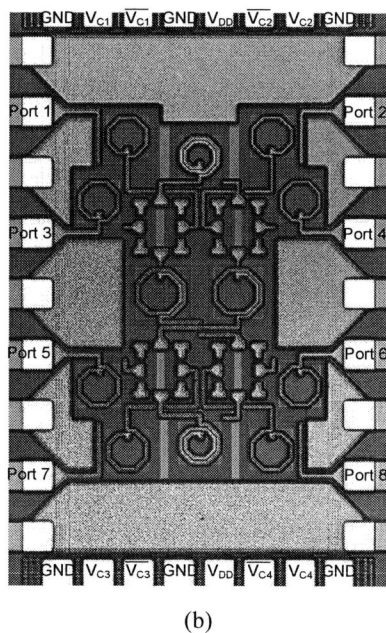

(b)

Fig. 4. Micro-photograph of 2x2 and 4x4 Switching Matrix

simulations.

The IBM model does not include accurate substrate resistance values which depend on the transistor layout, substrate contracts and deep trench isolation. These values were obtained using fitting with the measured data.

The fabricated 2x2 and 4x4 switching matrices occupy 0.66 and 1.44 mm^2, respectively, including pads. The chips do not consume any current and operate from a 1.5 V supply (Fig. 4).

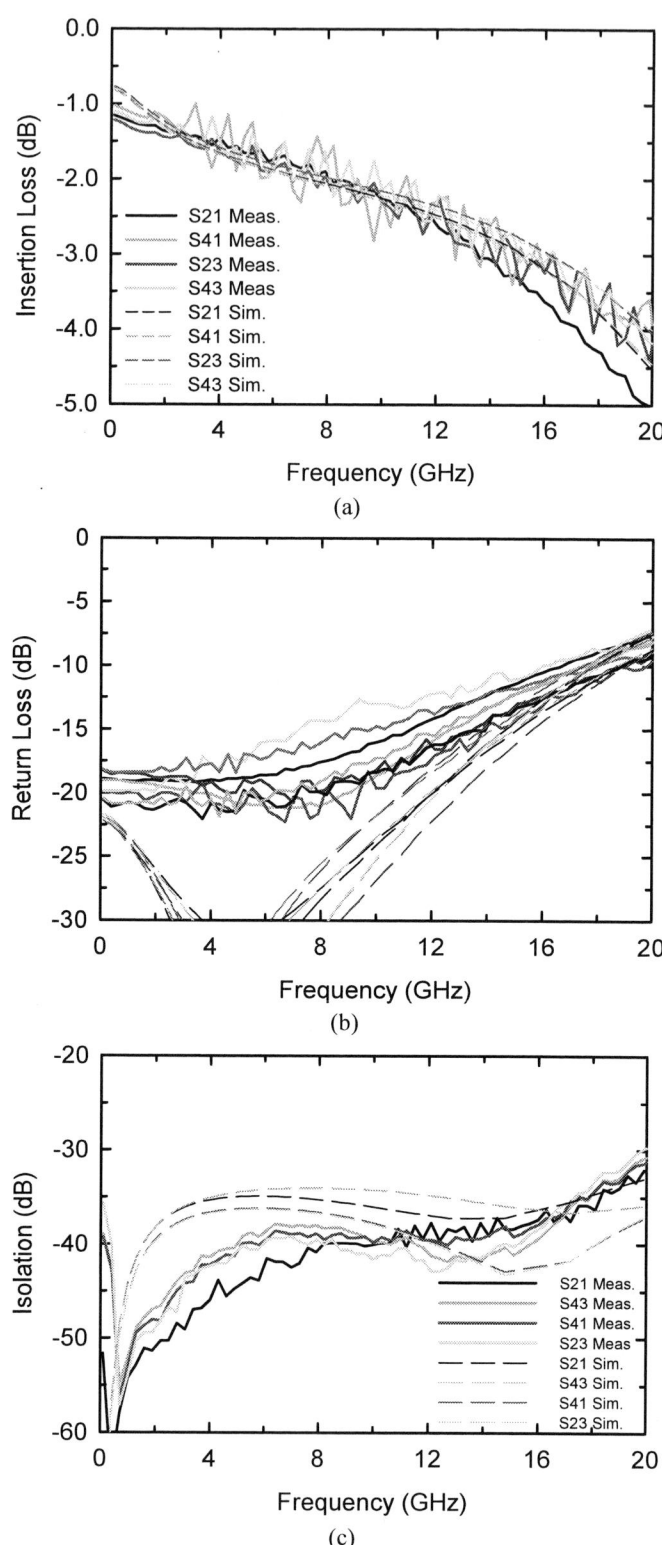

Fig. 5. Measured (solid) and simulated (dashed) insertion loss, return loss and isolation of 2x2 Switching Matrix

978-1-4244-5190-6/09 $25.00 © 2009 IEEE   94

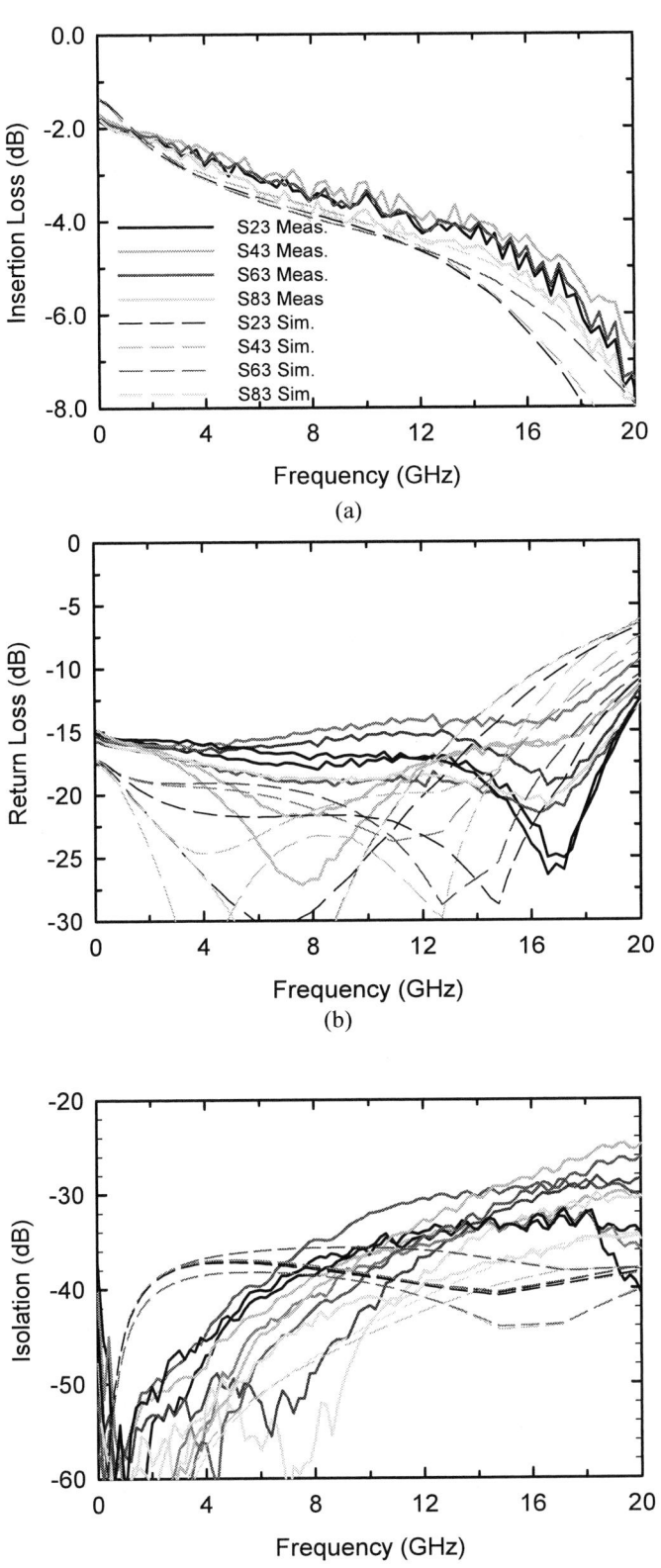

Fig. 6. Measured (solid) and simulated (dashed) insertion loss, return loss and isolation of 4x4 Switching Matrix

IV. MEASUREMENTS

The measured S-parameters of the 2x2 switching matrix are presented in Fig. 5. Measurements are done on-chip after a probe-tip SOLT calibration, and include the CPW input and output pad losses. The channels are well balanced and show an insertion loss and an isolation of < 2.5 dB and > 38 dB up to 12 GHz, and agree well with simulations. The measured Snn is < -15 dB up to 12 GHz, but the simulated dip in Snn at 4-8 GHz is not seen. This could be due to the finite substrate size and additional parasitic inductance.

The measured S-parameters of the 4x4 switching matrix are presented in Fig. 6. In this case, two out of the four ports at the input and output are terminated with a GSGSG probe, while the other ports are left open-circuited. This has no effect on the insertion loss or matching due to the high isolation of the 2x2 switching matrix cell, but does have a minor effect on the measured isolation, especially at high frequencies.

The measured insertion loss is 4.0 dB +/-0.3 dB over many different port combinations and agrees well with simulations. The measured Snn for port 3 when connected to ports 2, 4, 6, 8 are also shown. In general, Snn is < -12 dB up to 12 GHz for any connection between the input and output ports, with some being < -15 dB over the entire frequency range. The measured isolation is shown in Fig. 6 and is 35 +/-5 dB at 12 GHz. The variation is due to the non-terminated ports, coupling through the on-chip inductors, and coupling through the substrates. This is currently being investigated.

Finally, P1dB and IP3 measurements were done at 2, 6 and 12 GHz, and resulted in a P1dB and IP3 of 11+/-1 dBm and 31+/-1 dBm, respectively. This is similar to other series-shunt implementation in this technology.

V. CONCLUSION

This paper showed that 2x2 and 4x4 switching matrices are viable in CMOS switching networks and result in relatively low insertion loss and very high isolation up to 12 GHz. The 4x4 switching matrix can be improved in the future with further optimization.

ACKNOWLEDGMENT

This work was funded by the U.S. Army Research Office, Alfred Hung and Ed Viveiros, Program Monitors.

REFERENCES

[1] B. Min and G. M. Rebeiz, "Ka-band low-loss and high-isolation switch design in 0.13-μm CMOS," *IEEE Trans. Microw. Theory and Tech.*, vol. 56, no. 6, pp. 1364-1371, Jun. 2008.

[2] Y. Jin and C. Nguyen, "Ultra-compact high-linearity high-power fully integrated DC-20-GHz 0.18-μm CMOS T/R switch," *IEEE Trans. Microw. Theory and Tech.*, vol. 55, no. 1, pp. 30-36, Jan. 2007.

[3] M. Yeh, Z. Tsai, R. Liu, K. Lin, Y. Chang, and H. Wang, "Design and analysis for a miniature CMOS SPDT switch using body-floating technique to improve power performance," *IEEE Trans. Microw. Theory and Tech.*, vol. 54, no. 1, pp. 31-39, Jan. 2006.

[4] G. M. Rebeiz, *RF MEMS: Theory, Design and Technology* New York: Wiley, 2003.

[5] F.-J. Huang and K. K. O, "A 0.5-um CMOS T/R switch for 90-MHz wireless applications," *IEEE J. Solid-State Circuits*, vol. 36, no. 3, pp. 486-492, Mar. 2001.

[6] Y. Atesal, B. Cetinoneri, G. M. Rebeiz, "Low-loss 0.13-μm CMOS 50 – 70 GHz SPDT and SP4T switches," in *Proc. RFIC Symp.*, Boston, MA, June 2009, pp. 43-36.

[7] Sonnet, ver. 11.52, Sonnet Software Inc., Syracuse, NY, 1986-2007.

High-Resistivity SOI CMOS Cellular Antenna Switches

M. Carroll, D. Kerr, C. Iversen*, A. Tombak, J.-B. Pierres[#], P. Mason, and J. Costa

RF Micro Devices, 7628 Thorndike Road, Greensboro, NC USA
*RF Micro Devices, Denmark
[#]RF Micro Devices, Toulouse, France

Abstract — Results for cellular antenna switches using high-resistivity silicon-on-insulator (SOI) CMOS technology are presented. The performance of SOI RF switch FETs is presented and compared to a production GaAs pHEMT technology. Data from prototype high-resistivity SOI RF switch designs is presented and compared to pHEMT based designs.

Index Terms — RF CMOS, SOI, silicon-on-insulator, RF SOI, cellular switch, signal isolation

I. INTRODUCTION

Cellular handsets require low-loss, low-distortion, and low-cost antenna switches. Currently, GaAs pHEMT and silicon-on-sapphire (SOS) are the preferred technologies for RF switches [1,2]. Competitive RF switches have also been demonstrated using high-resistivity silicon-on-insulator (SOI) CMOS technologies [3-6]. High-resistivity SOI CMOS is now available as a foundry technology [7], leveraging the CMOS technology and manufacturing capability that has been developed over many years to support silicon ICs. In this work, we will show that high-resistivity SOI CMOS enables RF switch performance that is competitive with GaAs pHEMT technology, along with the integration capability and manufacturing scale available from CMOS technology.

II. CELLULAR RF SWITCH REQUIREMENTS

Cellular antenna switches must be capable of handling high RF power levels (up to +35dBm) along with widely varying antenna impedances. This translates to very high AC voltages across the switch during operation. The switch must also maintain high levels of linearity, and must be reliable for long term operation. Since the FETs available in most semiconductor technologies cannot handle the AC voltage in an RF switch without damage, several FETs must be stacked in series [8]. This stack of FETs is commonly referred to as a switch branch. If the impedance between the gate and body terminals of each FET is high enough, the AC voltage will divide nearly equally between the FETs in the switch branch at cellular frequencies (>700MHz). In SOI CMOS, a high resistivity handle wafer (>750 ohm-cm) and thick buried oxide (1μm) [7] are used to establish the required impedance between body terminals of the FETs.

III. SOI NFET PERFORMANCE

The 180nm technology used in this work offers 2.5V NFETs with gate oxide thickness of 5.2nm and gate length of 0.32μm [7]. For a single NFET, the typical on-resistance is 0.80 ohm-mm, and the off-capacitance is 310fF/mm. The resulting $R_{ON}*C_{OFF}$ product of 250fs is comparable to production GaAs pHEMT technologies. The 180nm design rules allow a much smaller contacted gate pitch than most GaAs technologies, resulting in very compact switch FETs. Table I shows a comparison of the key performance metrics of a SOI NFET switch branch from this work with a switch branch from a production pHEMT technology. Both of the switch branches in Table I are capable of handling up to +35dBm RF power under worst-case antenna impedance mismatch and temperature. The on-state insertion loss and resistance of the SOI switch branch is slightly higher than the pHEMT switch branch. However, the off-state isolation and capacitance of SOI switch branch are superior to the pHEMT switch branch. The resulting RF switch performance is similar between the two technologies, as demonstrated in section IV. The small-signal on-state insertion loss and off-state isolation of the series NFET stack is shown vs. frequency in Fig. 1. The insertion loss and isolation vs. RF input power at 1.8GHz are shown in Fig. 2. High isolation and low insertion loss are maintained up to +37dBm.

The maximum RF power handling capability (P_{MAX}) of the SOI switch branch in the off-state was studied to understand the number of FETs required in the series stack. Fig. 3 shows the measured P_{MAX} of the SOI switch branch vs. the number of FETs stacked in series, under worst-case antenna impedance mismatch and temperature. For RF power levels at or below P_{MAX}, no device degradation was observed during long-term reliability tests. For power levels above P_{MAX}, reduced isolation and permanent degradation of the FETs in the switch may occur.

IV. SOI RF SWITCH RESULTS

The results from a prototype single-pole, four-throw (SP4T) RF switch design in high-resistivity SOI are presented. Fig. 4 shows the RF switch die on an engineering evaluation board. The die has of four identical switch branches, with both series and shunt FETs included in each branch. The FET bias and

978-1-4244-5190-6/09 $25.00 © 2009 IEEE

control circuitry is included on the die. The insertion loss and isolation vs. frequency for the SOI RF switch die and a comparable pHEMT SP4T die are shown in Fig. 5. The insertion loss of the SOI and pHEMT die is similar below 1.5GHz, while SOI die shows a slower increase in insertion loss above 1.5GHz. The isolation of the SOI die is significantly higher than the pHEMT die. The improvement in isolation is partially due to the lower off-state capacitance and improved isolation of the SOI NFET. The shunt FETs on each branch of the SOI switch further improve the isolation. The 2nd and 3rd harmonic power vs. output power (P_{OUT}) at 900MHz is shown in Fig. 6 for both the SOI and pHEMT SP4T designs. The harmonic power data is from the RF1 port, and is the average of several die. Since the SP4T designs are symmetrical, the harmonic power from the other RF ports is nearly identical to the data in Fig. 6. Both designs meet GSM requirements for harmonic power. The pHEMT SP4T design shows superior 2nd harmonic power, except for very high output power levels where the two designs are comparable. The SOI SP4T design shows superior 3rd harmonic power.

V. TRANSMIT MODULE WITH SOI SWITCH

An SOI switch was also designed for an existing quad-band GSM/EDGE RF transmit module (TXM). The module contains two GaAs HBT power amplifiers, CMOS power management circuitry, and a double-pole, three-throw (DP3T) antenna switch. TXMs with SOI and GaAs pHEMT DP3T switches were measured as given in Figs. 7-9. In GMSK mode, measured P_{OUT} and PAE with SOI switches were approximately 0.3dB and 3% lower compared to TXMs with the pHEMT switches, respectively. The reason for this discrepancy was the use of a larger number of NFETs stacked in series in the TX switch branches than necessary. After design changes, equivalent performance is expected. The measured 2nd and 3rd harmonic power at the output are similar for pHEMT and SOI switches. The slight variation as a function of frequency is due to different impedances presented to the PAs from SOI and pHEMT switches, generating differing harmonic powers. In 8PSK mode, measured PAEs and ACPs are also very similar for both switches. Measured EVM was also very similar. The TXM performance at low band was also measured, and a similar performance difference was obtained as in high band.

VI. SILICON SUBSTRATE EFFECTS

Harmonic distortion of RF signals over high-resistivity silicon has been observed [9], and may be a concern for RF switches in high-resistivity SOI. Techniques for reducing harmonic distortion have been proposed, and significant improvements have been demonstrated for metal transmission lines over silicon [7,9]. As these techniques become available in production SOI processes, further improvement in RF switch performance may be possible.

VII. CONCLUSION

FETs from a production 180nm high-resistivity SOI process were shown to have comparable $R_{ON}*C_{OFF}$ product as FETs from a production GaAs pHEMT process. SOI FET switch branches were shown to offer low insertion loss and high isolation for cellular power levels. RF switch prototype designs using SOI were presented and shown to have comparable performance to pHEMT designs for insertion loss, isolation, and harmonic power levels.

REFERENCES

[1] H.C.Chiu, T.J. Yeh, Y.Y. Hsieh, T. Hwang, P. Yeh, C.S.Wu, "Low Insertion Loss Switch Technology Using 6-inch InGaP/AIGaAs/InGaAs pHEMT Production Process," proceedings of the Compound Semiconductor Symposium, (2004).

[2] D. Kelly, C. Brindle, C. Kemerling, M. Suber, "The state of the art of silicon-on-sapphire CMOS RF switches", proceedings of the IEEE Compound Semiconductor Symposium, (2005), pp 200-205.

[3] C. Tinella, O. Richard, A. Cathelin, F. Reaute, S. Majcherczak, F. Blanchet, D. Belot, "0.13 CMOS SOI Sp6T antenna switch for multi standard handsets," Silicon Monolithic Circtuis in RF Systems, 2006 Topical Meeting.

[4] J. Costa, M. Carroll, J. Jorgenson, T. McKay, T. Ivanov, T. Dinh, D. Kozuch, G. Remoundos, D. Kerr, A. Tombak, J. McMacken, M. Zybura, "A silicon RFCMOS SOI Technology for integrated cellular/WLAN RF TX applications", Proceedings of the IEEE MTS Microwave Symposium (2007), pp. 445-448.

[5] T. McKay, M. Carroll, C. Iversen, D. Kerr, Y. Remoundos, "Linear Cellular Antenna Switch for Highly Integrated SOI Front-End," 2007 IEEE SOI Conference, Oct. 2007, pp. 126-126

[6] T. McKay, M. Carroll, D. Kerr, J. Costa, "Advances in Silicon-on-Insulator Cellular Antenna Switch Technology," 2009 IEEE Topical Meeting on Meeting on Silicon Monolithic Integrated Circuits in RF Systems.

[7] A. Botula, A. Joseph, J. Slinkman, R. Wolf, Z-X. He, D. Ioannou, L. Wagner, M. Gordon, M. Abou-Khalil, R. Phelps, M. Gautsch, W. Abadeer, D. Harmon, M. Levy, J. Benoit, and J. Dunn, "A Thin-film SOI 180nm CMOS RF Switch Technology," 2009 IEEE Topical Meeting on Meeting on Silicon Monolithic Integrated Circuits in RF Systems

[8] M. Shifrin, P. Katzin, and Y. Ayasli, "Monolithic FET Structures for High-Power Control Component Applications," IEEE Trans. on Microwave Theory and Techniques, pp. 2134-2141, Dec. 1989.

[9] D. Kerr, J. Gering, T. McKay, M. Carroll, C. Neve, J.-P. Raskin, "Identification of RF Harmonic Distortion on Si Substrates and its Reduction Using a Trap-Rich Layer," 2008 IEEE Topical Meeting on Meeting on Silicon Monolithic Integrated Circuits in RF Systems, pp. 151-154.

Parameter	Unit	SOI	pHEMT
Insertion Loss (2GHz)	dB	0.34	0.24
Isolation (2GHz)	dB	20.2	15.0
Ron	Ohms	2.7	1.9
Coff	fF	92	147
Ron*Coff	fs	250	280

Table I - Comparison of FET RF switch branch parameters for SOI and pHEMT technologies.

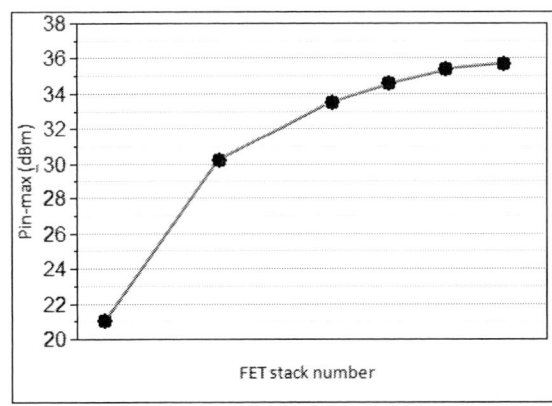

Fig. 3 – Worst-case maximum power handling capability (P_{MAX}) for a SOI RF switch branch vs. the number of FETs stacked in series.

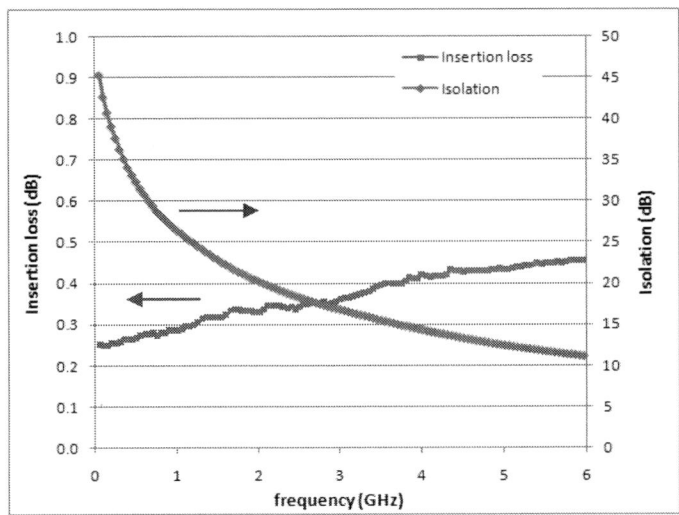

Fig. 1 – Small signal insertion loss and isolation vs. frequency for a SOI RF switch branch.

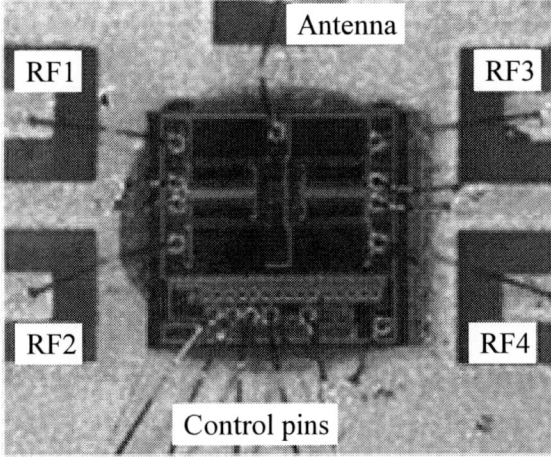

Fig. 4 – Photomicrograph of a SOI SP4T die on an engineering evaluation board.

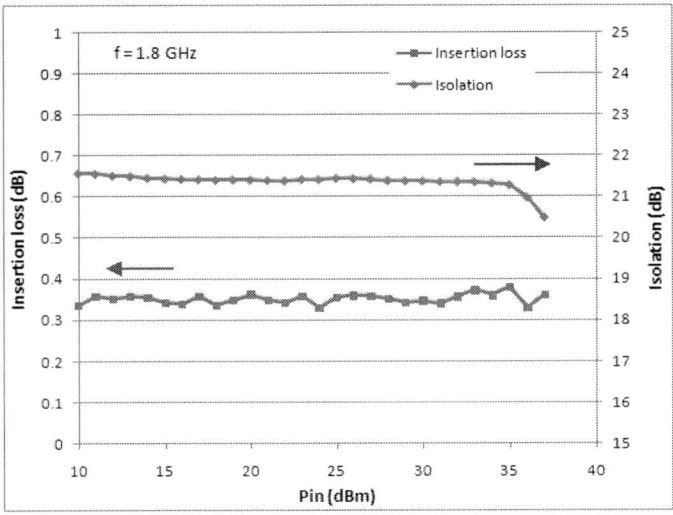

Fig. 2 – Insertion loss and isolation at 1.8GHz vs. RF input power for a SOI RF switch branch.

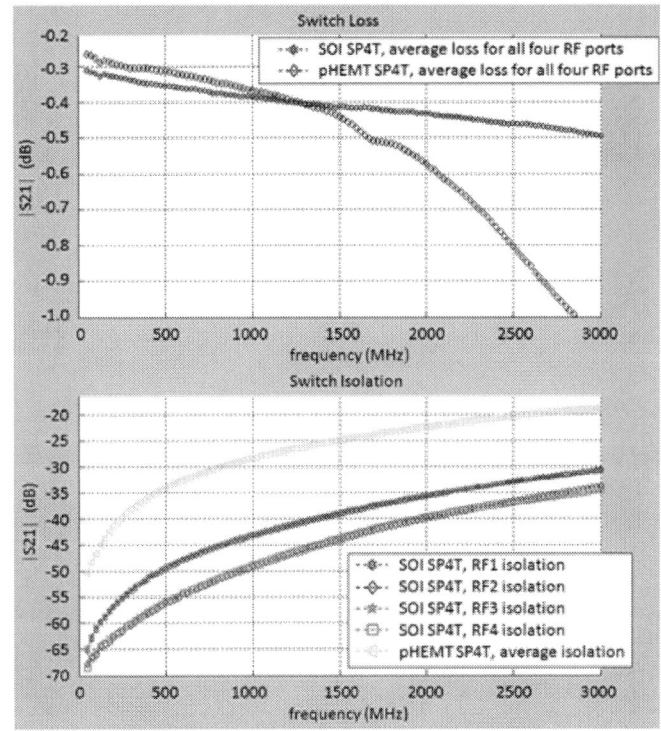

Fig. 5 – Insertion loss and isolation of SOI and pHEMT SP4T designs.

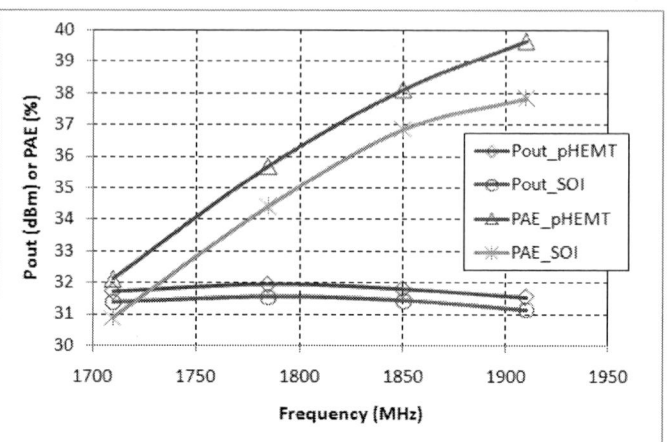

Fig. 7 – Output power and power added efficiency of transmit modules with SOI and pHEMT antenna switches.

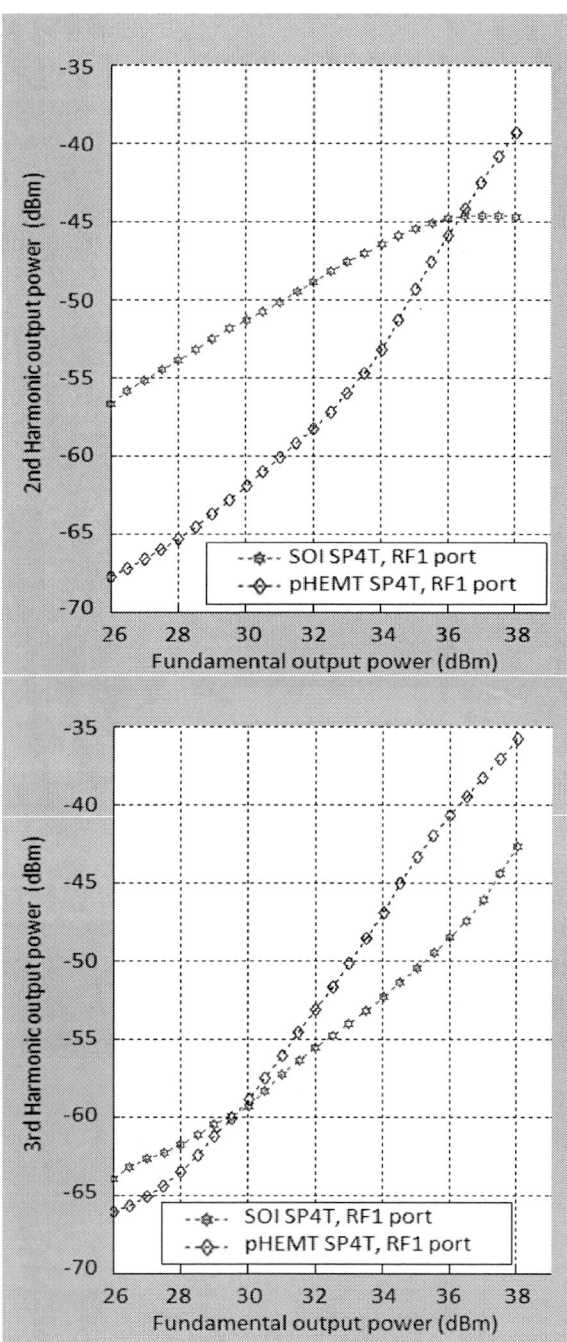

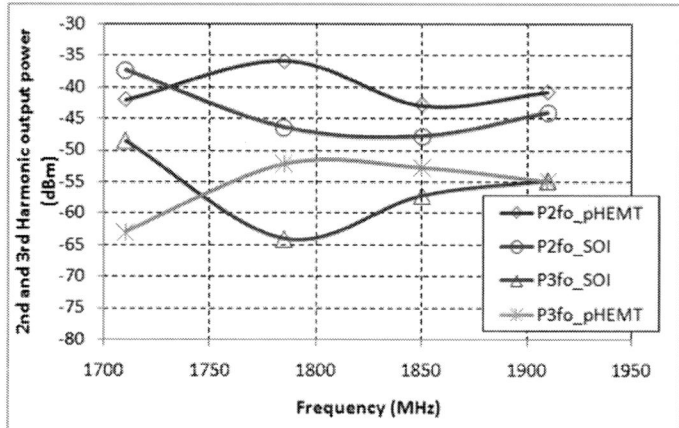

Fig. 8 – Harmonic power for transmit modules with SOI and pHEMT antenna switches operated in GMSK mode.

Fig. 6 – 2nd and 3rd harmonic power vs. output power at 900MHz for SOI and pHEMT SP4T designs.

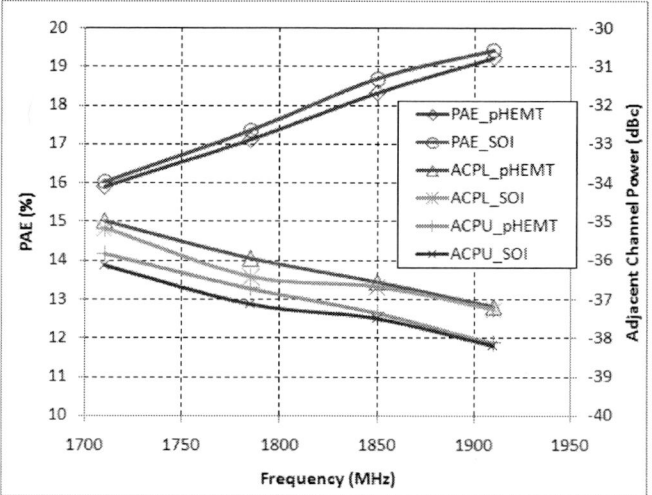

Fig. 9 – Power added efficiency and adjacent channel power for transmit modules with SOI and pHEMT antenna switches operated in 8PSK mode.

Advanced InP HBT Technology at Northrop Grumman Aerospace Systems

Augusto Gutierrez-Aitken, Cedric Monier, Pablo Chang, Eric Kaneshiro, Dennis Scott, Beckie Chan,
Matt D'Amore, Steven Lin, Bert Oyama, Ken Sato, Abdullah Cavus, Aaron Oki
Northrop Grumman Aerospace Systems
One Space Park, D1-1302J, Redondo Beach, CA 90278

Abstract— **Northrop Grumman Aerospace Systems (NGAS) has been developing InP-based heterojunction bipolar transistor technology for next generation high performance aerospace, defense and commercial applications. We present highlights and status of our production and advanced InP HBT technologies including ultra-high speed 0.25 micron emitter InP HBT.**

Keywords – InP; HBT; III-V semiconductors; high-speed microelectronics

I. INTRODUCTION

Indium Phosphide (InP) based heterojunction transistors (HBTs) technology has advanced significantly since the first InP HBT with a vertical epitaxial structure was demonstrated at Bell Labs by Malik et al in 1983 [1]. In the late 1990's the development of the InP HBT was pushed by the requirements of higher speed for optical fiber communication systems for 10 Gb/s and later 40 Gb/s. More recently there have been significant development efforts on this technology for applications in very high performance custom digital, mixed signal and RF circuits.

In the past few years, the need of higher performance and lower power has driven and guided the development of InP HBT technology at Northrop Grumman Aerospace Systems (NGAS), where we have several InP-based production and advanced processes to address the present and future high performance system needs. These technologies and some circuit examples demonstrated using these technologies are described below.

Several outstanding devices and circuits have been demonstrated to date in this technology. These include devices approaching THz operation, digital circuits operating above 170 GHz, amplifiers demonstrating gain above 200 GHz and oscillators with output signals above 300 GHz. InP HBTs are one of the highest performance semiconductor devices and are superbly suited for ultra-high speed and ultra-wide bandwidth digital, analog, mixed signal and RF applications. InP HBT technology, with its inherent electronic transport material advantages and flexibility in the device epilayer design, and emitter width scalability provides higher speed and higher breakdown voltage compared to other bipolar technologies. Fig. 1 shows a comparison of the HBT breakdown voltage (BV_{CE0}) as a function of f_T for all major bipolar technologies. As seen in this figure, the InP HBT technology demonstrates the highest performance.

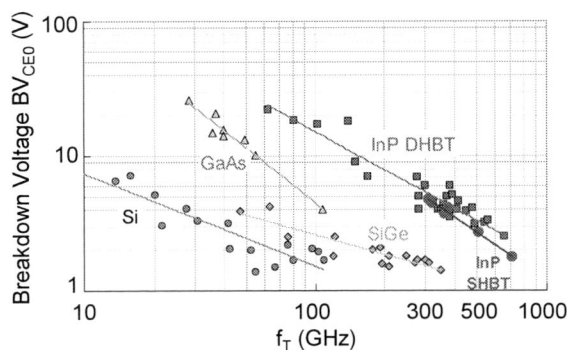

Figure 1. Comparison of the breakdown voltage (BV_{CE0}) as a function of f_T for all major bipolar technologies.

II. 0.8 um EMITTER InP HBT PRODUCTION TECHNOLOGY

A. Technology Description and Device Performance

The 0.8 um emitter width InP DHBT production process utilizes an InP/InGaAs/InP HBT structure, and has Schottky diodes, 2-metal gold interconnect, two values of precision NiCr thin film resistors (TFR): 20 and 100 ohm/sq, two values of metal-insulator-metal (MIM) capacitors: 130 and 430 pF/mm^2, and 75 um final substrate thickness with backside vias [2]. As shown in Fig. 2, this technology has excellent DC performance with $\beta > 70$, low V_{offset}, and low V_{knee}.

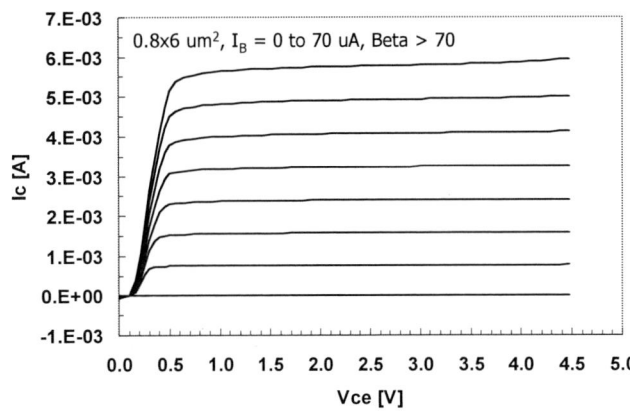

Figure 2. Measured common emitter output IV of the 0.8 um InP HBT production technology.

The forward Gummel plot of the 0.8 um technology, illustrated in Fig. 3, shows excellent characteristics such as low emitter resistance, low cross-over point and almost parallel lines to very low currents.

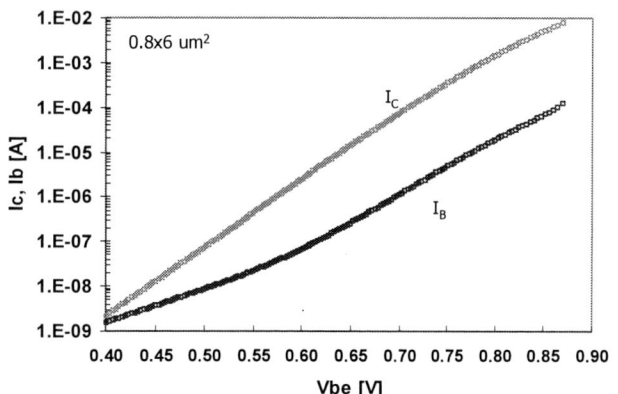

Figure 3.   Measured forward Gummel plot of the 0.8 um InP HBT production technology.

The peak performance of $f_T \approx 160$ GHz, $f_{MAX} \approx 200$ GHz and MAG ≈ 19 dB at a frequency of 20 GHz is in the region of 1 to 1.5 mA/um^2 emitter current density.

This process has a variation available for power applications based on a 1.6 um emitter width device and higher breakdown voltage. Several power amplifiers have been demonstrated with this process [3].

B.   0.8 um Demonstrated Circuit Examples

Several circuits have been demonstrated with the 0.8 um emitter production technology with complexities up to 5,000 HBTs. One example is a 2 Bit 20 Gsps ADC for digital equalization in high rate optical communications receivers [4]. The ADC provides 2 effective bits for low input frequencies and 1.8 effective bits at Nyquist when sampled at 20 Gsps. It consists of about 400 transistors, operates on a -3.3 V supply and dissipates 1.6 W of power. Fig. 4 shows the spur free dynamic range (SFDR) and the signal to noise ration (SNR) at Nyquist input as a function of sampling frequency. The SNR at 20 Gsps sampling rate varies from 13.5 dB, or 2.0 effective bits, at low frequencies, to 12.5 dB, or 1.8 effective bits, at Nyquist.

Another example demonstrated with this process is an ultra wide band DAC chip that integrates a 40-to-10 digital multiplexer with a 10-bit DAC and associated clock generation [5]. The chip has 5,000 transistors, is 9.25mm x 8.5mm in size, and dissipates 12.3W. Packaged parts were tested with approximately 20 different single-tone patterns distributed within Fs/4. These vectors were run at 1.5, 2.5, and 3.5 Gsps. Spectral data was captured via computer interface to the spectrum analyzer and stored in a text file. Analysis of the captured data did not reveal any noticeable pattern dependence of spur performance. Subsequent parts were run at 3.5 Gsps and data were only taken at three frequency locations (~Fs/100, Fs/10, and Fs/4). Fig. 5 shows the spectrum plot of the sampled data at 3.5 Gsps demonstrating an SFDR of 59 dBc

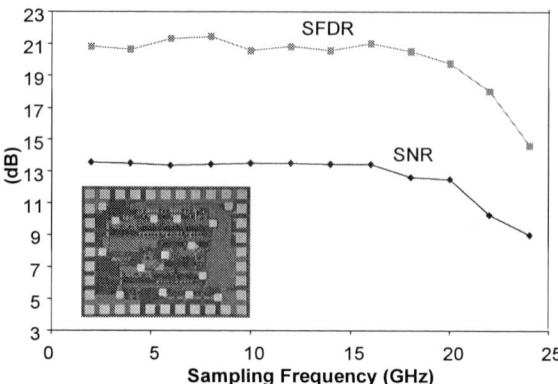

Figure 4.   Measured SFDR and SNR of the 2 Bit, 20 Gsps ADC. The inset shows a photo of the chip

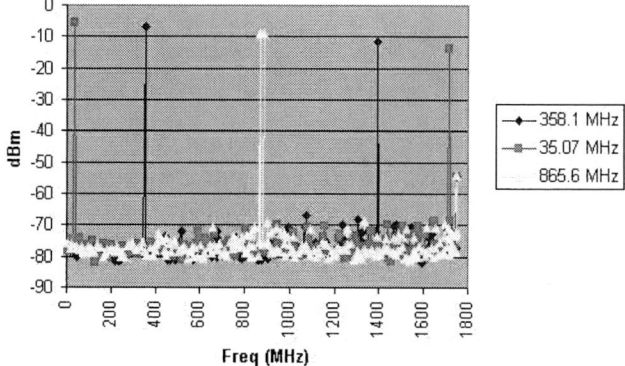

Figure 5.   Measured spectrum plot of the ultra wide band DAC chip with and SFDR of 59 dBc fabricated in the 0.8 um InP HBT production process.

III.   0.45 AND 0.6 um ADVANCED PRODUCTION InP DHBT TECHNOLOGY

A.   Technology Description and Device Performance

A more advanced InP DHBT process is in the baseline stage for production. This technology has also an InP/InGaAs/InP DHBT epitaxial structure but the emitter width has been reduced to a range of 0.45 to 0.6 um. The interconnect system has been significantly upgraded as well to a low-k planar 4-level metal interconnect. Similar to the 0.8 um technology, the advanced production technology includes Schottky diodes, 20 and 100 ohm/sq TFR, 190 pF/mm^2 MIM capacitors, and 75 um final substrate thickness with backside vias. In this technology there are two epitaxial profile alternatives. One is for high speed digital applications and the other is for high accuracy mixed-signal applications with low base-collector capacitance. The high-speed digital devices in this technology have peak performance in the region of 3 to 4 mA/um^2 emitter current density and demonstrate $\beta > 50$, $f_T \approx 250$ GHz and $f_{MAX} \approx 350$ GHz. The mixed-signal transistors have peak performance in the region of 1 to 1.5 mA/um^2 emitter current density and demonstrate $\beta > 50$, $f_T \approx 150$ GHz and $f_{MAX} \approx 400$ GHz [6].

The HBT structure has an InP emitter with an In-rich, n+ InGaAs cap for low emitter contact resistance. The InGaAs base layer is heavily doped p+ to obtain low base resistance, and the InP collector has an optimized doping level to support increased current density. The planar interconnect system has four plated Au metal layers and low-k Benzocyclobutene (BCB) as the interlayer dielectric (Fig. 6). In addition to this high yield front-side interconnect system, an InP backside process is available, which includes wafer thinning, through-wafer vias, and plated backside metal. The combination of front side and back side processes provides advantages in circuit speed, power dissipation, chip size, and design flexibility.

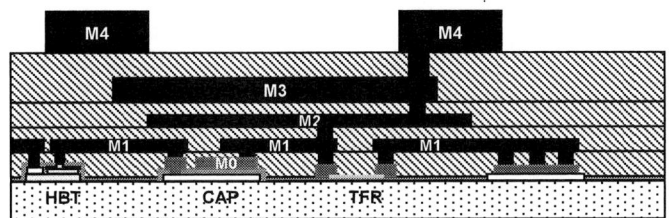

Figure 6.   Schematic cross-section of the advanced InP HBT technology, including 0.4 μm transistors, a 4 level low-k planar interconnect system, MIM capacitors and thin film resistors.

As shown in Fig. 7, the output I-V curves demonstrate well behaved analog properties such as low offset voltage, low knee voltage and a breakdown voltage in excess of 5 V. Typical peak current gain is over 50. The transistors deliver excellent and uniform RF performance with peak f_T and f_{MAX} of 250 and 350 GHz, respectively at 2.5 mA/μm² current density. Fig. 8 shows the RF data for a 0.6x6 μm² device for the digital profile.

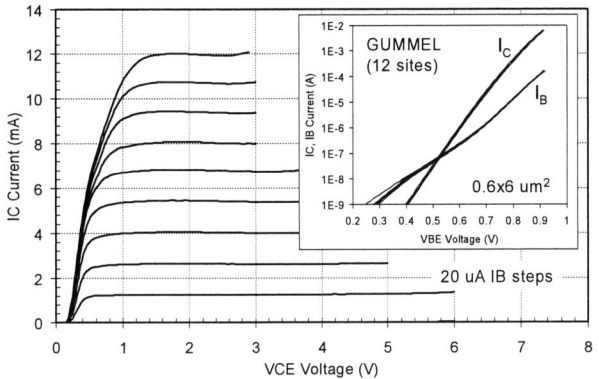

Figure 7.   Common-Emitter I.V. from a 0.6x6 μm² InP HBT. The inset shows Gummel plots measured from 12 different sites.

B.   0.45 and 0.6 um Demonstrated Circuit Examples

A large variety of digital and mixed-signal circuits have been fabricated in this technology with transistor counts ranging from 50 to > 4000 transistors. Digital building blocks, 40 Gb/s circuits (amplifiers, Demux, clock-data recovery) and mixed signal circuits have been demonstrated.

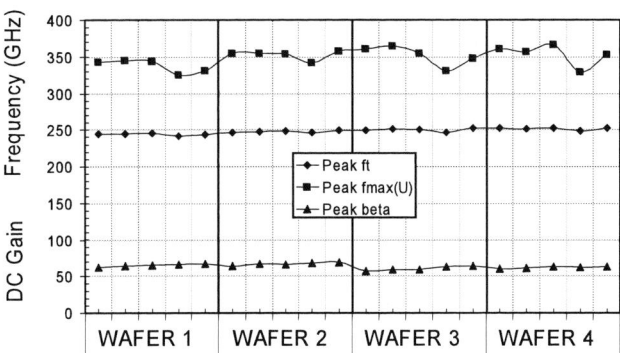

Figure 8.   Measured peak f_T, f_{MAX}, and DC gain of a 0.6x6 μm² InP HBTs showing very good uniformity.

As example, we include here an ultra-wideband 7-bit 5 Gsps ADC with 4,700 InP HBTs [7]. This monolithic ADC achieves 6 effective number of bits (EnoB) Nyquist performance at a sample rate of 5 Gsps. Fig. 9 shows the SINAD and SFDR of the ADC operating at 5 Gsps, as a function of analog input frequency. Note that 6 ENOB performance is maintained even in the 2nd Nyquist zone (2.5GHz to 5.0GHz), and >5.5 ENOB performance is maintained in the 3rd Nyquist zone (5.0 to 7.5 GHz).

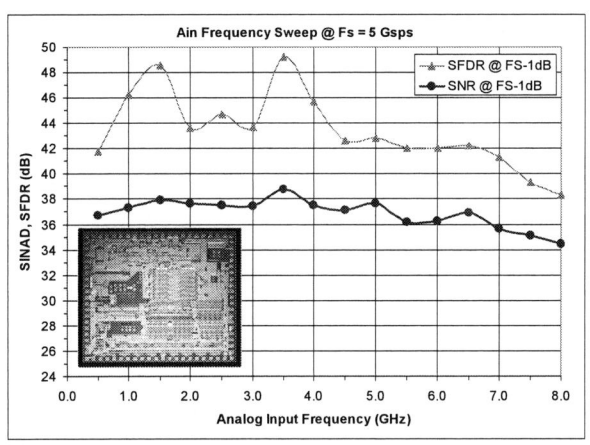

Figure 9.   Plot of Measured SINAD and SFDR of the ADC versus Analog Input Frequency operating at 5 Gsps.

IV.   0.25 um EMITTER ADVANCED INP HBT TECHNOLOGY

A.   Technology Description and Device Performance

To further increase speed and bandwidth and overall performance, NGAS has been developing scaled InP DHBTs with emitter widths of 0.25 um. A 0.25x4 um² emitter area device demonstrated β = 30, f_T = 530 GHz, f_{MAX} ≈ 650 GHz, and MAG/MSG = 15.5 dB at 100 GHz at an emitter current density of 13 mA/um² and V_{CB} = 0.4 V. A description of the process is included in [8]. An SEM cross-section illustrating all the components of this technology is shown below in Fig. 10. Measured common emitter output I-V curves and forward Gummel plots, including beta, are shown in Figs. 11 and 12, respectively.

978-1-4244-5190-6/09 $25.00 © 2009 IEEE

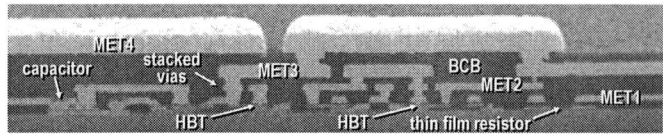

Figure 10. SEM cross-section of the complete four levels of interconnect including HBTs and passives for the advanced 0.25 μm InP HBT technology.

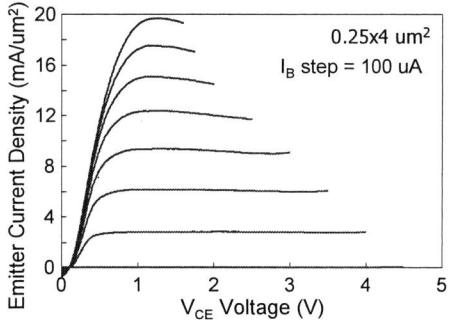

Figure 11. Measured common-Emitter I.V. characteristics from a 0.25x4um² InP DHBT.

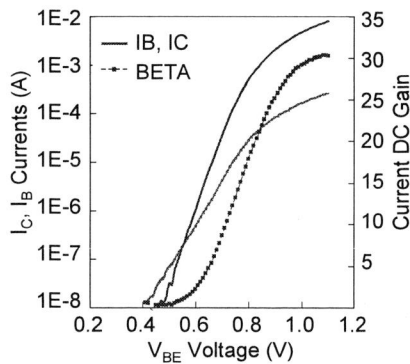

Figure 12. Meaured Gummel characteristics and DC gain Beta from a 0.25x4 um² InP DHBT.

B. 0.25 um Demonstrated Circuit Examples

A divide-by-two static digital divider circuit was implanted in this technology in a traditional master slave flip-flop topology with an inverted feedback loop. The circuit has ~ 40 transistors, all with 0.25 um emitter width. As shown in Fig. 13, the circuit demonstrated measured operation up to a maximum input clock frequency of 172.4 GHz [9].

VI. SUMMARY

At Northrop Gruman Aerospace Systems we have designed, fabricated and demonstrated a variety of state-of-the-art circuits in several InP HBT technologies with complexities up to ~ 5,000 HBTs per chip. In the advanced 0.25 um process we have produced a static divide-by-2 circuit operating up to input frequencies of 172.4 GHz. The HBT technologies presented here offer a combination of performance and integration suitable for high performance digital and mixed signal applications.

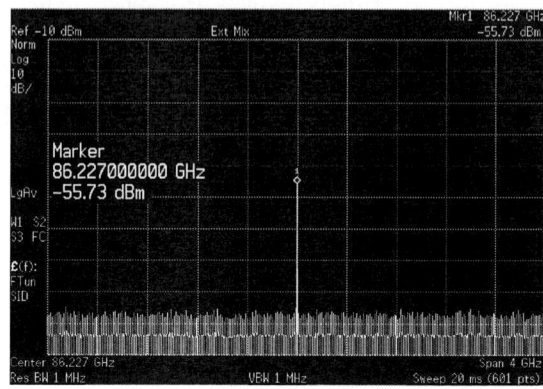

Figure 13. Measured spectrum of the divide-by-two circuit, corresponding to operatioon up to a clock input frequency of 172.4 GHz.

ACKNOWLEDGMENT

The authors would like to thank NGAS engineers and technicians in the microelectronics and mixed-signal centers for their effort and contributions to this work. The advanced development work was supported by the DARPA TFAST program under ONR contract No. N00014-02-C-0473.

REFERENCES

[1] Malik R. J. et al, "High-Gain Al0.48In0.52As/Ga0.53As Vertical n-p-n Heterojunction Bipolar Transistor Grown by Molecular-Beam Epitaxy." IEEE Electron Device Letters, Vol. 4, No. 10, Oct. 1983

[2] D. Sawdai, E. Kaneshiro, A. Gutierrez-Aitken, P.C. Grossman, K. Sato, W. Kim, G, Leslie, J. Eldredge, T. Block, P. Chin, L. Tran, A. Oki, D.C. Streit, "High performance, high yield InP DHBT production process for 40 Gbps applications," IPRM Conference Proceedings, May 2001.

[3] Aust, M.V.; Sharma, A.K.; Chau, A.T.; Gutierrez-Aitken, A.L.; "A High Efficiency and High Linearity 20 GHz InP HBT Monolithic Power Amplifier for Phased Array Applications," IEEE/MTT-S International Microwave Symposium, June 2007.

[4] Zheng Guo; D'Amore, M.; Gutierrez, A.; "A 2-bit 20 Gsps InP HBT A/D converter for optical communications," IEEE Compound Semiconductor Integrated Circuit Symposium, Oct. 2004.

[5] Ching, D.; Tsai, K.; Gutierrez, A.; Oyama, B.; "Ultra wideband digital to analog conversion based on advanced InP DHBT technology," IEEE Compound Semiconductor Integrated Circuit Symposium, 2005.

[6] Monier, C.; Scott, D.; D'Amore, M.; Chan, B.; Dang, L.; Cavus, A.; Kaneshiro, E.; Nam, P.; Sato, K.; Cohen, N.; Lin, S.; Luo, K.; Wang, J.; Oyama, B.; Gutierrez, A.; "High-Speed InP HBT Technology for Advanced Mixed-signal and Digital Applications, "IEEE International Electron Devices Meeting, Dec. 2007.

[7] Chan, B.; Oyama, B.; Monier, C.; Gutierrez, A.; "An Ultra-Wideband 7-Bit 5 Gsps ADC Implemented in Submicron InP HBT Technology," IEEE Compound Semiconductor Integrated Circuit Symposium, Oct. 2007.

[8] Scott, D.W.; Chang, P.C.; Sawdai, D.; Dang, L.; Wang, J.; Barsky, M.; Phan, W.; Chan, B.; Oyama, B.; Gutierrez-Aitken, A.; Oki, A.; "Sub-Micrometer InP/InGaAs Heterojunction Bipolar Transistors with fT = 400 GHz and fmax > 500 GHz," IPRM Conference Proceedings, May 2006.

[9] C. Monier, M. DAmore, D. Scott, A. Cavus, S. Lin, E. Kaneshiro, L. Dang, P.C. Chang, K. Sato, M. Truong, P. Nam, D. Li, B. Chan, R. Sandhu, J. Wang, B. Oyama, A. Gutierrez,. and A. Oki. "172 GHz Divide-by-Two Circuit Using a 0.25-um InP HBT Technology, International Conference on IPRM Conference Proceedings, May 2009.

V-band Amplifier MMICs using Multi-finger InP/GaAsSb DHBT Technology

Jean Godin, Virginie Nodjiadjim,
Muriel Riet, Philippe Berdaguer,
Stéphane Piotrowicz, Olivier Jardel,
André Scavennec

Alcatel-Thales III-V Lab,
Marcoussis, France
email: jean.godin@3-5lab.fr

Jean-Christophe
Nallatamby

XLim,
Brive,
France

Christophe Gaquière

IEMN,
Villeneuve d'Ascq,
France

Matthieu Werquin

MC² Technologies,
Villeneuve d'Ascq,
France

Abstract — **We report on the development of a multi-finger InP/GaAsSb DHBT technology, optimized for the fabrication of RF MMICs. Geometry (number of fingers) and structure optimization has allowed to get $F_T \sim 200$ GHz and $F_{MAX} > 300$ GHz, to suit microwave applications. Special attention has been paid to critical thermal behavior. Key devices have been modeled using a modified Gummel-Poon model. Based on these models, and a passive devices library, RF amplifiers have been designed to operate at 60 GHz, fabricated using the developed process, using 2-finger devices, and measured. 1-stage amplifier delivers 5.5 dB, and 2-device 2-stage amplifier achieves up to 10 dB gain.**

HBT, InP, GaAsSb, MMIC, Amplifier

I. INTRODUCTION

Thanks to its unique capability to provide both very high speed and large breakdown voltage, InP HBT technology is a clear contender, studied by various teams, for demanding applications, be it high bit-rate optical communications (100 Gbit/s and above [1]) or RF MMICs operating at higher frequency bands (E, W, G, i.e. 80-200+ GHz [2-5]).

Alcatel-Thales III-V Lab has previously reported on a submicron InP/InGaAs DHBT process for high-speed high-swing mixed-signal ICs [6]. This process has been tuned to address very high frequency microwave applications, such as amplifiers and oscillators: the structure has been optimized to get higher F_{MAX}, GaAsSb has been used for the base [7] and multi-finger devices were assessed for power applications.

II. DHBT STRUCTURE

A. Material choice

As we target high speed operation, the devices have to operate at high collector current, where Kirk-like effects are detrimental. Base-Collector junction has to be optimized to limit these effects. In classical InGaAs-based DHBTs, 'type I' discontinuity has to be smoothed, and various approaches have been studied. In GaAsSb-based DHBTs, on the other hand, 'type II' heterojunction allows simpler structure. In [8], we have compared these different approaches, and concluded that "type II" DHBTs provide a larger current range over which high performances are achieved, as well as a higher F_{MAX}.

Following developments were done using GaAsSb as Base material.

B. Structure epitaxy

The epitaxial material was grown by Picogiga International, using MBE on a 3" SI-InP wafer, based on optimization reported in [8]. The layer structure includes a carbon doped (1.5×10^{20} cm^{-3}) GaAsSb base and a thick (230 nm) and relatively low doped (5×10^{16} cm^{-3}) collector to ensure a high breakdown voltage. As thermal effects is another key point for amplifier operation, InP is used for the major part of the subcollector for thermal resistance reduction.

III. DEVICE FABRICATION AND PERFORMANCE

A. Device Fabrication

For the device fabrication, we have modified our triple-mesa self-aligned process, described in [6] to take into account the different Base material.

The emitter contact width is 1 µm and the self-aligned base contact extends 0.5 µm on each side of the emitter. Both contacts are defined by electron beam lithography to get the needed alignment accuracy, while other steps rely on optical lithography. Benzocyclobutene (BCB) is used for passivation, isolation and planarization.

The process also includes three interconnection metal levels, as well as resistors and MIM capacitors.

B. Multi-finger layout

To provide the amplifier designer with the more relevant devices, various multi-finger configurations were studied, based on an hexagonal finger layout to minimize the modification to our one-finger process; layouts with finger number ranging from 2 to 8 were compared and reported in [9]. For the higher frequency applications, 2-finger devices were found to be best suited, and selected for subsequent work.

Figure 1 shows a microphotograph of the 2-finger 1x15 µm² device, with plug base contact, labeled 2x1M15B5R10.

978-1-4244-5190-6/09 $25.00 © 2009 IEEE

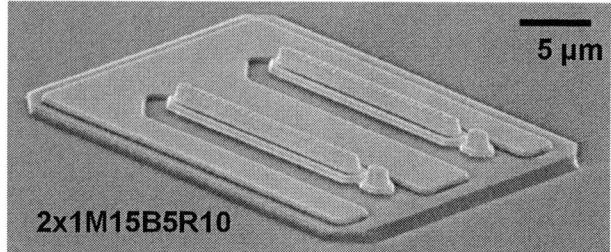

Figure 1: two-finger (2 x 1μm x 15μm) device microphotograph

C. Device performances

DC characteristics are presented in Figure 2 for eleven different 2-finger 1x15 μm² devices.

Gummel plot (top) shows a base current ideality coefficient of 1.5, typical of GaAsSb base-emitter heterojunction; I_C (V_{CE}) curves (bottom) demonstrate fair uniformity, with DC current gain about 30. Measured break-down voltage BV_{CE0} is about 7 V and $V_{knee} = 0.95$ V @I_C=84 mA.

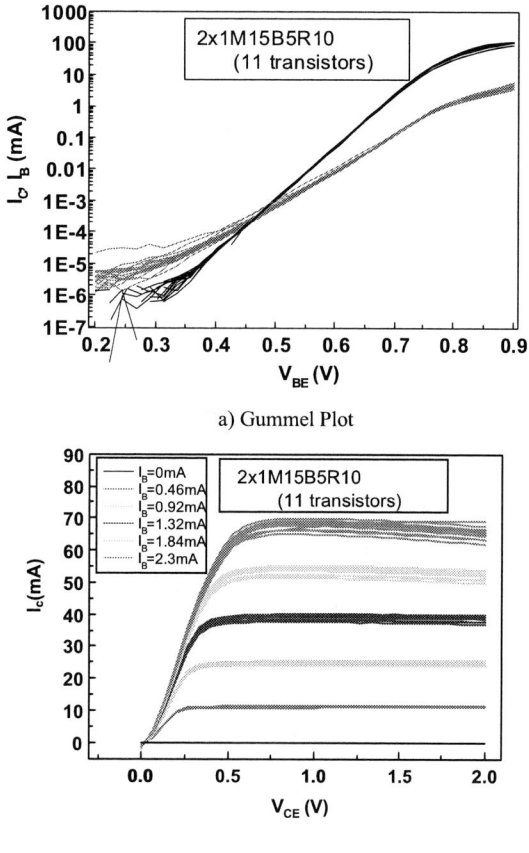

a) Gummel Plot

b) $I_C(V_{CE})$

Figure 2: DC characteristics for eleven 2x1x15 μm² devices

Frequency characteristics for the same devices are shown for V_{CE}=1.6 V, in Figure 3: F_T~200 GHz and F_{MAX} > 300 GHz are achieved, with a relatively good uniformity, on a significant current range.

As Maximum Available Gain is the more relevant data for amplifiers, it is reported on Figure 4, which shows 11 dB at 60 GHz.

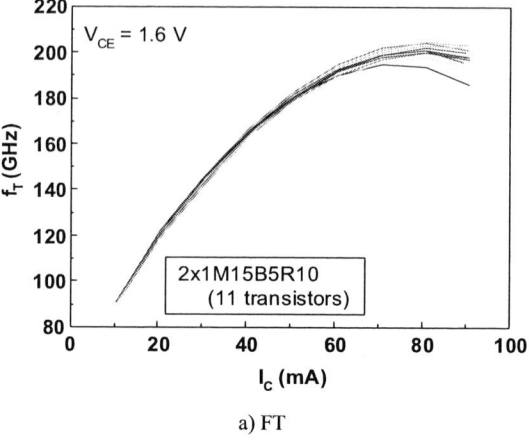

a) FT

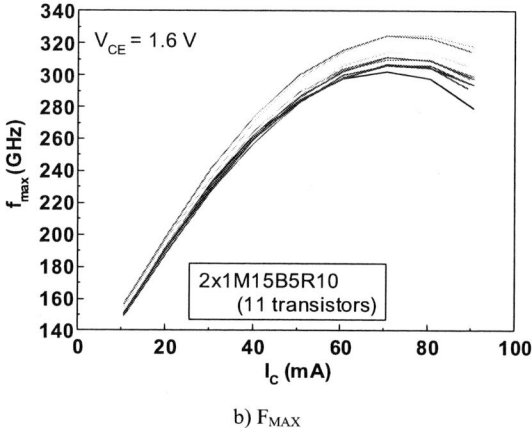

b) F_{MAX}

Figure 3: AC characteristics for eleven 2x1x15 μm² devices

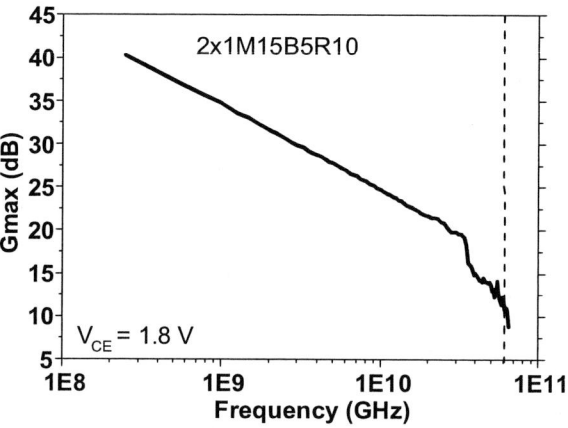

Figure 4: MAG for 2x1x15μm² device

D. Multi-finger devices comparison

While first assessment (reported in [9]) has led us to focus on 2-finger devices, we have in parallel improved the layout of multi-finger devices in order to reduce thermal effects impact.

Figure 5 shows the AC performances (at V_{CE}=1.6 V) for various number of fingers (up to 8) for 1 μm (top) and 0.7 μm (bottom) emitter width. With these modifications, similar or even improved performances are achieved for up to six fingers. The larger current range for which high performances are achieved bodes well for the amplifier design in the future.

978-1-4244-5190-6/09 $25.00 © 2009 IEEE

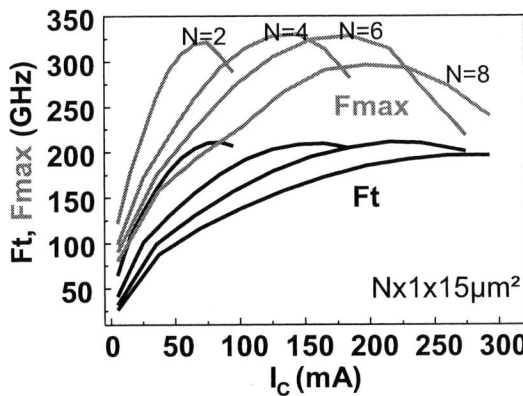

a) 1x15 μm² multi-finger DHBT w.r.t number of fingers

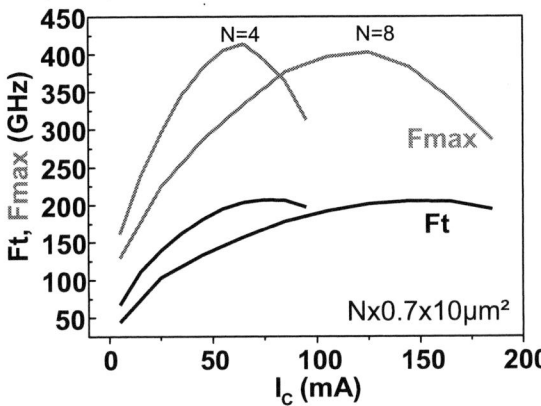

b) 0.7x15 μm² multi-finger DHBT w.r.t number of fingers

Figure 5: improved multi-finger DHBT performances

IV. MODELING

For accurate amplifier circuit design, precise modeling of both passive and active devices is needed.

A comprehensive mask-set (Figure 6) of needed CPW passive elements has been fabricated and characterized up to 110 GHz. Scalable models have been extracted to build the needed library.

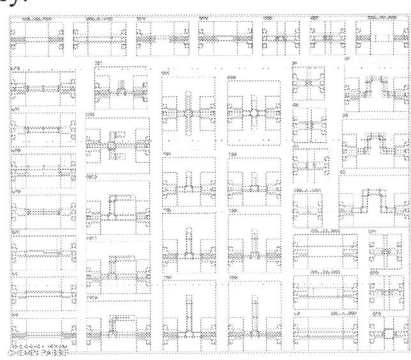

Figure 6: passive devices mask-set

For active devices, a slightly modified Gummel-Poon model is used, including a two-cell thermal sub-circuit to take thermal effects into account.

Modeling is based on various measurement [I(V), S-parameters] done for different temperatures. This allows to extract the thermal impedance Z_{TH}.

The methodology is described with more details in [10].

Load-pull measurements have been carried out over the Ka band (26-40 GHz) for various biasing and impedance conditions, and compared with simulations based on developed model.

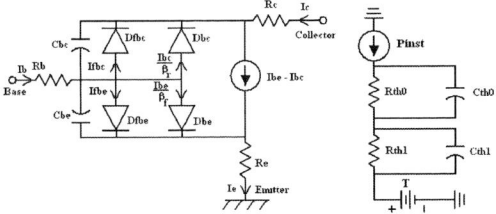

Figure 7: π GP model with thermal sub-circuit

Figure 8 shows one such comparison (for I_C=40 mA, I_B=1.8 mA, Z_{LOAD}=30+j6 Ω) for gain (top) and PAE (bottom); the good agreement achieved for operating conditions in the range of interest validates the model and its suitability for the design of amplifiers.

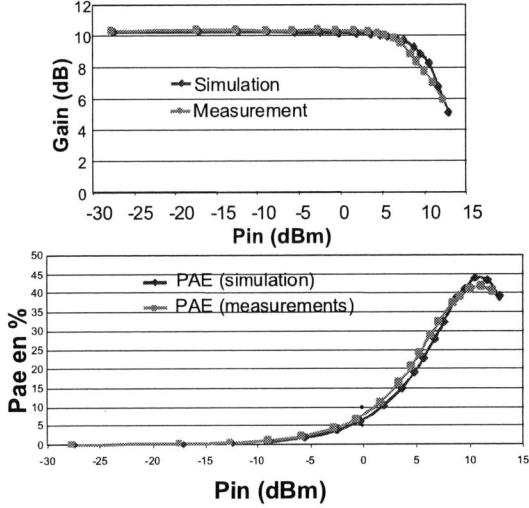

Figure 8: Load-pull measurement compared to simulation

V. MMIC DESIGN & CHARACTERIZATION

For assessment and validation purpose, 60-GHz amplifiers have been designed and characterized. 2 versions were studied, based on 2-finger 1x15 μm² devices: a 1-stage and a 2-stage amplifiers. The device biasing point is chosen as a trade-off between gain, junction temperature and breakdown voltage.

A. 1-stage amplifier

One-stage amplifier has been designed to assess modeling accuracy and device potential. Figure 9 shows small signal measurement vs. simulation comparison, with 5 dB gain at 60 GHz; $S_{11} \approx$ -15 dB and $S_{22} \approx$ -10 dB are also achieved.

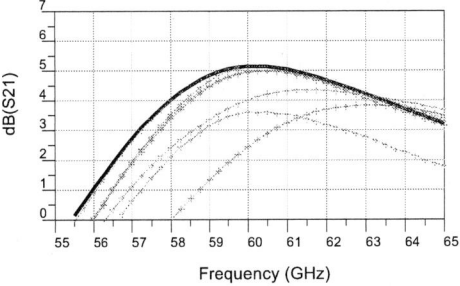

Figure 9: small-signal S_{21} measurements (dash) vs. simulation (solid)

Large signal measurements (gain=5.5 dB, P_{outMAX}=12.25 dBm and PAE=6.8%) confirm the good correlation obtained with simulations.

B. 2-stage amplifier

Figure 10 shows a microphotograph of the 2-stage 3.9x2.3 mm² amplifier.

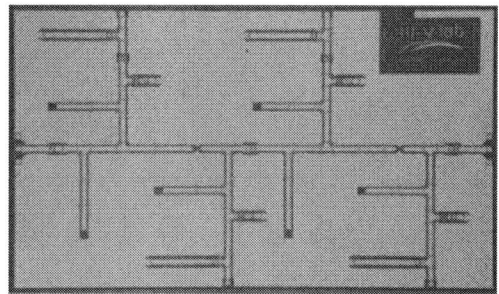

Figure 10: 3.9x2.3 mm² 2-stage amplifier µphotograph

Small signal measurements where carried out for various biasing conditions. Figure 11 shows S_{21} measurement for varying base currents, with a comparison with the simulation for $I_{B1,2}$ = 2.5 mA. A slight frequency shift is observed, as well as a 1 dB/stage gain improvement, attributed to minor epitaxy difference between modeled and final devices.

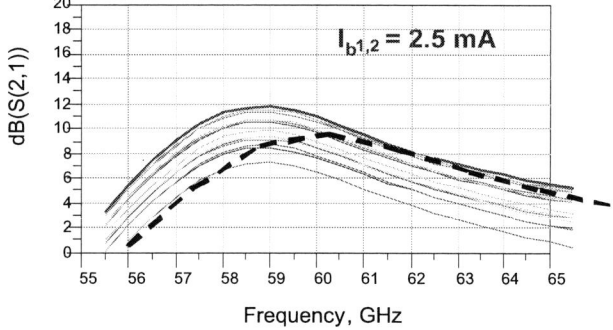

Figure 11: S_{21} measurements compared to simulation

While large signal simulations had let us expect >75 mW output power at P_{-3dB}, our input RF source delivering 8 dBm maximum limited our ability to reach 3 dB compression during the measurements. Results are shown on Figure 12, for biasing conditions: V_{CE1}=3 V, I_{C1}=35 mA, V_{CE2}=3 V, I_{C2}=55 mA, in a 50 Ω/50 Ω measurement setup.

For 2.4 dB compression, the measured linear gain is 9.8 dB, $Pout_{MAX}$=15.4 dBm and PAE_{MAX}=11%, with a good low level stability. We note a good correlation between measurements and simulation, with a low level gain difference of 0.5 dB.

VI. CONCLUSION AND PERSPECTIVES

A multi-finger InP/GaAsSb DHBT process has been developed, with frequency characteristics suitable for RF applications. Thanks to accurate modeling, MMIC amplifiers operating at 60 GHz have been designed and characterized. Further optimized vertical structure allowed to get >400 GHz F_{MAX} for 4-finger 0.7x10 µm² devices. These devices will allow improving amplifier performances, and are relevant for various RF applications, including E-band transmissions.

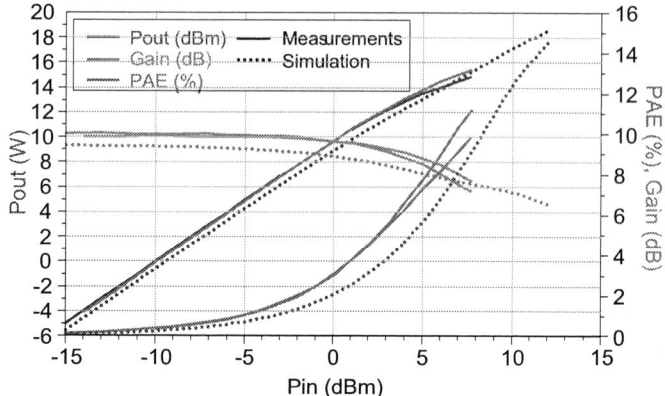

Figure 12: Large signal 60 GHz 2-stage amplifier measurements and simulation

ACKNOWLEDGMENT

Support from French National Research Agency (*Agence Nationale pour la Recherche*) through ANR/pNano ATTHENA program is gratefully acknowledged. GaAsSb based DHBT structures were grown by Picogiga International SAS.

REFERENCES

[1] RE Makon, R. Driad, K. Lösch, J. Rosenzweig, and M. Schlechtweg, "A Fully Integrated InP DHBT-Based CDR/1:2 DEMUX IC operating at 100 Gbit/s", CSICS '08, Oct. 08

[2] Z. Griffith, Yingda Dong, D. Scott, Yun Wei, N. Parthasarathy, M. Dahlstrom, C. Kadow, V. Paidi, M.J.W. Rodwell, M. Urteaga, R. Pierson, P. Rowell, B. Brar, Sangmin Lee, N.X. Nguyen, C. Nguyen, "Transistor and circuit design for 100-200-GHz ICs", IEEE Journal of Solid-State Circuits, Volume 40, Issue 10, Oct. 2005, pp. 2061-2069

[3] V. Radisic, D. Sawdai, D. Scott, W. R. Deal, L. Dang, D. Li, A. Cavus, A. Fung, L. Samoska, R. To, T. Gaier, R. Lai, "Demonstration of 184 and 255-GHz Amplifiers Using InP HBT Technology", MWCL-18, n° 4, April 2008, pp. 281-283

[4] Y. Baeyens, N. Weimann, V. Houtsma, J. Weiner, Y. Yang, J. Frackoviak, P. Roux, A. Tate, Y.K. Chen, "Submicron InP D-HBT single-stage distributed amplifier with 17 dB gain and over 110 GHz bandwidth", IMS 06 Digest, June 2006, pp. 818-821

[5] R.-E. Makon, R. Driad, K. Schneider, R. Aidam, M. Schlechtweg, G. Weimann, "Fundamental W-Band InP DHBT-Based VCOs With Low Phase Noise and Wide Tuning Range", IMS 07 Digest, 3-8 June 2007, pp. 649-65

[6] J. Godin, V. Nodjiadjim, M. Riet, P. Berdaguer, O. Drisse, E. Derouin, A. Konczykowska, J. Moulu, J.-Y. Dupuy, F. Jorge, J.-L. Gentner, T. Johansen, V. Krozer and A. Scavennec, "Submicron InP DHBT technology for high-speed high-swing mixed-signal ICs", CSICS '08, Oct. 08

[7] A. Konczykowska, M. Riet, P. Berdaguer, P. Bove, M. Kahn and J. Godin, "40 Gbit/s digital IC fabricated using InP/GaAsSb/InP DHBT technology", IEE Elec. Letters, Vol. 41 No. 16, 4th August 2005, pp. 905-906

[8] V. Nodjiadjim, M. Riet, A. Scavennec, P. Berdaguer, J.-L. Gentner, J. Godin, P. Bove, M. Lijadi, "Comparative collector design in InGaAs and GaAsSb based InP DHBTs", IPRM 2008, May 2008, paper WeB1.7

[9] M. Riet, V. Nodjiadjim, A. Scavennec, O. Drisse, E. Derouin, J. Godin, P. Bove, M. Lijadi, "GaAsSb/InP multifinger DHBTs for power applications", IPRM 2008, May 2008, paper WeP64

[10] S. Laurent, AA Lisboa de Souza, JC Nallatemby, M. Prigent, M. Werquin, C. Gaquière, M. Riet, P. Berdaguer, "Non-linear InP/GaAsSb model validated by 40 GHz load-pull measurement", presented at JNM 2009, May 2009

Sub-mW Operation of InP HEMT X-band Low-Noise Amplifiers for Low Power Applications

C. H. Lin, X. B. Mei, Y. C. Chou, L. S. Lee, J. M. Yang, M. Y. Nishimoto, P. H. Liu, R. To, A. Cavus, R. Tsai, M. Wojtowicz and R. Lai

Northrop Grumman Corporation, One Space Park, Redondo Beach, CA 90278

Abstract — For the first time, the sub-mW operation of InP HEMT X-band low-noise amplifiers on 4-inch InP wafer was demonstrated. With optimized non-alloyed ohmic contact, gate recess profile and epitaxial layer design, the InP HEMT achieves average peak transconductance of 1150 mS/mm at V_{DS}=0.3 V. The mean current cut off frequency is above 150 GHz at V_{DS}= 0.3 V and I_{DS}= 150 mA/mm. The developed low power InP HEMTs enables the manufacturing of low-noise amplifiers for low power applications. The superior performance of X-band low-noise amplifiers was also demonstrated. Operating at a supply voltage of 0.25 V and drain current of 3.75 mA with DC power of 0.937 mW, the low noise amplifier exhibits noise figure of 1.6 dB and gain of 11 dB at frequencies from 6 to 12 GHz.

Index Terms — InP HEMT, X-band, Low-noise amplifiers.

I. INTRODUCTION

Recently, low-noise amplifiers (LNAs) for low power operation have been developed to reduce the total power consumption in a large-aperture phased-array system. A decrease of DC power dissipation for each component would result in DC power reduction on the order of kW for a large phased-array system with millions of LNAs. This is important for systems requiring small size, light weight and low cost. As a result, device technologies of antimonide-based compound semiconductor (ABCS) high-electron-mobility-transistor (HEMT) and SiGe-based heterojunction bipolar transistor (HBT) are being developed for low power operations [1-2].

On the other hand, the InP-based HEMTs have been proven to be a mature technology for high-speed and low-noise applications [3]. However, its potential for low power LNAs is still not explored. In this study, superior performance of InP HEMT LNAs operating at low V_{DS} ($\leq$0.3 V) was demonstrated. The achieved X-band LNAs exhibit sub-mW operation with an associated gain of 11 dB and noise figure of 1.6 dB, respectively

II. DEVICE DESIGN AND FABRICATION

Figure 1 shows the cross-section of the epitaxial and device structures used in this study. The epilayer structure was grown by molecular-beam epitaxy (MBE) on 4-inch semi-insulating InP (100) substrates. The epitaxial profile consists of an InAlAs buffer, an $In_{0.60}Ga_{0.40}As$ channel, an InAlAs spacer, a Si delta-dope layer, an InAlAs schottky, and an n^+ InGaAs/InAlAs cap layer. The sheet charge density and

mobility measured from a capless epi-structure are 3.5 x10^{12} cm^{-2} and 10500 cm^2/V-sec, respectively. Device isolation was performed by wet etching and ion implantation. The evaporated ohmic contact metals form a non-alloyed ohmic contact [4]. The T-shaped gate was defined uisng electron beam lithography with gate length of 0.1 µm. The gate recess was achieved using a wet etching technique. The devices and circuits were fully passivated with silicon nitride deposited by plasma-enhanced chemical vapor deposition (PECVD) technique. The passive elements of the MMIC consist of 300 pF/mm^2 nitride based metal-insulator-metal capacitors, 100 Ω/sq thin film resistor and two levels metal interconnects.

III. DEVICE PERFORMANCE

Figure 2 shows the drain I-V characteristics of a low power InP HEMT in this study with gate voltage (V_{GS}) swept from –1 V to 0.4 V. The knee voltage (V_{KNEE}) is the drain-source voltage (V_{DS}) for the transition of drain current (I_{DS}) from linear to saturation region. For devices to reach saturated I_{DS} of 400 mA/mm, it was observed that V_{KNEE} is around 0.3V for the low power InP HEMT. The maximum drain current (I_{MAX}), defined as the I_{DS} at V_{GS}= 0.4V for V_{DS}= 1V, is higher than 700 mA/mm for the developed low power InP HEMT. The low V_{KNEE} and high I_{MAX} allow the transistor to operate at lower

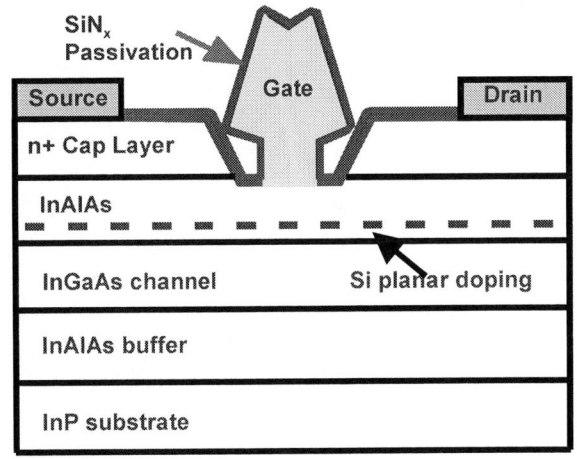

Figure 1: The schematic cross-section of a low power InP HEMT.

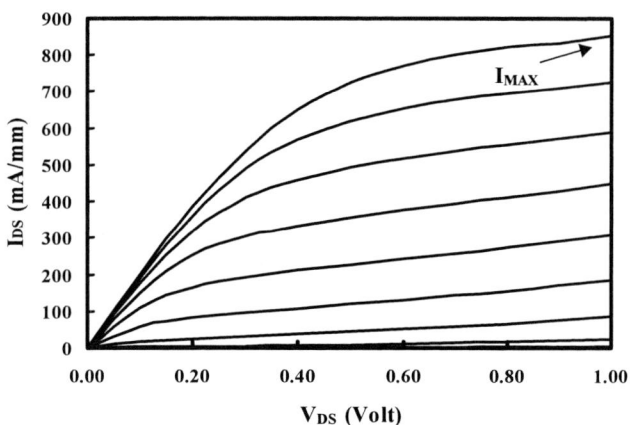

Figure 2: The drain I-V characteristics of a low power InP HEMT (V_{GS} of top curve= 0.4V, V_{GS} step = 0.1V).

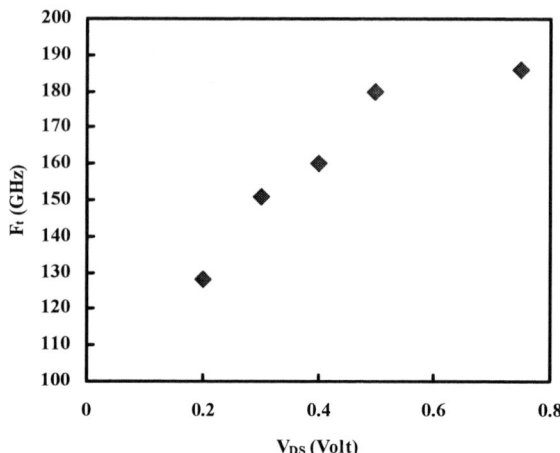

Figure 4: F_t versus V_{DS} at I_{DS}= 150 mA/mm of the low power HEMT.

drain voltages. These improvements were realized by optimizing non-alloy ohmic contact, gate recess process, and epitaxial structure to achieve low source and drain resistances (R_s and R_D). The average ohmic contact resistance (R_c) of the low power InP HEMTs is approximately 0.04 ohm-mm. Typical R_s and R_D values are below 0.3 ohm-mm for the low power InP HEMT.

The extrinsic voltage (V_{DS}) across the source and drain pads and the potential across channel (V_{DS}') have the relation of $V_{DS}= V_{DS}'+ (R_s + R_D) \times I_{DS}$. Lower R_s and R_D can reduce the power consumed by the parasitic elements and also bias the channel more efficiently for an applied V_{DS}. In addition, the extrinsic transconductance (G_M) is related to the intrinsic transconductance of the transistor (G_{M0}) according to G_M =

$G_{M0}/(1+R_sG_{M0})$. Therefore, reduction of R_s and R_D effectively improves not only V_{KNEE} but also G_M at low drain voltage. Figure 3 shows the I_{DS}-G_M characteristics of 36 low-power InP HEMTs biased at V_{DS}= 0.3 V across a 4-inch wafer. The average peak G_M value is approximately 1150 mS/mm with standard deviation of 67 mS/mm while the average V_{GS} value at peak G_M is around 0.14 V with standard deviation of 60 mV.

Moreover, the radio frequency (RF) characteristics of the low power InP HEMT was also characterized on 4-fingers 200 μm-wide transistors at various V_{DS} and I_{DS} conditions. Figure 4 shows the current gain cut-off frequency (F_t) with respect to V_{DS} at fixed I_{DS} of 150 mA/mm. The F_t is approximately 153 GHz for V_{DS}= 0.3 V. The F_t values increase with increasing V_{DS}, which is attributed to the higher G_M values at higher V_{DS}.

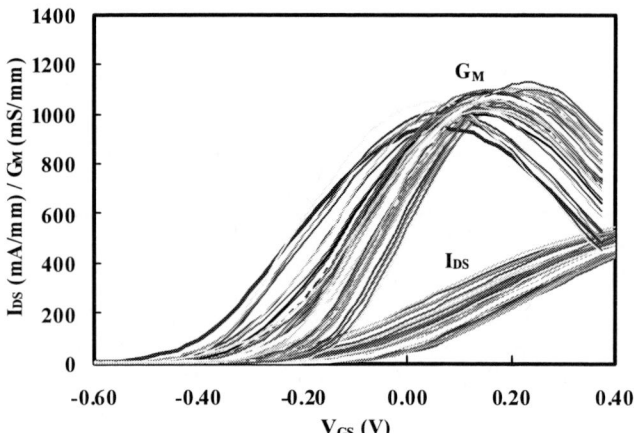

Figure 3: The I_{DS}-G_M versus V_{GS} characteristics at V_{DS} of 0.3 V on 200 μm transistors across a 4-inch wafer.

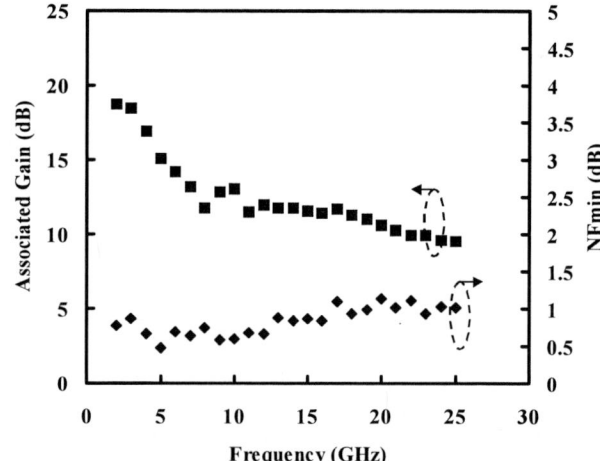

Figure 5: The NF and associated gain of a low power InP HEMT biased at V_{DS}= 0.3 V, I_{DS}= 75 mA/mm.

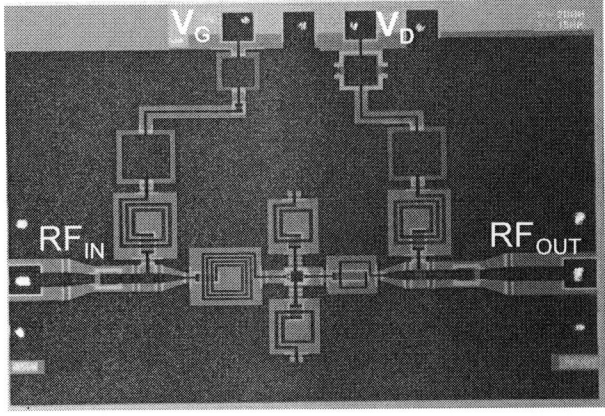

Figure 6: Photograph of a single stage X-band LNA used in this study.

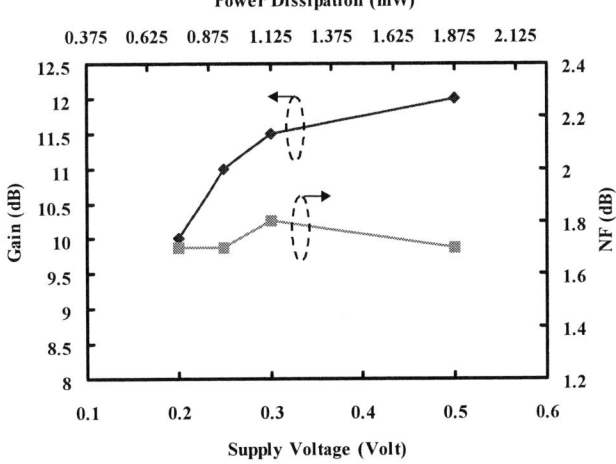

Figure 8: Gain and NF versus supply voltage and DC power dissipation for a InP HEMT X-band LNA at 10 GHz.

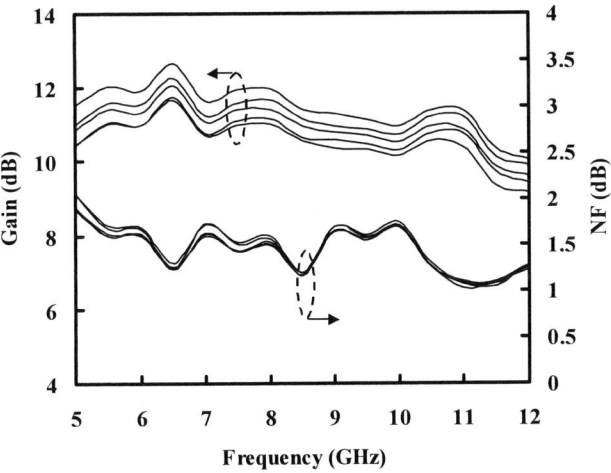

Figure 7: The NF and gain performance of single stage X-band LNAs operating at 0.25 V and 3.75 mA

The F_t values are around 180 GHz for V_{DS} higher than 0.5 V and exhibit less dependence on V_{DS}. The average maximum available gain (MAG) at 26 GHz is 12.4 dB for V_{DS}= 0.3 V and I_{DS}= 150 mA/mm. The noise figure (NF) performance at V_{DS}= 0.3 V and I_{DS}= 75 mA/mm of a 4-fingers 120 μm-wide transistor is also shown in Figure 5. It is noticed that the minimum NF is smaller than 1 dB with associated gain of 11.5 dB at 12 GHz for the HEMT operating at DC power of 22.5 mW/mm.

IV. LNA RF PERFORMANCE

RF performances of monolithic single-stage X-band LNAs (gate periphery of 80 μm as shown in Figure 6) with coplanar design were also evaluated. The S-parameter and NF were measured using an on-wafer probe station. As shown in Figure 7, the gain is around 11 dB and the NF is approximately 1.6 dB at frequencies from 6 GHz to 12 GHz at supply voltage (V_{DD}) of 0.25 V and total drain current (I_{DD}) of 3.75 mA. The total dc power consumption of the LNA is 0.937 mW. For the first time, the InP HEMT LNAs with sub-mW operation was demonstrated.

Furthermore, performance of the LNA over various bias conditions has also been evaluated. Figure 8 shows the performance of the low DC power LNA with respect to the DC supply power for various V_{DD} with constant I_{DS}. It was observed that the NF is below 1.8 dB and is not sensitive to the V_{DD}, while the gain increases from 10 dB to 12 dB when V_{DD} increases from 0.2 V to 0.5 V. By trading off performance against DC power, the LNA also demonstrates the capability of operating at 0.75 mW. These results indicate that the technology provides enough margin for the sub-mW operation.

Table 1 summarizes recently reported state-of-art low DC power X-band LNAs from several device technologies. It was noticed that the noise performance of the LNAs is superior to that of the SiGe-based HBT and is comparable with ABCS-based HEMT. Evidently, the developed InP HEMT technology here offers an alternative device technologies for low power applications.

Table I: Summary of state-of-art low power X-band LNAs.

Technology	Supply Voltage (Volt)	Current (mA)	Gain @ 10 GHz (dB)	NF @ 10 GHz (dB)	DC Power (mW)	Ref
InP HEMT (this work)	0.25	3.75	11	1.6	0.937	
SiGe HBT	1.2	0.7	10	3.6	0.8	[2]
ABCS HEMT	0.15	9.2	18	1.8	1.38	[1]
ABCS HEMT	0.25	3.93	12.2	N/A	0.98	[5]

V. CONCLUSION

A low power InP HEMT technology with good performance was demonstrated. The achieved X-band LNAs exhibit operation at DC power of 0.937 mW with NF and gain of 1.6 dB and 11 dB, respectively. The low power InP HEMT technology demonstrated here offers an alternative device technologies for large-aperture phased-array applications requiring millions of LNA elements.

REFERENCES

[1] W. R. Deal, R. Tsai, M. D. Lange, J. B. Boos, B. R. Bennett, and A Gutierrez, "A low power/low noise MMIC amplifier for phase-array applications using InAs/AlSb HEMT" 2006 IEEE MTT-S Int. Microwave Symp. Dig, pp. 2051-2054, June 2006.

[2] P. Roux, Y. Baeyens, J. Weiner and Y. K. Chen, "Ultra-low-power X-band SiGe HBT low noise amplifier" 2007 IEEE MTT-S Int. Microwave Symp. Dig, pp. 1787-1790, June 2007.

[3] R. Lai, Y. C. Chou, L. J. Lee, P. H. Liu, D. Leung, Q. Kan, X. Mei, C. H. Lin, D. Farkas, M. Barsky, D. Eng, A. Cavus, M. Lange, P. Chin, M. Wojtowicz, T. Block, A. Oki,"High performance and high reliability of 0.1 um InP HEMT MMIC technology on 100 mm InP substrates" 2007 IEEE Intl. Conf. InP and Related Materials Proc., pp. 63-66, May 2007.

[4] X. B. Mei, D. Farkas, W. B. Luo, C. H. Lin, J. Lee, W. Liu, P. H. Liu, A. Cavus, and R. Lai, "InGaAs/InAlAs/InP power HEMT with an improved ohmic contact and an extremely high operating voltage" 2009 IEEE Intl. Conf. InP and Related Materials Proc., pp. 204-206, May 2009.

[5] J. B. Hacker, J. Bergman, G. Nagy, G Sullivan, C. Kadow, H. K. Lin, A. C. Gossard, M. Rodwell and B. Brar, " An ultra-low power InAs/AlSb HEMT X-band low noise amplifier and RF switch" Proceeding of CS Mantech Conference, pp. 239-242, 2006.

Accurate HEMT Switch Large-Signal Device Model Derived from Pulsed-Bias Capacitance and Current Characteristics

Shinichiro Takatani[1] and Cheng-Duan Chen

WIN Semiconductors Corp.
No. 69, Technology 7th Rd., Hwaya Technology Park, Kuei Shan Hsiang, Tao Yuan Shien, Taiwan 333
[1]takatani@winfoundry.com

Abstract—**Large-signal operation of HEMT (High Electron Mobility Transistor) switch device is found to be much affected by trap-induced slow dynamic effects. Off-state gate and drain capacitances derived from pulsed S-parameter measurements with a large negative gate quiescent voltage are less voltage-dependent than capacitances measured by a continuous bias. Two-dimensional device simulation suggests that surface traps near the gate edge are responsible for the observed dynamic effect. A new large-signal device model is developed that takes both C-V and I-V pulsed dynamic behavior into account. The new model is shown to accurately predict harmonics generated from off- and on-state switches.**

Keywords—*HEMT; switch; pulsed measurement; device model*

I. INTRODUCTION

There is an increasing demand of accurate large-signal device model for HEMT (High Electron Mobility Transistor) switch. RF switch circuits require design of insertion loss, isolation as well as nonlinear properties. For antenna switch in hand phone power amplifier modules, nonlinear properties such as harmonics [1], [2] and inter-modulation distortion [3] must be carefully designed to meet the specifications of communication systems. There are some efforts in improving HEMT/MESFET device model for switch design [4]. However, there still needs improvement to accurately design nonlinear behaviours of switch circuits.

On the other hand, pulsed measurement technique is frequently used to characterize thermal and low-frequency dispersive effects in HEMT/MESFET, and device models incorporating those effects have been developed for amplifier design [5], [6], [7]. However, there has been no report on device modeling based on pulsed-bias measurements for switch application.

In this paper, new pulsed dynamic effects in HEMT and a device model incorporating them for designing switch are reported. Pulsed S-parameter measurements show significantly different voltage dependence in high-frequency capacitance at off-state, responsible for nonlinearity of off switch, compared to those obtained by a nonpulsed measurement. This effect is shown to be important in accurately simulating nonlinearity of off-switch.

II. PULSED DYNAMIC EFFECT IN HEMT SWITCH DEVICE

The device investigated in this study is a 0.5μm gate double heterostructure InGaAs channel pseudomorphic HEMT on GaAs. Devices with 75μm and 125μm finger lengths with a total gate width from 150 to 1000μm are used for device characterization, modeling and model verification.

A. Pulsed S-parameter and I-V Measurements

In the pulsed S-parameter and I-V measurement, the gate quiescent voltage (Vgsq, Vgdq) is varied from on state (=0.5V) to a large negative voltage covering voltages at which a HEMT operates in a switch circuit. Drain quiescent voltage (Vdsq) is kept at zero. Pulse width is 1μsec, and pulse period is 1msec. S-parameters are measured up to 10GHz.

Fig. 1 shows gate capacitance-voltage characteristics obtained from nonpulsed and pulsed measurements. Cgs is derived from Y-parameters as Cgs=(Im(Y11)+Im(Y12))/ω. C-V curves are significantly different for nonpulse and pulse modes. At pulse mode, off-state Cgs is less voltage dependent, i.e. slope is less steep, than that measured at nonpulse mode. The voltage dependence decreases with increasing negative gate quiescent voltage. Furthermore, the threshold voltage at which capacitance rises shifts toward less negative voltage at pulse mode. The shift becomes larger as the gate quiescent voltage becomes more negative. Cgd behaves similarly to Cgs.

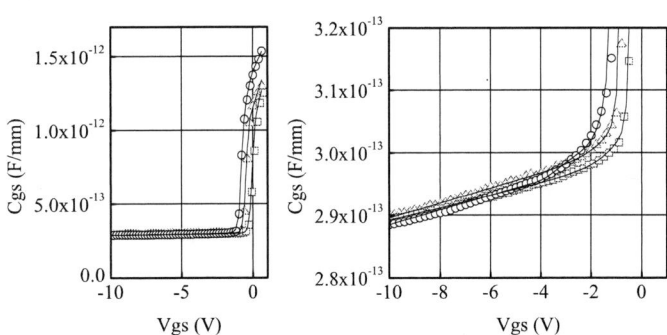

Fig. 1 Cgs derived from nonpulsed and pulsed S-parameter measurements at 10GHz. Magnified off-state characteristics are show on the right. Circle: nonpulsed, triangle: pulsed at Vgsq=-3V, square: pulsed at Vgsq=-5V. Vdsq=0 for both cases. Lines are modeled.

Drain capacitance, Cds, depends on gate voltage. The dependence is small, but is large enough to contribute to harmonic generation from an off-switch. Fig. 2 shows Cds (=(Im(Y22)+Im(Y12))/ω) at Vds=0 as a function of Vgs for pulse and nonpulse modes. Similarly to Cgs, capacitance is less voltage dependent at pulse mode than nonpulse mode.

A pulsed dynamic behavior is also observed in I-V characteristics (i.e. gate lag). Fig. 3 shows Id-Vgs curves measured at different Vgsq. Threshold voltage shifts to less negative voltage for more negative quiescent voltage in accordance with the shift observed in C-V curves.

B. Physical Model of Capacitance Pulsed Dynamic Effect

The pulsed dynamic behavior observed in this study is different from the low-frequency capacitance dispersion reported by other researchers [8]. In [8], excess input capacitance is observed at a low frequency. In this study, however, capacitances are measured at microwave frequencies. High-frequency capacitance in semiconductor is determined by the distance between gate electrode and carrier depletion front in the semiconductor. At an off state, the position of carrier depletion front is affected by trapped charge at recess surface. We examined the effect of surface traps employing two-dimensional device simulation. Deep levels consisting of a pair of donor and acceptor [9] were introduced on the recess surface near the edge of gate electrode, as shown in the inset of Fig.4 (a). Simulated C-V curves at 1GHz for devices with and without surface traps are shown in Fig. 4 (a). The device without surface traps shows a large capacitance slope at voltage below threshold. However, the capacitance slope becomes smaller for the device with surface traps. The traps charge negatively, and the amount of negative charge increases as the gate is biased more negatively. The depletion front moves away from the gate, giving low and flat capacitance curve.

C-V curves measured for actual devices are closer to the one simulated with the surface traps. Presence of surface traps favorably reduces voltage dependence of gate capacitance,

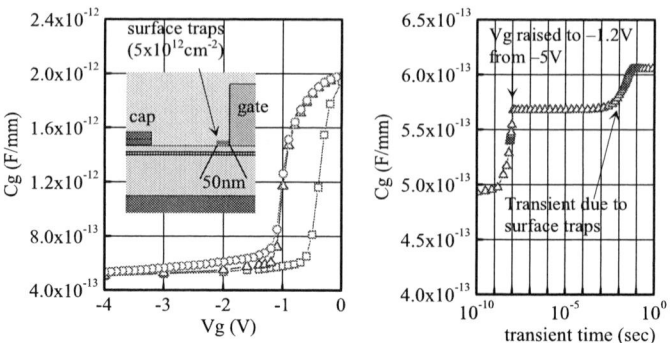

Fig. 4 (a) Gate capacitance (Cgs+Cgd) at 1GHz calculated by two-dimensional device simulation. Circle: no surface trap, triangle: with surface traps, square: pulsed-bias with surface traps. Inset shows the device structure near the gate edge. (b) Transient response after Vg is raised in 10nsec from -5V to -1.2V.

thereby reducing nonlinearity of off-switch.

Even more interestingly, C-V curve is further flattened at pulse operation. Fig. 4 (b) shows the transient response of gate capacitance after the gate voltage is raised in 10nsec from –5V to –1.2V. A transient with a millisecond time constant determined by the physical parameters of surface traps is seen. Capacitances before the transient are evaluated for various pulse gate voltages, and the obtained pulsed-bias C-V curve (Vgsq=-5V) is shown in Fig. 4 (a). The capacitance is even less voltage-dependent, and the threshold voltage shifts to a less negative voltage, reproducing what is observed in the pulsed bias measurement. The trapped charge at surface freezes at pulse operation, i.e. the measurement is in an isotrapping condition. The large negative charge created at quiescent bias keeps the depletion front away from the gate, giving capacitance even smaller and less voltage-dependent. The effect is low-frequency dispersive in nature, although the effect is observed in high frequency capacitance.

Assuming that there is no further dispersion in the time faster than the pulse used in the measurement (1μsec), it is the C-V curve measured at pulse mode that is actually experienced by large microwave signal applied to off-state FET in a switch circuit.

III. LARGE-SIGNAL DEVICE MODEL

A. Model Formulation

Analytical expressions for current and capacitance used for device modeling are based on Angelov model [10], [11], but some modifications are introduced to improve accuracy of characteristics key for switch applications.

1) Gate Capacitance: To improve the accuracy of voltage-dependent gate capacitance Cgs at off-state, a polynomial function ΔCgs shown below is added to the gate capacitance given in Angelov model.

$$\Delta Cgs = \Sigma Cgs_n (Vgs - Vgs0)^n \qquad (1)$$

The polynomial is expanded at a large negative voltage, Vgs0. The value is chosen so that RF signal does not swing beyond Vgs0 in a switch circuit to be designed. In this study, Vgs0 is set at –10V. Minimum number of terms is chosen to fit the measured C-V curves to avoid ripple that is often created in

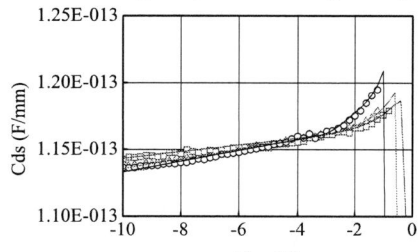

Fig. 2   Cds derived from nonpulsed and pulsed S-parameter measurements at 10GHz. Circle: nonpulsed, triangle: pulsed at Vgq=-3V, square: pulsed at Vgq=-5V. Vds=0 and Vdsq=0. Lines are modeled.

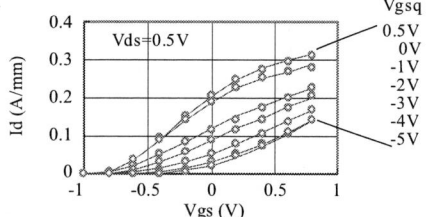

Fig. 3 Pulsed-bias Id-Vgs characteristics. Symbols are measured, and lines are modeled. Vgsq is varied from 0.5V to –5V. Vdsq=0 in all cases.

polynomial fitting with many terms. The curves in Fig.1 are actually fitted with only 1st and 11th polynomials.

Vds dependence given in the original Angelov model is neglected, so that Cgs and Cgd depend only on their terminal voltages. This simplification is acceptable for a switch model. The same equations and parameter values are used for Cgs and Cgd for symmetry required for a switch model. In the actual model, the gate capacitance is implemented using a charge formulation.

Using the model, Pulsed-bias C-V characteristics taken at various gate quiescent voltages are fitted using a circuit simulator, and model parameters are extracted. Good fit, particularly at off state, is obtained as shown in Fig.1.

2) Drain Capacitance: Cds is also expressed by polynomial functions as shown below:

$$Cds = \Sigma Cds_n(Vgeff-Vgs0)^n$$
$$Vgeff = (Vgs+Vgd+sqr((Vgs-Vgd)^2+\delta^2))/2. \quad (2)$$

The model is symmetric with respect to source and drain. The parameters at various Vgsq are extracted for Vds=0 and Vdsq=0. A good fit is obtained as shown in Fig. 2. Again, only 1st and 11th polynomials are used. δ in (2) is adjusted to fit Cds at nonzero Vds.

The drain capacitance model is implemented using direct capacitance formulation instead of charge formulation to avoid introduction of transcapacitance.

3) Drain Current: Angelov model symmetric with respect to source and drain is used for drain current formulation. The model fits fairly well at voltage range for switch operation. The fitting results are shown for Id-Vgs in Fig. 3 for various quiescent voltages. Fig. 5 shows pulsed Id-Vds characteristics for Vgsq=0.5V and the fitting results. The model fits well at on state up to Vds=0.8V and Vgs=0.8V.

4) Gate Current: An accurate gate Schottky model is required for switch design, because the reverse and forward currents affect the effective gate bias voltage of both on and off FETs. Angelov model accurately fits forward characteristics. For improving the accuracy in reverse characteristics, an exponential term is added. The whole gate current is formulated as (see [11] for parameter name definitions),

$$I_{GS}=IJ*(exp(PG*tanh(2*(Vgs-VJG)))-exp(PG*tanh(-2*VJG)))$$
$$- IJR*(exp(R1*(-Vgs)+R3*(-Vgs)^3)-exp(0)), \quad (3)$$

where parameters IJR, R1, and R3 in the additional term have positive values. Good fit to measured curves is obtained as shown in Fig. 6.

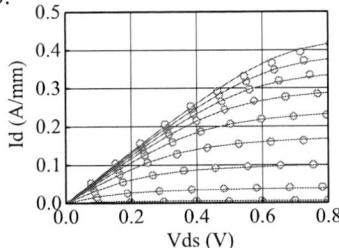

Fig. 5 Pulsed Id-Vds characteristics measured at Vgsq=0.5V (symbol). Lines show modeled curves. Vgs is varied from 0.8 V with –0.2V step.

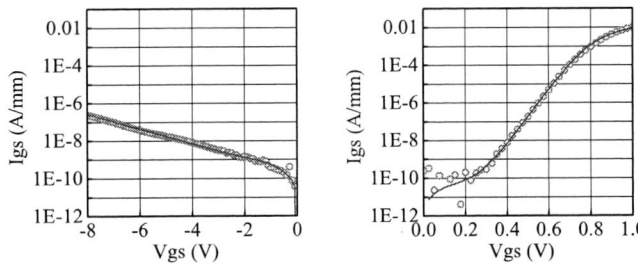

Fig. 6 Measured (symbols) and modeled (lines) gate I-V characteristics

B. Modeling of Dynamic Behavior

The component models described in the previous section are implemented into an intrinsic FET equivalent circuit shown in Fig. 7. RC reference circuits [6], [7] are introduced at gate-source and gate-drain to separate the DC voltage component (Vgsq, Vgdq) from the total voltage Vgs and Vgd. It simulates trap dynamics with a simplification that net charging and discharging processes have a same time constant. Note that in the present model the capacitance in the RC circuit is not intended to contribute to a low frequency response, so that a negligible value is chosen for C with an RC time constant characteristic of gate lag phenomena of the device.

For gate/drain capacitance and drain current, the sets of model parameter values extracted from pulsed-bias characteristics at various Vgsq are brought into the equivalent circuit as a look-up table. The parameter values are checked if they vary fairly smoothly with gate quiescent voltage. The parameter values are read from the table using interpolation function built in a commercial circuit simulator. For Cgs and Cgd, parameter values are read at Vgsq and Vgdq, respectively, in the equivalent circuit shown in Fig. 7. For Cds and Id, parameter values are read at an average value of Vgsq and Vgdq to avoid using different parameter set at positive and negative Vds, avoiding discontinuity at Vds=0V. This simplification does not affect simulation results if the switch is operated with no drain DC bias.

This table-based approach makes modeling simple and yet accurate. Since only parameter values are tabulated, scalability provided by the analytical models is maintained.

IV. MODEL VERIFICATION

A. Harmonic Distortion

Harmonic distortions generated from off-FET and on-FET are measured for device in shunt and through configuration, respectively, with the gate DC bias (VgsDC) fed through 10kΩ

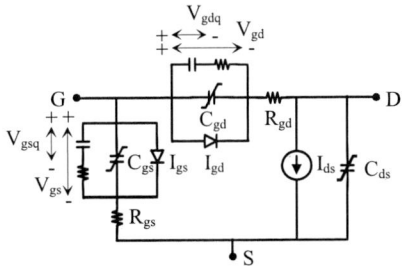

Fig. 7 Large-signal equivalent circuit of intrinsic FET.

resistor. The fundamental frequency is 0.9GHz.

Fig. 8 shows the input power dependence of measured and simulated 2nd and 3rd harmonic distortions (2HD, 3HD) generated from off-FET at various VgsDC. Improvement in accuracy for the model derived from pulsed characteristics, particularly for 3HD, is remarkable. HD's at low input power due to the slope of off-state C-V curve and their sharp rise at high input power due to C-V and I-V characteristics around threshold voltage are both accurately simulated.

Fig. 9 shows the measured and simulated 2nd and 3rd HD's from on-FET at VgsDC=0 and 0.5V. Fairly good fit is obtained. Improvement in simulation accuracy is not significant for the on-state harmonics. It is, however, noted that, incorporating pulsed dynamic effect in I-V characteristics, the model is made capable of simulating transient phenomena due to the gate lag in a voltage swing from off-state to on-state.

B. Insertion Loss and Isolation

Fig. 10 (a) shows measured and simulated small-signal insertion loss of a device in the through configuration. The error in simulation is less than 0.03dB. Fig 10 (b) shows the transfer characteristics from on-state to deep off-state. S21 at off-state (i.e. isolation) is fitted well at every bias point owing to accurate modeling of voltage-dependent Cgs, Cgd and Cds.

V. CONCLUSIONS

Large-signal operation of HEMT switch device is much affected by trap-induced slow dynamic effect. A model derived from pulsed-bias S-parameter and I-V characteristics

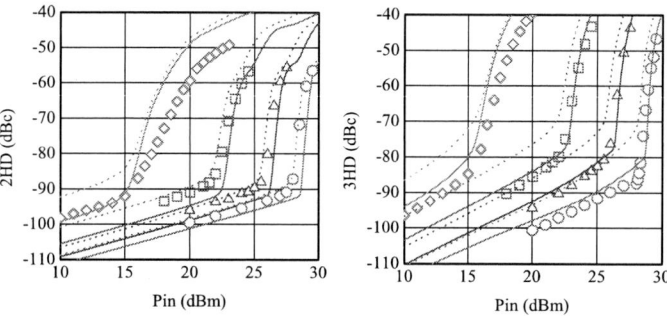

Fig. 8   2nd and 3rd harmonic distortion generated from off-state FET (Wg=125x8μm). Fundamental frequency is 0.9GHz. Circle: VgsDC=-5V, triangle:VgsDC=-4V, square: VgsDC=-3V, diamond: VgsDC=-2V. Solid lines: simulation with model extracted from pulsed characteristics. Dotted lines: simulation with model extracted from nonpulsed characteristics.

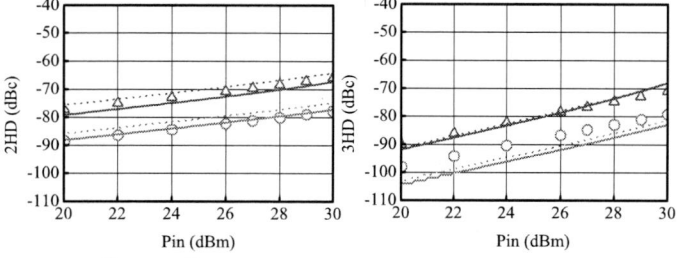

Fig. 9   2nd and 3rd harmonic distortion generated from on-state FET (Wg=125x8μm). Fundamental frequency is 0.9GHz. Circle: VgsDC=0.5V, triangle:VgsDC=0. Solid lines: simulation with model extracted from pulsed characteristics. Dotted lines: simulation with model extracted from nonpulsed characteristics.

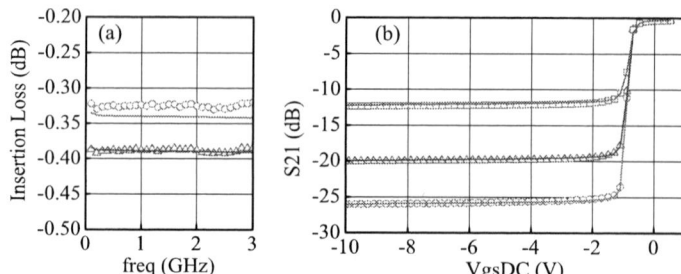

Fig. 10 (a) Small-signal insertion loss of Wg=125x4μm device measured at VgsDC=0.5V (circle) and 0V (triangle). Lines are simulated. (b) Transfer characteristics of the same device as a function of VgsDC. Circle: 1GHz, triangle: 2GHz, square: 5GHz. Lines are simulated.

successfully simulates the device behavior. The model is expected to enable accurate nonlinear design of RF switch such as antenna switches for hand phone and WLAN modules.

ACKNOWLEDGMENT

Authors thank Prof. Angelov of the Chalmers University for useful discussions on model formulation. They also thank J. Liu of WIN for his help in harmonics measurements, R. Kuo of WIN for his help in RF measurements, S. J. Li of WIN for his help in 2D device simulation, and National Nano Device Laboratories in Taiwan for their help in pulsed-bias measurements.

REFERENCES

[1]   D. Prikhodko, Y. Tkachenko, S. Sprinkle, R. Carter, S. Nabokin, and J. Chiesa, "Design of a low VSWR harmonics, low loss SP6T switch for GSM/Edge applications," in Proc. 2nd European Microwave Integrated Circuits Conference, 2007, p. 32

[2]   Michael Roberts, Luffi Albasha, Wolfgang Bosch, Damian Gotch, James Mayock, Pallavi Sandhiya, Ian Bisby, "Highly linear low voltage GaAs PHEMT MMIC switches for multimode wireless handset applications," in Proc. Radio and Wireless Symposium, 2001, p.61

[3]   T. Ranta and J. Ella, and H. Pohjonen, "Antenna switch linearity requirement for GSM/WCDMA mobile phone front-ends," in Proc. The European Conference on Wireless Technology, 2005, p. 23.

[4]   M. A. Holm, and D. M. Brookbanks, "Advanced meander gate p-HEMT model for accurate harmonic modeling of switch MMIC designs," in GAAS 2005, p. 317.

[5]   T. Fernandez, Y. Newport, J. M. Zamanillo, and A. Tazon, "Extracting a Bias-Dependent Large Signal MESFET Model from Pulsed I/V Measurements," IEEE Trans. Microwave Theory and Techniques, Vol. 44, No. 3, pp..372-378, March 1996

[6]   G. Rafael-Valdivia, R. Brady, and T. J. Brazil, "Single function drain current model for MESFET/HEMT devices including pulsed dynamic behavior," IEEE MTT-S International Microwave Symposium Digest, 2006, p.473.

[7]   G. Rafael-Valdivia, R.Brady, and T. J. Brazil, New drain current model for MESFET/HEMT devices based on pulsed measurements," in The 1st European Microwave Integrated Circuit Conference, 2006, p.289

[8]   J. Graffeuil, Z. Hadjoub, J. P. Fortea, and M. Pouysegur, "Analysis of Capacitance and Transconductance frequency dispersions in MESFETs for Surface Characterization," Solid-State Electronics, Vol. 29, No.10, pp.1087-1097, 1986

[9]   Y. Mitani, D. Kasai, and K. Horio, "Analysis of Surface-State and Impact-Ionization Effects on Breakdown Characteristics and Gate-Lag Phenomena in Narrowly Recessed Gate GaAs FET," IEEE Trans. Electron Devices, Vol. 50, No. 2, pp.285-291, February 2003

[10] I. Angelov, L. Bengtsson, and M. Garcia, "Extension of the Chalmers Nonlinear HEMT and MESFET Model," IEEE Trans. Microwave Theory and Techniques, Vol. 44, No. 10, pp. 1664-1674, October 1996.

[11] Agilent Technologies, "ADS Documentation, Nonlinear Devices", August 2005, P. 3-7.

An Image Reject Mixer for High-Speed E-band (71-76, 81-86 GHz) Wireless Communication

Marcus Gavell[1,2], Mattias Ferndahl[1,2], Sten E. Gunnarsson[1,2], Morteza Abbasi[1,2], Herbert Zirath[1,2]

[1]Chalmers University of Technology, Microwave Electronics Laboratory, MC2, Göteborg, 412 96, Sweden
[2]Gotmic AB, Göteborg, 417 03, Sweden
E-mail: marcus.gavell@gotmic.se

Abstract— **In this paper, the design and characterization of a broadband image reject mixer for the next generation of point-to-point microwave links is presented. The manufacturing has been made in a commercially available 0.15 μm gate length GaAs mHEMT technology. The measured performance demonstrates a conversion loss of 9 dB and an image rejection ratio of 25 dB on average across the full E-band (71-76 and 81-86 GHz). Performance peaks at 77 GHz with conversion loss of 7 dB and image rejection of 40 dB. Furthermore, these results have been achieved with a LO power requirement of 4 dBm. To the best of the authors' knowledge this is the first reported image reject mixer suitable for the full E-band.**

Keywords-component; E-band, image reject mixer, single side-band, MMIC, mHEMT, GaAs

I. INTRODUCTION

In October 2003 the Federal Communications Commission (FCC) released 13 GHz of yet unused spectrum in the frequency bands 71-76, 81-86 and 92-95 GHz, dedicated for high speed wireless communications in the United States [1][2]. The two former bands are now usually referred to as the E-band. By now, many countries have followed FCC's example by licensing the E-band for communication purposes. The E-band has thus become very interesting for the next generation of point-to-point microwave backhaul links all over the world. The E-band has attracted a lot of interest because it is possible to achieve multi-Gb/s in full duplex even with simple binary modulation. Over the past few years data traffic in the wireless mobile networks has increased continuously, reaching the maximum capacity of today's microwave backhauls. In spite of this emerging challenge, little has been published on this subject.

A receiver system benefits from an Image Reject Mixer (IRM) in terms of noise as well as a relaxation of filter requirements. The noise from the image frequency will be downconverted to the same Intermediate Frequency (IF) increasing the overall system noise unless the image is cancelled by filtering. However, even with filtering, the noise performance is degraded due to losses in the filter. More importantly, high order filter introduces distortion which is highly unwanted in broadband signals.

In this paper a suitable and flexible IRM for the E-band with state-of-the-art performance in terms of Conversion Loss (CL) and Image Rejection (IR) is presented.

II. CIRCUIT DESIGN

The design of the IRM is implemented in a GaAs 0.15 μm gate length mHEMT (metamorphic high electron mobility transistor) technology from WIN semiconductors. The simplified schematic of the IRM is shown in Fig. 1. The IRM consists of two resistive mixers, each a 2×50 μm transistor with LO at the gate and RF and IF at the drain [5]. The transistors work in the linear region with gate voltage close to threshold and drain voltage at zero, making it a passive circuit. The RF is split with a Wilkinson divider to the drains and the LO is divided into 0 and 90 degree signals with a Lange coupler to pump the gates. The gates are equally biased through the LO port and with a resistor in parallel with the Lange coupler. A coupled line connected between the Wilkinson divider and drain is used to block the IF signal from the RF port. A quarter wavelength stub in addition with a compact lowpass filter with broad stopband characteristics [3] is used in the IF path to block the RF and LO signals. The IF is extracted at the drains in a 0 (I) and 90-degree (Q) quadrature signals. The advantage with this solution is that the quadrature signals are accessible from the IF output, alternatively an off-chip IF hybrid can be used. An off-chip IF-hybrid is flexible and at the same time cost effective. Depending on the direction of the off-chip IF-hybrid, the mixer can be used as Lower Side Band (LSB) or Upper Side Band (USB) mixer. Special care was taken to make the design symmetric to enhance the broadband characteristics. A chip photo is shown in Fig. 2.

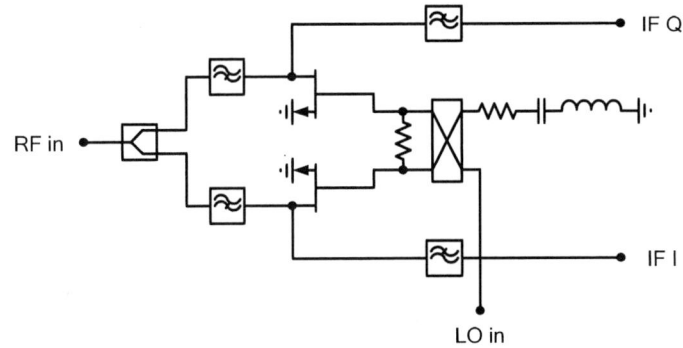

Figure 1. Schematic representation of the image reject mixer

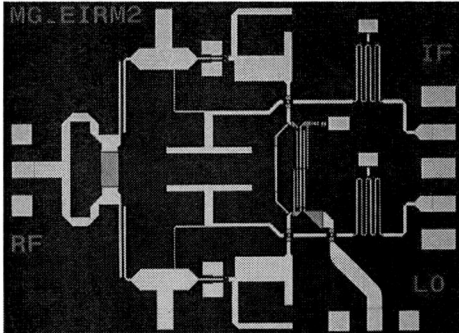

Figure 2. Chip photo of the image reject mixer. The chip size measures 2×1.5 mm².

III. CIRCUIT CHARACTERISATION

The IRM described in this paper demonstrates a CL of 9 dB and an IR of 25 dB on average across the full E-band. The measurements were carried out on wafer with coplanar GSG-probes and an off-chip 90-degree 1-12 GHz IF-hybrid. An off-chip bias-T at the LO port was used for gate biasing. Millimeter-wave source modules were used as signal generators and a spectrum analyzer was used to measure the power levels at the IF. All losses prior and post the MMIC has been calibrated for.

In Fig. 3 and Fig. 4 contour plots of the CL vs. LO and IF frequencies in LSB and USB configuration are shown respectively. The roll-off characteristic between the LSB and USB is due to the lack of data below 1 GHz. The lower frequency limit was determined by the IF-hybrid and not the IRM. The IRM is capable of signals close to DC. The CL maxima on the image side in the contour plots in Fig. 3 and Fig. 4 are due to a minimum in the phase and amplitude mismatch in the overall IRM, including the IF-hybrid.

In Fig. 5 a 3D plot of the CL vs. LO and IF frequency is plotted as a complement to the contour plots in Fig. 3 and Fig. 4.

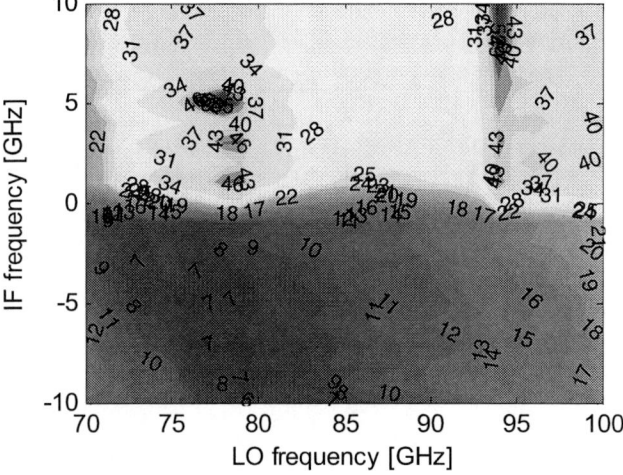

Figure 3. A contour plot of the CL with respect to LO and IF frequencies. The LO power is fixed at 4 dBm. The mixer is measured in LSB configuration.

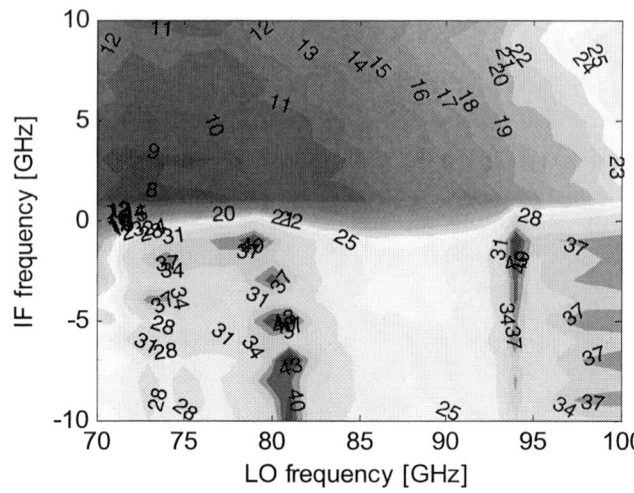

Figure 4. A contour plot of the CL with respect to LO and IF frequencies. The LO power is fixed at 4 dBm. The mixer is measured in USB configuration.

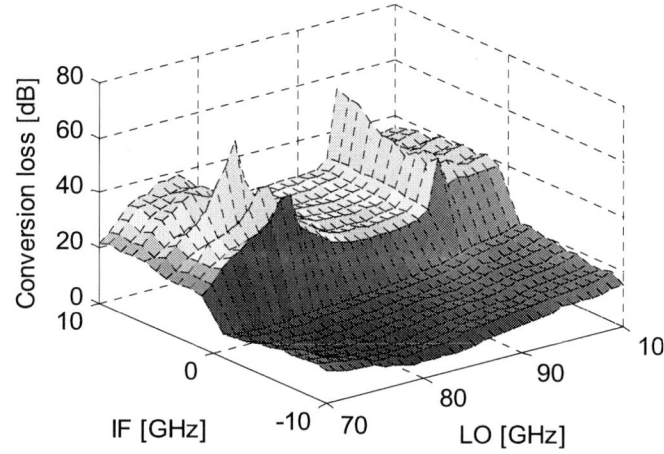

Figure 5. A 3D-plot of the CL with respect to LO and IF frequencies. The image frequency is on the USB.

In Fig. 6 and Fig. 7 CL vs. fixed LO is plotted such that the RF fall in the two E-bands (71-76 and 81-86 GHz). The ripple in CL is less than 1 dB at the same time that the IR is on average 25 dB.

In Fig. 8 the CL vs. a fixed IF frequency of 1 GHz is plotted. The 3 dB RF bandwidth covers the full E-band.

978-1-4244-5190-6/09 $25.00 © 2009 IEEE   118

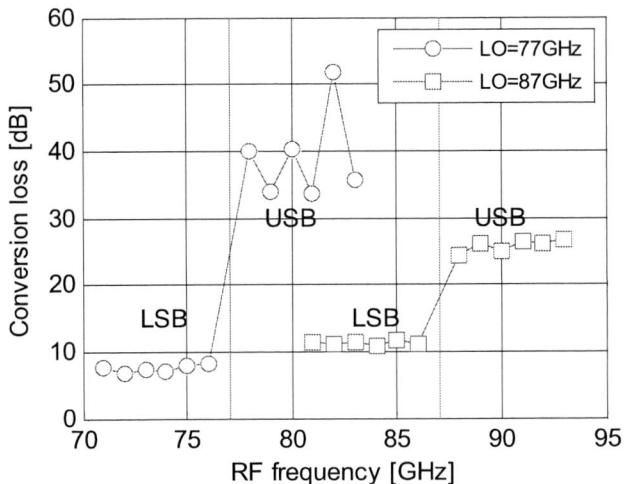

Figure 6.   CL vs. fixed LO in LSB configuration.

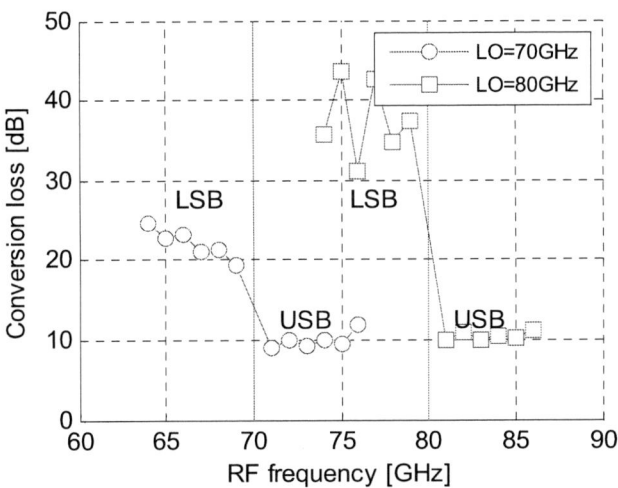

Figure 7.   CL vs. fixed LO in USB configuration. IF bandwidth= 6 GHz.

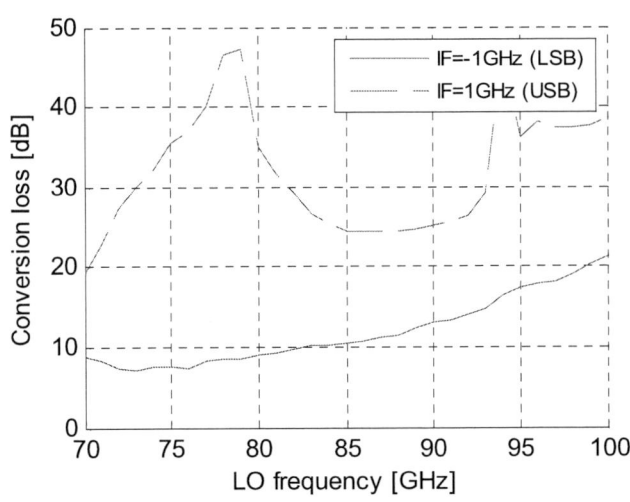

Figure 8.   CL vs. fix IF.

The LO power requirement and linearity are two very important parameters for a mixer. A low LO power requirement simplifies the power buffer amplifier in the LO-chain and the high linearity is important for modulation formats such as QAM. The CL as a function of LO power is plotted in Fig. 9. Due to the limitations in output power from the millimeter wave source modules, 4 dBm was the maximum output power that could be generated.

In Fig. 9 the CL vs. LO power is illustrated. The optimum LO power for best CL lie just beyond 4 dBm. If power generation is limited the IRM can be backed off. Even at -1 dBm the increase in CL is only 2 dB.

In Fig. 10 the linearity of the mixer is plotted. The 1 dB compression point (P1dB) occurs with an RF input power of 1 dBm.

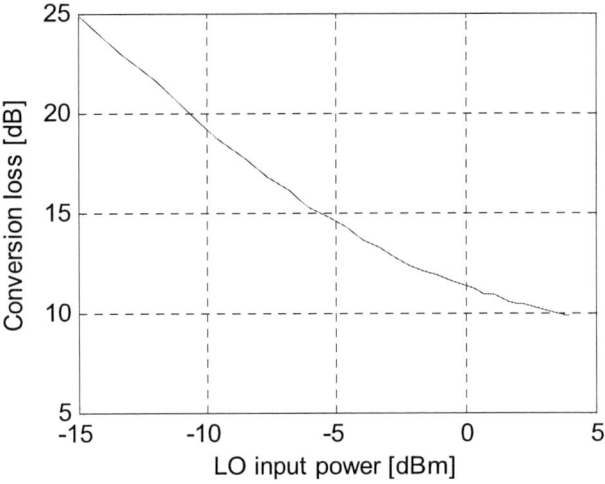

Figure 9.   CL vs LO power. The LO frequency is 80 GHz and the IF frequency is 1 GHz.

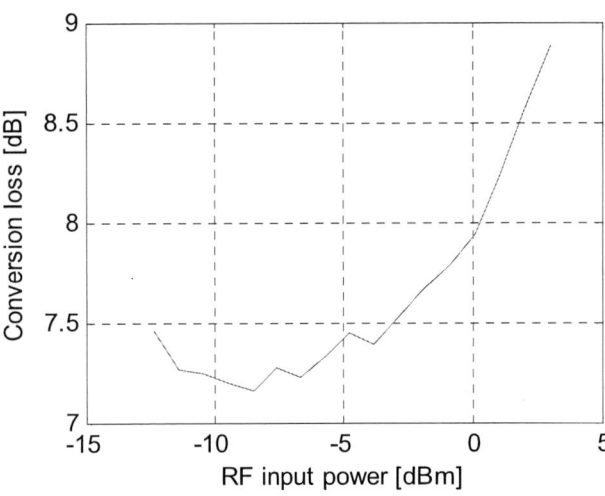

Figure 10. CL vs. RF power. The LO is fixed at 76 GHz and 4 dBm. IF frequency is 1 GHz.

978-1-4244-5190-6/09 $25.00 © 2009 IEEE

IV. SUMMARY AND COMPARISON

A summary of the key results measured with the IRM in this paper is shown in TABLE I. The performance of the IRM has been thoroughly measured in both LSB and USB. Coversion loss vs. IF frequency, LO frequency, LO power and RF power has been measured and plotted. Contour plots of CL vs. IF and LO frequency has been plotted to give a good overview of the IRM's performance.

TABLE I.   SUMMARY OF THE KEY FIGURE OF MERITS

Parameter	Value
Conversion loss (avg)	9 dB
Image Rejection (avg)	25 dB
IF bandwidth	0-10 GHz
LO bandwidth	70-90 GHz
Optimum LO power	4 dBm
Input P1dB	1 dBm

As previously mentioned few publications are found on E-band MMICs in the open literature. Therefore a direct comparison with E-band mixers cannot be made. Instead a comparison with mixers in the vicinity to the E-band is done in TABLE II. The overall performance from the IRM presented in this paper stands out in this comparison. CL, IF- and LO bandwidth, IR and LO power are all among the best.

TABLE II.   COMPARISON WITH OTHER IRMs

Work	Conv. loss [dB]	Image Suppr ession [dB]	IF BW [GHz]	LO power [dBm]	LO BW [GHz]	Technology
This work	9	25	10	4	70-90	mHEMT
[4]	10	30	2.2	1	55-65	pHEMT
[6]	11	16	0.2	10	94	Diode HEMT
[7]	10	20	0.14	5	55-62	pHEMT
[8]	3	12	0.14	5.8	58-62	pHEMT
[9]	10	19	5	6	50-69	pHEMT
[10]	20	18	5	15	38.75-49.25	Subharmonic pHEMT

V. CONCLUSION

A resistive Image Reject Mixer for the E-band has been designed and fabricated in a commercially available GaAs 0.15 μm gate length mHEMT technology. The measurements show conversion loss (CL) of 9 dB (avg) and image rejection (IR) of 25 dB (avg) across the E-band with an LO power of 4 dBm in both lower and upper side band. Its best performance is at 77 GHz where the CL is 7 dB and IR 40 dB. The P1dB for the RF occurs at 1 dBm. The measured results make this circuit suitable for high-speed communications in the E-band.

ACKNOWLEDGMENT

Kjell Håkan Närfelt, Ann-Mari Fineman, and Marie Wall at VINNOVA, The Swedish Governmental Agency for Innovation Systems, are acknowledged for the support of this work through the VINN-VERIFIERING-project '60 GHz MMIC'.

REFERENCES

[1] J.A. Wells, "MultiGigabit wireless technology at 70, 80 and 90 GHz", R.F. Des. pp. 50-58, May 2006

[2] J.A. Wells, "Faster than fiber: The future of Multi-Gb/s wireless", IEEE Microwave Magazine, vol. 10, No.3, pp. 104-112, May 2009

[3] Rui Li, Dong Il Kim, "A New Compact Low_Pass Filter with Broad Stopband and Sharp Skirt Characteristics", 2005 Asia-Pacific Microwave Conference, p 3 pp., 2006

[4] Sten E. Gunnarsson, Dan Kuylenstierna, Herbert Zirath, "A 60 GHz MMIC pHEMT Image Reject Mixer with Integrated Ultra Wideband IF Hybrid and 30 dB of Image Rejection Ratio", 2005 Asia-Pacific Microwave Conference, p 4 pp., 2006

[5] Stephen A. Maas, "A GaAs MESFET mixer with very low intermodulation", IEEE Transactions on Microwave Theory and Techniques, v MTT-35, n 4, p 425-9, April 1987

[6] T.N. Ton, et al.,"A W-band Monolithic InGaAs/GaAs HEMT Schottky Diode Image Reject Mixer", Gallium Arsenide Integrated Circuit (GaAs IC) Symposium, 1992. Technical Digest 1992., 14th Annual IEEE , 4-7 Oct. 1992, Pages:63 – 66

[7] K. Nishikawa, K. Kamogawa, R. Inoue, K. Onodera, T. Tokumitsu, M. Tanaka, I. Toyoda, and M. Hirano, "Miniaturized millimeter-wave masterslice 3-D MMIC amplifier and mixer," IEEE Transactions on Microwave Theory and Techniques, vol. 47, pp. 1856, 1999.

[8] T. Saito, et al., "60-GHz MMIC image-rejection downconverter using InGaP/InGaAs HEMT", Gallium Arsenide Integrated Circuit (GaAs IC) Symposium, 1995. Technical Digest 1995., 17th Annual IEEE , 29 Oct.-1 Nov. 1995, Pages:222 – 225

[9] K. Fujii, et al., "A 60 GHz MMIC chipset for 1-Gbit/s wireless links", IEEE MTT-S International Microwave Symposium Digest, 2002, Vol. 3, 2-7, June 2002, Pages: 1725 – 1728.

[10] John W. Archer, "A 80–100 GHz Image reject passive HEMT mixer", Microwave and optical technology letters, vol. 48, No. 12, December 2006, Pages: 2429 – 2433.

Robust AlGaN/GaN Low Noise Amplifier MMICs for C-, Ku- and Ka-band Space Applications

E.M. Suijker[1], M. Rodenburg[1], J.A. Hoogland[1], M. van Heijningen[1], M. Seelmann-Eggebert[2], R. Quay[2], P. Brückner[2], F.E. van Vliet[1]

[1]*TNO Defence, Security and Safety*
Oude Waalsdorperweg 63, 2597 AK, Den Haag, The Netherlands
Erwin.Suijker@tno.nl, Tel. +31(0)70 3740498, Fax. +31(0)70 3740653

[2]*Fraunhofer Institute of Applied Solid State Physics*
Tullastraße 72, 79108 Freiburg, Germany

Abstract—**The high power capabilities in combination with the low noise performance of Gallium Nitride (GaN) makes this technology an excellent choice for robust receivers. This paper presents the design and measured results of three different LNAs, which operate in C-, Ku-, and Ka-band. The designs are realized in 0.25 μm and 0.15 μm AlGaN/GaN microstrip technology. The measured noise figure is 1.2, 1.9 and 4.0 dB for the C-, Ku-, and Ka-frequency band respectively. The robustness of the LNAs have been tested by applying CW source power levels of 42 dBm, 42 dBm and 28 dBm for the C-band, Ku-band and Ka-band LNA respectively. No degradation in performance has been observed.**

I. INTRODUCTION

Due to the current trends in the global security scenario the interest in secure and robust satellite communication systems is increasing. A space-born receiver is one of the most important, but also one of the most sensitive components in the satellite communication chain. These receivers must also be functional during severe jamming and no degradation is allowed due to high input powers from hostile electromagnetic attacks.

Currently enabling technologies like GaN and SiC are being developed for both military as well as governmental and commercial applications. GaN is expected to improve the performance regarding robustness together with microwave capabilities comparable to currently used technologies like GaAs, InP and Si. Due to the combination of high power and low noise operation GaN is very suitable for the realization of secure robust RF front-end (RFFE) modules. In addition, the overdrive capability and survivability levels are exceptionally high. The development of both GaN HEMTs as well as MMIC processes makes it possible to realize GaN low noise receivers at millimeter wave frequencies. Such receivers are very attractive for secure communication systems due to the changing trade-off between performance and cost. Therefore GaN based receivers offer important potential for next generation satellite communication systems.

This work aims at technology demonstrations at C-, Ku- and Ka-band by designing, building and testing GaN low noise

amplifier MMICs, as part of the receiver front-end. Section II describes the RFFE and improvement in robustness by using GaN components. Section III describes the AlGaN/GaN technology used for the MMICs. A detailed description of the design of the three LNAs for each frequency band is given in section IV. Section V shows the small signal as well as the large signal measurement results. Finally conclusions are given regarding the low noise performance and survivability levels.

II. ROBUST RECEIVERS

The advantages of GaN enable the improvement of the receiver performance. The high power capabilities allow the receiver to handle high input power levels without any degradation in RF performance. The combination of high power and low noise allows optimizing the trade-off between NF and IP3. Figure 1 shows a simplified block diagram of a RFFE receiver. As can be seen in the figure the proposed front-end modules based on GaN MMICs do not include a separate limiter, which improves the overall noise figure.

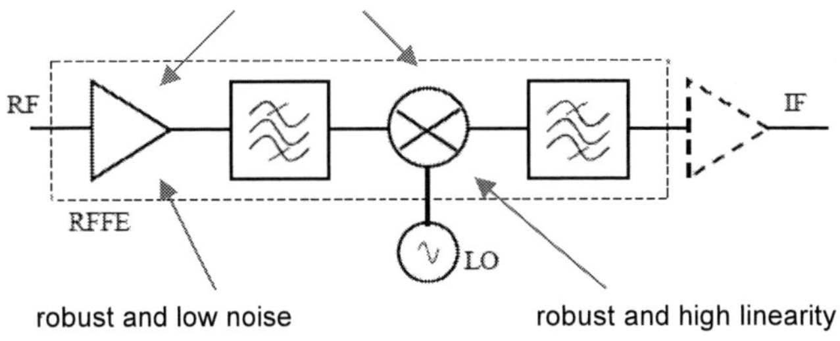

Fig 1. Simplified block diagram of a RFFE receiver.

The survivability capability of GaN LNAs has already been proven by others [1]-[7]. State-of-the-art performance for the input power capability is 37 dBm for CW mode and 41 dBm in pulse mode. For this work, the target performance for survivability of the C-band RFFE is 40 dBm CW power and 37 dBm CW power for the Ku- and Ka-band RFFE.

III. ALGaN/GaN PROCESSING TECHNOLOGY AND MODELLING

The LNA MMICs have been processed in the AlGaN/GaN microstrip transmission line technology of Fraunhofer IAF. The C- and Ku-band designs have used 0.25 µm gate length and the Ka-band design 0.15 µm gate length.

A. AlGaN/GaN HEMT MMIC Technology description

AlGaN/GaN single heterostructures are grown on s.i. SiC by MOCVD in a 12×3-inch multi-wafer reactor. Three-inch wafers are processed for a combined 0.25 and 0.15 µm gate length HEMT technology. The gate modules comprise optimized field plates by which the two-terminal breakdown voltages of the produced power HEMTs are increased above ≥100 V. The MMIC technology yields a transit frequency f_T of 33 GHz and 50 GHz at V_{DS}= 28 V for gate lengths of 0.25 µm and 0.15 µm respectively.

Covering transistors from 21 wafer cells, a load-pull power mapping is routinely performed at V_{DS}= 28 V in CW-mode for a device with a gate width of 0.48 mm. Robustness, yield and power performance are assessed as part of a reproducibility check. An average high-gain performance is achieved with ≥15 dB linear gain at 10 GHz with a yield of about 90%.

After the front side processing the full three-inch s.i. SiC wafer is thinned to 100 µm thickness and the backside is processed for via-hole formation, in particular for contacting the source electrodes.

B. Device and MMIC Modeling

A large data base was established by measurement of noise- and S-parameter measurements in the frequency range from 2 to 26 GHz for drain voltages from 5 V to 15 V and drain current densities from 100 mA/mm to 200 mA/mm for several transistor geometries. From this data base compact small signal models were built comprising noise sources with bias dependent noise temperatures. Large-signal performance can be assessed by a two-dimensional voltage-lag model to accurately describe thermal effects and low-frequency dispersion. For the MMIC design a library of passive microstrip components is available, including all technology specific elements like MIM capacitors, resistors, and inductors.

IV. LNA DESIGNS

The C-band LNA is a three stage amplifier. All three stages use a 8x60 µm transistor with 0.25 µm gate length. The large FET generates sufficient gain at C-band and the larger gate length and gate periphery will improve the robustness of the amplifier. All three stages are biased at 10 V drain-source voltage and a drain current density of 150 mA/mm, which is about 15% of Idss. The matching of each stage is optimized for noise to improve the overall noise figure and still fulfill the gain specification. A photograph of the C-band LNA is depicted in figure 2.

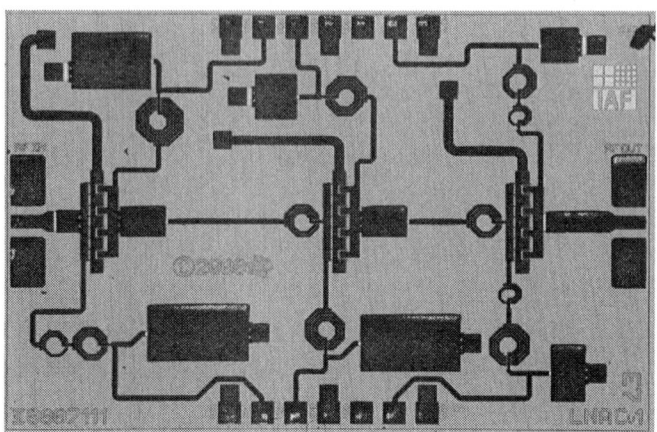

Fig. 2 Microscope photograph of the C-band LNA.

The Ku-band LNA is a three stage amplifier as well. The gate length of each stage is 0.25 µm, which provides sufficient gain at 14 GHz and improves the robustness of the LNA. The first two stages use a 2x50 µm transistor the get sufficient gain. The last stage is a 8x60 µm transistor to achieve an output power higher than 20 dBm at P_{1dB}. The first stage is matched for optimum noise figure and for the last two stages the matching is a compromise between gain and noise. All three stages are biased at 10 V drain voltage and 150 mA/mm drain current density. Figure 3 depicts a photograph of the Ku-band LNA.

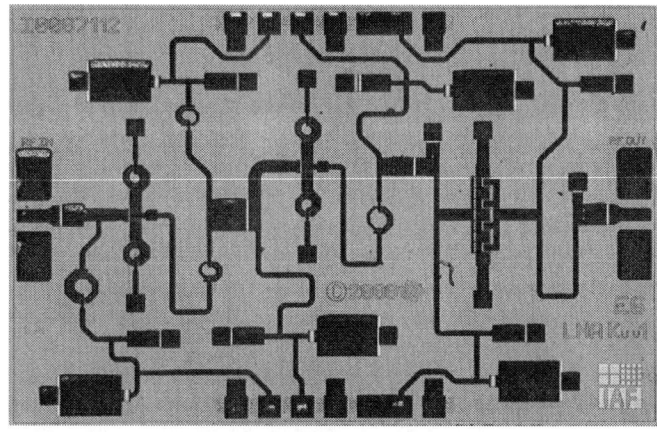

Fig. 3 Microscope photograph of the Ku-band LNA.

At Ka-band frequencies four stages are required to achieve an overall gain higher than 19 dB. The gate length of the used transistors is 0.15 µm to increase the gain at these high frequencies. The gate periphery is 2x50 µm and is equal for all four stages. All stages are biased at V_{DS}= 10 V and a drain current density of 200 mA/mm. The higher current results in more gain without sacrificing the noise performance. All stages are matched for optimum noise figure. A photograph of the Ka-band LNA is depicted in figure 4.

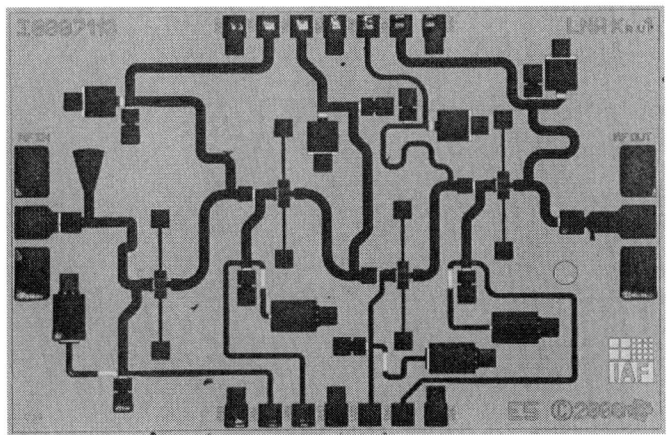

Fig. 4 Microscope photograph of the Ka-band LNA.

For each frequency band the performance of the LNA will be demonstrated within an RF frontend (RFFE) module. To reduce the complexity of fabrication of each module the dimensions of the MMIC LNAs are made equal and are 3x2 mm^2.

V. MEASUREMENT RESULTS

Measurements on the three LNA versions are performed and compared to the simulated results. First the complete wafer was measured (21 MMIC's for each LNA version) to determine the known good dies. Five of the known good dies have been mounted on a CuMo carrier, which works as a heat sink. Besides the mounted samples two bare dies of each LNA version have been characterized for mounting in the RFFE module. To stabilize the LNAs, the gate connections are decoupled with 10 pF low pass networks. For each LNA version, the following measurements are performed: S-parameters, noise figure, output power compression, third order intermodulation and electrical stress. Table 1 lists a summary of all measured parameters.

Table 1: Specifications of all three LNAs

Parameter	C-band		Ku-band		Ka-band	
	Sim	Meas	Sim	Meas	Sim	Meas
Freq [GHz]	6	6	14	14	27.5-28.5	28
NF [dB]	1.0	1.2	1.7	1.9	<2.7	4.0
Gain [dB]	23.3	>21	19.7	>19.8	21.9	18
P1dB [dBm]	25	>28	24	>28	-	>12.5
RL$_{IN}$ [dB]	-15	<-11.5	-10.7	<-6	-11.2	-12
RL$_{OUT}$ [dB]	-23	<-12	-18.2	<-13	-9.6	<-6.5

Based on functional S-parameters the overall yield of the C-band LNA is 86%. The S$_{21}$ is higher than 18.8 dB and the nominal value is 21.8 dB. The noise figure is better than 1.6 dB and the nominal NF is 1.2 dB at 6 GHz. The output power at the 1 dB compression level is higher than 28 dBm, which is even higher than simulated. The noise figure and small-signal gain are depicted in figure 5.

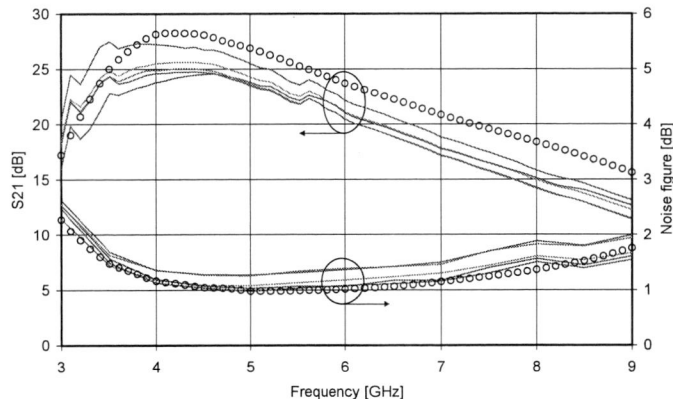

Fig. 5 Measured S$_{21}$ and noise figure of the C-band LNA. The circles represent the simulations.

The yield of the Ku-band LNA is 100% based on functional S-parameters. The noise figure and small-signal gain of the mounted samples are depicted in figure 6. The measured noise figure varies from 1.9 dB to 2.4 dB and the gain varies from 21 to 27 dB. The output power at 1 dB compression is higher than 28 dBm at 14 GHz.

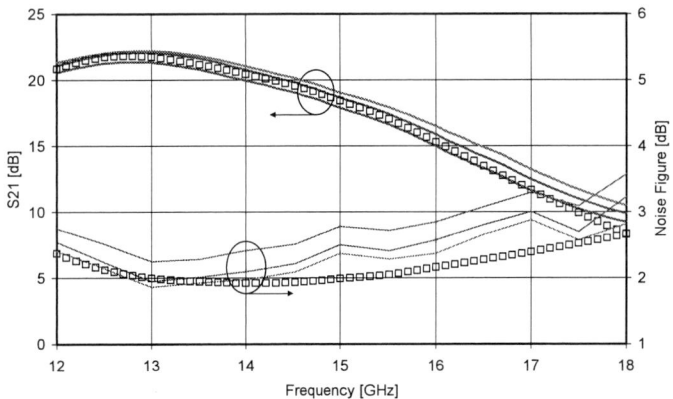

Fig. 6 Measured S$_{21}$ and noise figure of the Ku-band LNA. The squares represent the simulations.

The measured S-parameters of the Ka-band LNA show the least variation of the three designs. One sample shows an unexpected behavior for the output return loss, which results in an overall yield of 95%. The gain at 28 GHz varies over all samples from 18.9 to 22.5 dB. The noise figure varies between 4 and 5 dB, which is quite high compared to the simulation. New extracted noise models of the transistors have different optimum noise impedance and simulations with these models predict a noise figure of 4 dB. The noise figure and small-signal gain are depicted in figure 7. The output power at the 1 dB compression level is 12.5 dBm. To improve the compression behavior the last stage will be biased at higher V$_{DS}$.

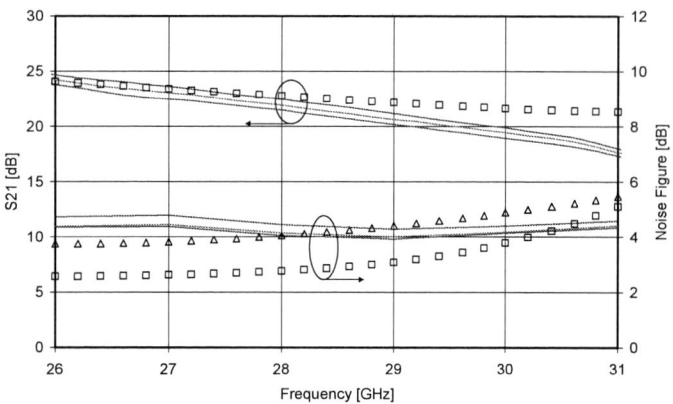

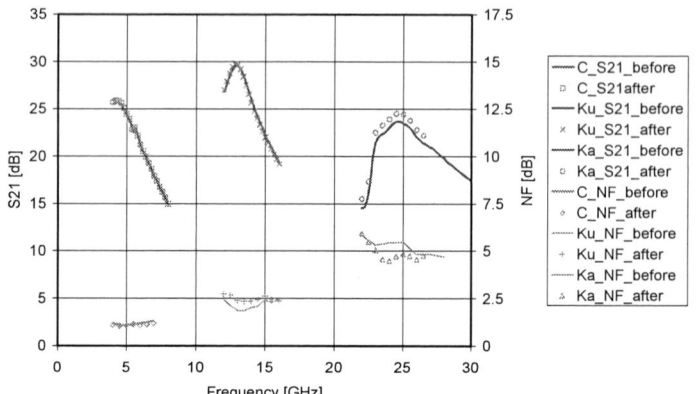

Fig. 7 Measured S_{21} and noise figure of the Ka-band LNA. The squares represent the simulations. Triangles represent a noise figure simulation with updated noise model.

Fig. 9 Small-signal gain and noise figure of each LNA measured before and after the devices have been stressed with high source powers.

The survivability performance of each LNA has been measured by monitoring gate currents, drain currents, output power and gain as a function of source power. The gate current can be limited by using a series resistor in the gate bias network. The resistor and gate current result in a voltage drop and decrease the gate bias voltage. Therefore the device is pinched-off and the output power drops at high input power levels, see figure 8.

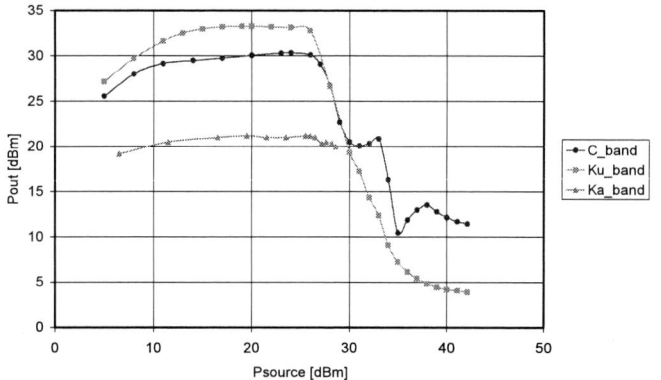

Fig. 8 Output power versus source power for each LNA versions.

Figure 9 shows the small-signal gain measured before and after the devices have been stressed with high input power levels. As can be seen no degradation in gain can be observed. The noise figure measured after the devices have been stressed is 1.1, 2.3 and 4.8 dB for the C-, Ku- and Ka-band LNA respectively. This means a variation of 0.1, 0.3 and 0.1 dB in noise figure for these selected devices. This variation is within the noise figure measurement uncertainty and also partly caused by the use of different measurement equipment. The C-band and Ku-band LNA can survive source power levels of 42 dBm without degradation in gain or noise figure. The survivability of the Ka-band LNA has been tested up to 28 dBm, which is limited by the capabilities of the measurement equipment.

VI. CONCLUSIONS

Three LNA GaN MMICs have been designed, realized and measured. With a yield around 90% it is demonstrated that sub quarter-micron GaN MMIC technology is increasingly mature. The nominal noise figure is 1.2, 1.9 and 4.0 dB for the C-, Ku- and Ka-band respectively. Due to the absence of a limiter GaN will be an excellent choice for low-noise receivers. The noise figure of the C-band and Ka-band LNA are state-of-the-art results. The robustness for receiver applications is demonstrated by applying CW source powers of 42 dBm, 42 dBm and 28 dBm for the C-, Ku- and Ka-band LNA respectively without degradation in small-signal gain or noise figure. The survivability figures are also state-of-the-art results.

ACKNOWLEDGMENT

This work has been supported by European Space Agency project "GaN Technology for Robust Communication Receivers" (ITT-5322). The research program is executed in co-operation with Thales Alenia Space in Toulouse, France, United Monolithic Semiconductors in Ulm, Germany and Fraunhofer Institute of Applied Solid State Physics in Freiburg, Germany.

REFERENCES

[1] D. Krausse, et al., "Robust GaN HEMT Low-Noise Amplifier MMICs for X-band Applications", 12th GAAS Symposium – Amsterdam, 2004.
[2] S. Cha, et. al, "Wideband AlGaN/GaN HEMT Low Noise Amplifier For Highly Survivable Receiver Electronics", IEEE MTT-S Digest 2004.
[3] J.C. deJaeger, et al., "Noise Assessment on AlGaN/GaN HEMTs on Si or SiC substrates:Application to X-band Low Noise Amplifiers", 13th GAAS Symposium – Paris, 2005.
[4] M. Rudolph, et al., "Analysis of the survivability of GaN Low-noise Amplifiers", IEEE Trans on Microwaves Theory and Techniques, vol. 55, January 2007.
[5] M. Micovic, et al., "Robust Broadband (4-16GHz) GaN MMIC LNA", CSICS 2007.
[6] J.P.B. Janssen, et al., "X-band AlGaN/GaN MMICs with over 41 dBm Power Handling", CSICS 2008.
[7] M. Rudolph, et al., "Highly Rugged 30 GHz GaN Low-Noise Amplifiers", IEEE Microw. Wireless Compon. Lett., vol. 19, no. 4, Apr. 2009.

978-1-4244-5190-6/09 $25.00 © 2009 IEEE

Compact and Broadband Millimeter-Wave Mixer Based on the New Phase Relationship

Yu-Ann Lai, Shih-Han Hung, and Yeong-Her Wang*

Institute of Microelectronics, Department of Electrical Engineering,
National Cheng-Kung University,
Tainan, 701 Taiwan, R.O.C.
*E-mail: YHW@eembox.ncku.edu.tw

Abstract—**A millimeter-wave monolithic mixer with new phase relationship is demonstrated and fabricated through a 0.25 μm PHEMT process. This mixer is composed of two 90⁰ hybrids with new phase relationships, which provides the function like the 180⁰ hybrids in the double balanced mixer structure. And it does not only provide flexible layout design, but also offer wideband performance. The measured results demonstrate a conversion loss of 10.2~14.5 dB over 18-40 GHz operation bandwidth, and the 1-dB compression power of 6 dBm at 27 GHz. Moreover, the LO-IF and RF-IF isolation are better than 20dB from 28-40 GHz, and this proposed configuration leads to a die size of less than 0.48 × 0.77 mm².**

Keywords-Phase relationships, double balanced mixer (DBM), hybrid, PHEMT

I. INTRODUCTION

Recently, the interests in the use of millimeter-wave integrated circuits for satellite communication systems and wireless cable television are increasing. In these systems, the most important component of conversion base-band data is the mixer [1]. Similar to the previous mixer designs, it is obvious that the ring or star double balanced structures are often used. The double balanced mixers (DBM) have similar requirements for those of communication systems that use millimeter-wave: good port-to-port isolation, broad bandwidth and low conversion loss.

These DBM are used alone or built in the system and the use of baluns at the RF/LO port is the most common realization of the ring and the star mixer, as shown in Figs. 1. The ring mixer also requires a rather complex IF extraction circuit; accordingly, this circuit increases IF parasitic inductance and narrows IF bandwidth. Therefore, the star mixer is used as a better configuration to overcome the problems of the ring mixer. Four Schottky diodes are connected to the common node, and RF/LO signals pass through the two Marchand dual balun to produce the IF signal [2-3]. The advantage of the star mixer's IF port is low IF parasitic, which provides wide IF bandwidth. However, the traditional ring or star mixer have the same problem, they need two rather large half-wavelength Marchand baluns.

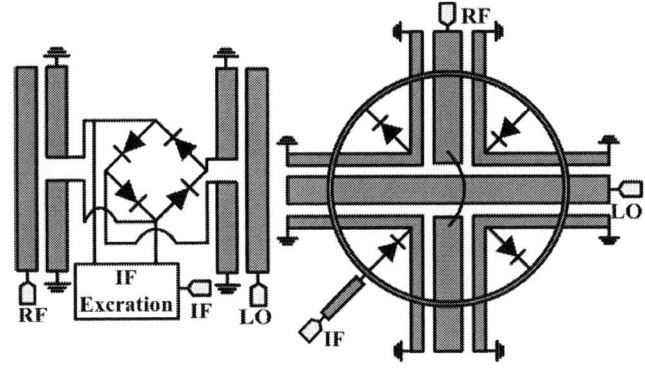

Figure 1. Conventional double-balance mixer layout: (a) ring mixer, (b) star mixer.

To construct the hybrid easily, the DBM utilized the two identical RF/LO Marchand dual baluns with a plastic packaged diode at S-band [4]. However, the circuits still need half-wavelength baluns to lead to large areas. In order to make a DBM more compatible with conventional MMICs, a multi-coupled line dual balun and compensated capacitors were used [5]. Furthermore, reports about DBM utilize spiral transformer baluns to achieve smaller chip size were announced [6]. Thus the configurations of the previous mixers are based on two Marchand balun-likes structure, which may still have large circuit dimensions [1]-[6]. Although the 180⁰ hybrids circuit still need further study in DBM, but emphasizing on improving the efficiency and simplifying the circuits of the mixer product are essential.

In this paper, a novel configuration of the mixer is proposed to further reduce the DBM size for K- to Ka-band applications. This proposed structure adopts the new phase relationship to realize the MMIC mixer, which achieves wideband performance and small chip size. The mixer has the best conversion loss of 10.2 dB with a chip size of 0.48 × 0.77 mm².

II. DESIGN OF THE PROPOSED MIXER CONFIGURATION

The configuration of a mixer is proposed as shown in Fig. 2. The feature of the proposed structure is the use of two 90⁰

978-1-4244-5190-6/09 $25.00 © 2009 IEEE

hybrids, which makes the mixer design easier. According to the mixing mechanism of the conventional DBM, the new phase relationships has served mixer reject spurious responses and perform an IF current. Furthermore, this mixer replaced the lumped-element structure with traditional 90^0 hybrids, which makes this design more compact [7]. This hybrid structure was used the high-pass and low-pass elements to realized the equal out power with 90-degree phase difference. Besides, it used the transmission line as the inductor itself to reduce the hybrid area. The symmetrical plane configuration of the network can transfer a four-port network into a two-port one, and analyze it by means of the transmission (ABCD) matrix method.

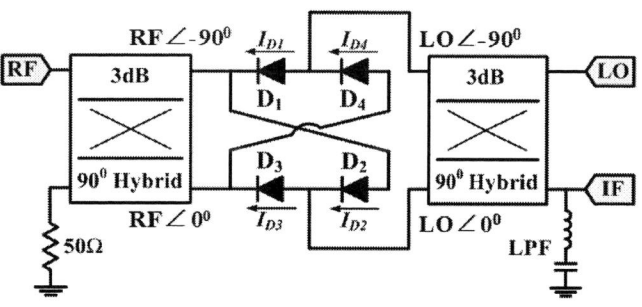

Figure 2. The configuration of the proposed mixer

Hence, the phase relationships of the diodes, the mixer products can be conveniently analyzed by means of the Fourier series expansion. Due to the non-linearity of the diode, the RF voltage across the diode will contain the harmonic terms of the RF frequency when RF signal is applied to the diode. In addition, the transconductance has a time-varying behavior which varies at the rate of the LO frequency. The current waveform of $I_D(t)$ generated by the product of the signal voltage, $V_{RF}(t)$, and transconductance, $g(t)$, can be represented as follows [3]:

$$I_D(t) = V_{RF}(t) \times g(t) = \sum_{m=-\infty}^{\infty} V_m e^{jm\omega_{RF}t} \times \sum_{n=-\infty}^{\infty} g_n e^{jn\omega_{LO}t} \quad (1)$$

TABLE I
SIGNAL PHASE RELATIONSHIPS IN DIODE CURRENTS

Diode current	RF signal phase	LO signal phase
I_{D1}	$\angle -90^0$	$\angle -90^0$
I_{D2}	$\angle -90^0$	$\angle 0^0$
I_{D3}	$\angle 0^0$	$\angle 0^0$
I_{D4}	$\angle 0^0$	$\angle -90^0$

As Table I shows, the new phase relationships of RF and LO signal can be realized by two 90^0 hybrids. Consequently, a comparison of the relative phase difference is made between diode's anode and cathode, and the current of D_3 is determined

to the reference current as I_{REF}. And the total current for the proposed mixer can be expressed in terms of I_{REF} as follows:

$$I_{TOTAL} \quad (2)$$
$$= I_{D1} + I_{D2} + I_{D3} + I_{D4}$$
$$= \left[(-j)^m (-j)^n + (-j)^m (1)^n + (1)^m (1)^n + (1)^m (-j)^n \right] I_{REF}$$

This represents the resulting total current where m and n are the order of RF/LO harmonics, respectively. When the (m, n) equals (1, -1) is the desired down conversion IF current, where $I_{TOTAL} = 2I_{REF}$ can be obtained. The resulting total current also shows spurious mixing products corresponding (m, n) = (1, 2), (m, n) = (2, 1), and (m, n) = (2, 2) are suppressed. In contrast with the conventional DBM, the mixing mechanism operation is in the quarter of the LO cycle. It not only produces available IF current in all Schottky diodes, but also suppress the spurious mixing products effectively. In addition, the proposed configuration can extract IF current from the isolation port of the 90^0 hybrids easily, and it doesn't require the additional air-bridge and microstrip lines for the IF extraction circuit. In the IF port, the low-pass filter (LPF) is used as a microstrip line and capacitor to separate the RF/LO signals. It uses the microstrip line instead of the inductor to minimize the number of circuit elements. Also, it improves the isolation and the desired small size. The proposed new phase relationship can be easily employed to obtain the IF current and suppressed spurious mixing products in the mixer.

Figure 3. Photograph of the fabricated mixer MMIC.

III. MIXER FABRICATION

This mixer was fabricated by the low-noise GaAs PHEMT process on a 50.8 μm thick GaAs substrate. Various techniques, such as wide-head 0.25 μm T-shaped gate, air bridges, via-hole etching, and back side processing were all used during the device fabrication. To design the simple mixer, the entire layout was verified using the Agilent Advanced Design System and Zeland IE3D software. And four simple 2×25 μm^2 diode's models with the low-junction series

978-1-4244-5190-6/09 $25.00 © 2009 IEEE

resistance and junction capacitance were used for each diode. As shown in Fig. 3, the chip size of the fabricated mixer is 0.48 × 0.77 mm².

IV. MIXER PERFORMANCE

This mixer was attached on carrier plates for testing. Measurements of the signals were provided by coplanar GSG and GSGSG on-wafer probes measurement systems based on the Agilent E4446A spectrum analyzer, which was calibrated by an E44198 power meter. The losses of the probes and cables were also calibrated by a PNA E8364A network analyzer.

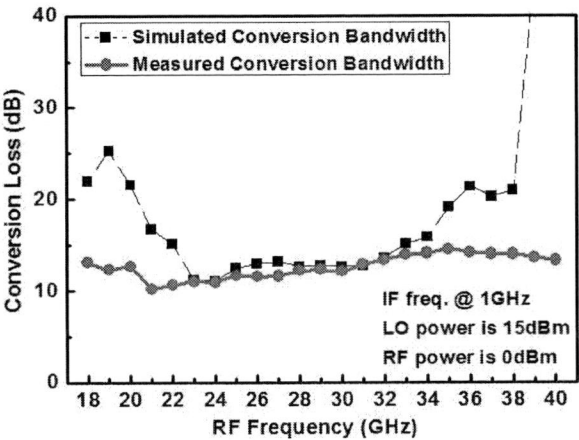

Figure 4. Conversion loss vs. RF frequency when IF frequency fixed at 1GHz.

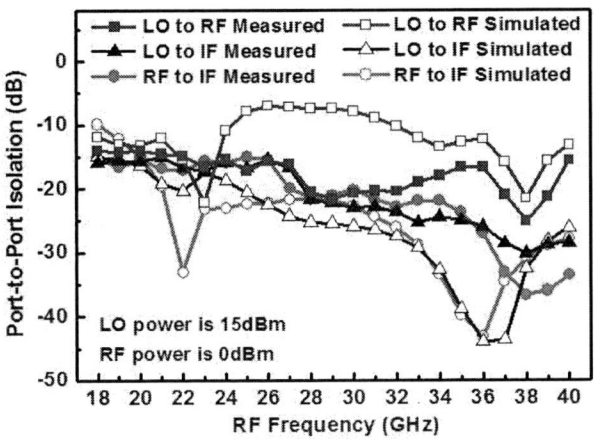

Figure 5. Isolation of the mixer at the 15 dBm LO level, at 1 GHz IF frequency.

Figure 4 shows the measured and simulated conversion loss as a function of RF frequency. The measurements were performed with an LO power level of 15 dBm. The measured conversion loss is 10.2 to 14.5 dB from 18 to 40 GHz with the IF fixed at 1 GHz. Although the deviations can be seen from the simulated and measured results, but these results are still promising. Figure 5 shows the measured and simulated LO-to-

RF, LO-to-IF, and RF-to-IF isolations as functions of RF frequency from 18 to 40 GHz. The measured LO-to-IF and RF-to-IF isolations are about 15 dB at the lower end of the band, and a better value than 20.2 dB at the high end of the band ranging from 28 to 40 GHz can be achieved. This indicates that the low pass filter can provide superior isolation between the IF port and LO/RF ports. But the simulated results are not similar to measured results, the frequency response of the mixer was caused by additional capacitor effect and discontinue junction from the lumped-element hybrid. Nevertheless, the LO-to-RF isolation is limited to the inherent condition by the proposed structure; it consequently resulted in 14 to 25.1 dB at the same frequency.

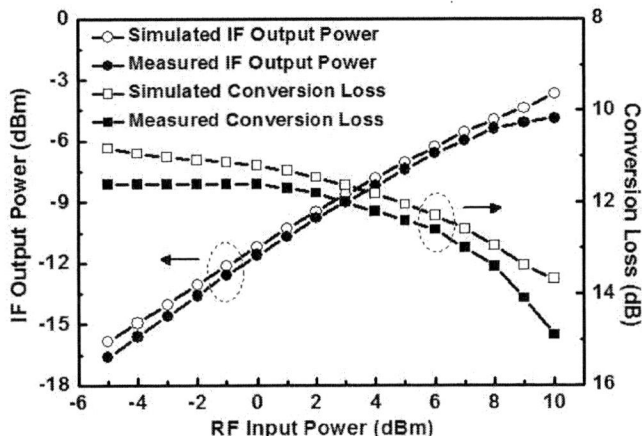

Figure 6. Measured conversion loss and IF output power versus RF input power at LO input power of 15 dBm.

TABLE II
COMPARISON OF DOUBLE BALANCED RING AND STAR MIXERS

Ref.\ Parameters	[2] #	[5]	[8]	[9]	[10]	This Work
RF Freq. (GHz)	30~45	52~68	28~36	16~44	50~75	18~40
LO power (dBm)	4	12	16	15	13	15
LO-to-IF Isolation (dB)	N/A	>18	N/A	>27	N/A	15 ~ 30.2
CL* (dB)	10~14	8.6 ~ 12.5	6~10	11~14	13~18	10.2 ~ 14.5
Chip size (mm2)	4	0.4	6.72	0.64	2.25	0.37

CL*: conversion loss, #: with DC supply

Figure 6 shows the conversion loss and the IF output power against RF input power with the RF fixed at 27 GHz and the IF fixed at 1 GHz. The measured results exhibit a low conversion loss of 11.6 dB, and a 1dB compression point of 6 dBm. It can be found that the measured data is in close agreement with the simulated results, which signifies that the new configuration is workable. However, all simulations have limited accuracy, especially these difficult simulations at millimeter-wave frequencies, and inaccuracies are not necessarily only the EM

978-1-4244-5190-6/09 $25.00 © 2009 IEEE   127

simulation. Therefore, more electrical parameters should be considered in the design process, such as the dielectric constant, metal loss, discontinue junction and the parasitic capacitive with inductance of Schottky diode's model. Table II summarizes the comparisons of the proposed mixer with the reported double balanced ring and star mixers [2, 5, 8-10]. As Table II indicated, this work certainly demonstrates the smallest size with good performances compared to the existing publications.

V. CONCLUSION

In this paper, a compact K- to Ka-band mixer based on new phase relationship has been demonstrated. The proposed mechanism eliminates the use of two 180^0 hybrids structure for application in the conventional double balanced mixer as well as provides the suppress function for spurious mixing products. As the measured results show, this mixer achieves the low conversion loss of 10.2 dB, and good input 1-dB compression point of 6 dBm. The LO-to-IF and RF-to-IF isolation is higher than 21.7 and 20.2 dB from 28 to 40 GHz, respectively. The proposed configuration provides not only a novel current mixing mechanism in designing a MMIC mixer, but also a compact size, which is very attractive for RF front end applications.

ACKNOWLEDGMENT

This work was supported in part by the National Science Council of Taiwan under contract NSC NSC95-2221- E-006-428-MY3 and the Foundation of Chen, Jieh-Chen Scholarship

of Tainan, Taiwan.

REFERENCES

[1] C.H. Lin, J.C. Chiu, C.M Lin, Y.A. Lai and Y.H. Wang, "A variable conversion gain star mixer for Ka-Band applications," IEEE Microw. Wireless Compon. Lett., vol. 17, no. 11, pp. 802-804, Nov. 2007.

[2] M. Yu, and R. H. Walden, A. E. Schmitz, and M. Lui, "Ka/Q-band doubly balanced MMIC mixers with low LO power," IEEE Microw. Guided Wave Lett., vol. 10, no. 10, pp.424-426, Oct. 2000.

[3] S. A. Maas, Microwave Mixers, 2nd ed. Norwood, MA: Artech House, 1993.

[4] S. S. Kim, J. H. Lee and K. W. Yeom, "A novel planar dual balun for doubly balanced star mixer," IEEE Microw. Wireless Compon. Lett., vol. 14, no.9, pp. 440-442, Sep. 2004.

[5] T.Y. Yang, W.R. Lien, C.C. Yang and H.K. Chiou, "A compact V-Band star mixer using compensated overlay capacitors in dual baluns," IEEE Microw. Wireless Compon. Lett., vol. 17, no. 7, pp. 537-539, July. 2007.

[6] C. J. Trantanella, "Ultra-small MMIC mixers for K- and Ka-band communications," in IEEE MTT-S Int. Microw. Symp. Dig., vol. 2, pp. 647-650, 2000.

[7] J.A. Hou and Y.H. Wang, "A compact quadrature hybrid based on high-pass and low-pass lumped elements," IEEE Microw. Wireless Compon. Lett., vol. 17, no. 8, pp. 595-597, Aug. 2007.

[8] S. A. Maas and K.W Chang, "A broadband, planar, doubly balanced monolithic Ka-band diode mixer," IEEE Trans. Microw. Theory Tech., vol. MTT-41, no. 12, pp. 2330-2335, Dec. 1993.

[9] C.M. Lin, H.K. Lin, C.F. Lin, Y.A. Lai, C.H. Lin and Y.H. Wang, "A 16-44 GHz Compact Doubly Balanced Monolithic Ring Mixer," IEEE Microw. Wireless Compon. Lett., vol. 18, no. 9, pp. 620-622, Sep. 2008.

[10] K. W. Yeom and D. H. Ko, "A novel 60-GHz monolithic star mixer using gate-drain-connected pHEMT diodes," IEEE Trans. Microw. Theory Tech., vol. 53, no. 7, pp. 2435 - 2440, July 2005.

X/Ku-Band SiGe BiCMOS Phased Array Chips with Simultaneous 2- and 4-Beam Capabilities

Dong-Woo Kang, Gabriel M. Rebeiz
Electrical and Computer Engineering
UCSD
San Diego, CA, USA
dwkang@ucsd.edu, rebeiz@ece.ucsd.edu

Kwang-Jin Koh
Technology and Manufacturing Group
Intel Corporation
Hillsboro, OR, USA
kjinkoh@gmail.com

Abstract— **This paper presents RFIC phased array receive chips capable of formation 2- and 4- simultaneous beams from the same antenna input with 4-bit amplitude and phase control. The design is based on a SiGe BiCMOS process (Jazz SiGe18Hx) and results in excellent isolation between the different beams. The 2-beam chip results in a gain of 5-6 dB per channel at 14-15 GHz, a noise figure of 10.0 dB, a P1dB of -13 dBm per channel (IIP3 of -6 dBm), and an RMS phase and gain error of < 12° and 1.5 dB, respectively. A gain control of 17 dB is available at each channel. Most important, the on-chip isolation between the channels has been fully characterized and is > 30 dB at 11-15 GHz. The beams can operate over an instantaneous bandwidth of > 1 GHz at any frequency between 11 and 15 GHz, and both beams can be at the same frequency if required. The 4-beam chip is currently being tested and the results will be reported at the conference.**

Keywords- SiGe BiCMOS, phased array, phase shifter

I. INTRODUCTION

Phased arrays based on silicon RFICs emerged as a practical solution for phased array back-ends due to their high integration density, yield, and functionality on a single chip [1-3]. Several silicon-based phased arrays with 4, 8 and even 16-elements have been demonstrated at X, Ku and even Q-band frequencies and with excellent uniformity [4-9]. This not only reduces the number of chips to be assembled in the phased array but also simplifies the control routing in large arrays. A phased array design for high performance will therefore use III-V components for low-noise amplification and RF power generation together with silicon RFICs for the back-end functions (phase and gain control, power combining, digital control and processing, etc.).

A natural progression from the early work on phased arrays with transmit- or receive-only capabilities is transmit/receive and multi-beam phased arrays on a single-chip [1]. The multi-beam design presents special challenges since high isolation is typically required between the beams. Also, the beams should be able to operate at different frequencies and with wide bandwidth. As will be seen in this work, this can be implemented using silicon RFICs and with excellent performance.

This work was supported by a SPAWAR contract under funding from DARPA, MTO, Mark Rosker, Program Monitor.

II. SYSTEM LEVEL AND CIRCUIT DESIGN

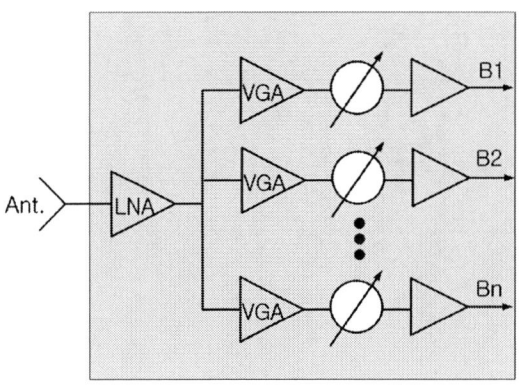

Figure 1. Phased-array chip with one antenna input and N ouput beams.

Fig. 1 presents the system-level diagram of a multiple-beam phased array chip. The RF port is split into N different paths (N=2-4) and each path contains amplitude and phase control circuitry which determines the location of the corresponding phased-array beam. There are N outputs per chip, one for each beam. The interaction between the different paths (beams) should be ideally zero, and the phase and amplitude setting in one beam cannot affect the performance of the other beams. In the implementation shown in Fig. 1, the beams from antenna 1 are added to the beams from antenna 2 using off-chip power combiners. In the future, and for multiple-input designs, the beam combiners can also be integrated on the silicon RFIC.

The circuit level implementation for the 2-beam chip is shown in Fig.2. Starting from the antenna port, the input LNA/Balun is a cascade design with an inductive load and de-Q resistors for wideband operation and followed by a differential emitter follower stage (V_{DD}= 3.3 V, I=20 mA). The LNA/Balun results in a simulated gain and noise figure of ~ 10 dB and 5.4 dB, respectively, at 14-15 GHz, and with wideband operation. The output signal is then using a short passive transmission-line network to two cascade differential amplifiers to form the power dividing network. A cascade design is chosen so as to result in high isolation between the two channels. These amplifiers also have a 1-bit gain control for 0/-10 dB using current steering, and a low output impedance (25 Ω) so as to drive an all-pass I/Q network. The simulated gain is -6 dB at 14-15 GHz.

978-1-4244-5190-6/09 $25.00 © 2009 IEEE

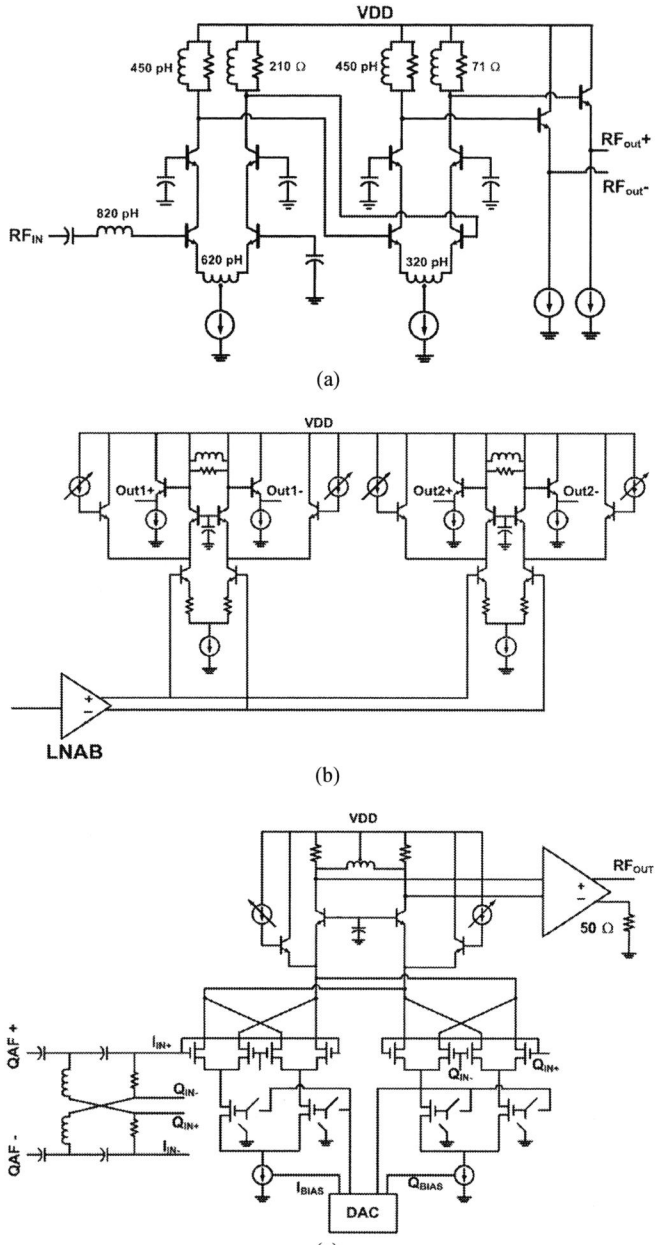

(a)

(b)

(c)

Figure 2. (a) Circuit schematic for LNA/Balun (b) divider block diagram and circuit level description (c) vector moduleator and output stage.

The phase shifter is based on an I/Q vector modulator with 4-bit DAC control. It is a similar design to the work reported in [2], but with an additional 3-bit amplitude control using current steering. Again, inductive loads are used with de-Q resistors for wideband operation. The outputs are passed by a differential common-emitter stage and one part of the output signal is terminated internally by 50 Ω (this results in a 3 dB loss in the signal). The phase shifter consumes 12 mA of current and results in a gain of 5 dB at 14-15 GHz.

The 2-beam array is implemented in the Jazz SBC18XL process with 6 metal layers. This process has 0.18μm SiGe

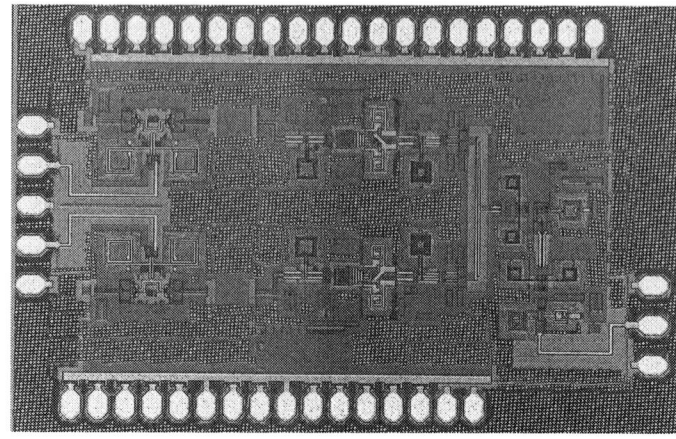

Figure 3. Microphtograph of the 2-beam phased-array chip. The dimensions are 2.4 mm x 1.5 mm.

transistors (f_T of 155 GHz) and 0.18μm CMOS transistors (f_T of 50-60 GHz). A 50 Ω microstrip line is realized with top metal (Metal 6) as a signal line and Metal 4 as a ground plane. Standard Jazz transistor cells and models are used, and a ground metal barrier stack formed from Metal 1 to Metal 6 is employed to reduce substrate coupling between the two channels. Jazz ESD protection diodes were placed at the RF ports (1.6 kV, 1.1 A) and larger ESD diodes are placed on the digital control pots (3 kV, 2 A). Full electromagnetic modeling is done on the inductors and transmission-lines using Sonnet[1]. A total of 100 pF de-coupling capacitors are placed on-chip between the VDD and ground to also enhance the isolation between the channels. A microphotograph of the 2-beam chip is shown in Fig. 3.

III. MEASUREMENT RESULTS

The 2-beam phased array was measured on-chip after a standard probe-tip SOLT calibration. The control inputs applied to the chips are supply voltage, the biases of current mirror of each stage, and data bits (4-bit for amplitude and 4-bit for phase). No calibration is done on the chip.

A. S-Parameters

The measured S-parameters for a single channel (channel1) are shown in Fig.4. The input and output ports are well matched and agree well with simulations. The measured gain is 3-5 dB lower than simulated (depending on frequency) and this is currently being investigated. The measured NF agrees well with simulations at 14-15 GHz, but is higher at 11-14 GHz due the lower channel gain. A 17 dB gain control is achieved per channel (simulations is 20 dB). The rms phase and gain error at < 12° and 1.5 dB, respectively, up to 15 GHz. The measured linearity results in an input P1dB of -13 dBm and an input IIP3 of -6 dBm at maximum gain setting. The linearity is determined by the divider stage/I-Q network junction due to the low impedance of the I-Q network and the current driving

---

[1] Sonnet, ver. 11.52, Sonnet Software Inc., Syracuse, NY, 1986-2007

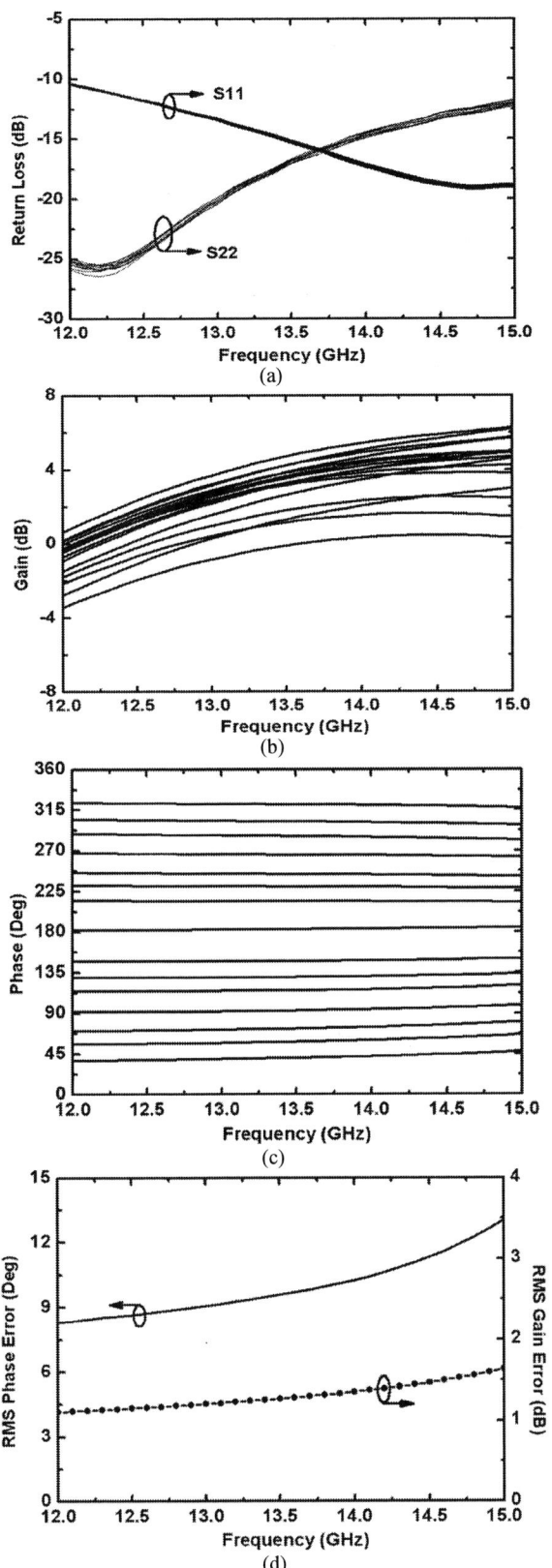

Figure 4. Measured S-Parameter of Channel 1 for the 16 different phase setting : (a) S11 and S22 (b) gain (c) phase and (d) RMS gain and phase error.

capability of the divider stage. Both channels results in

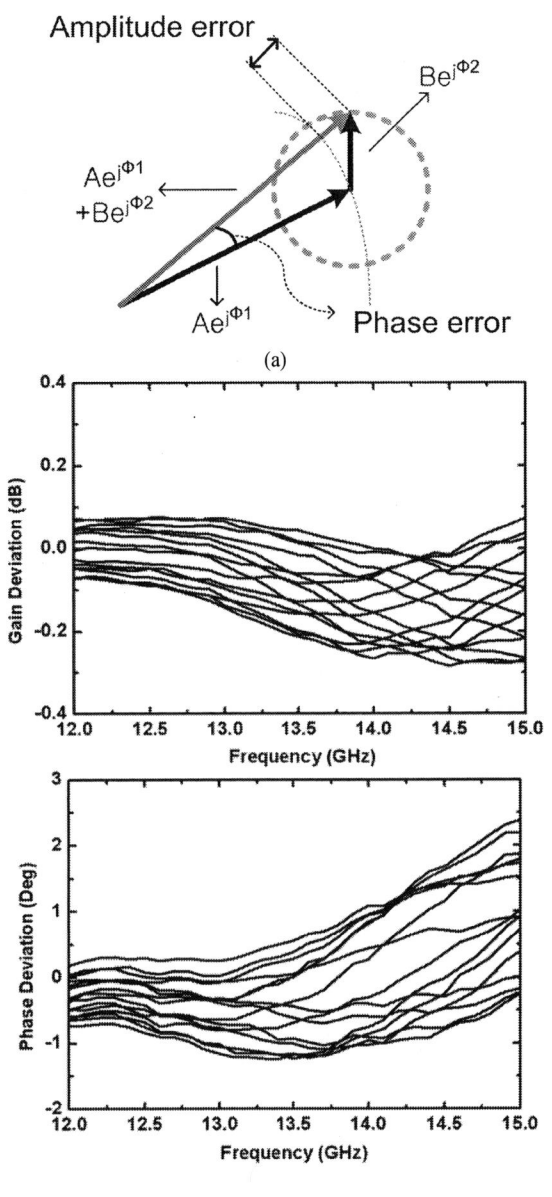

Figure 5. (a) Vectorial representation of coupling between Channel 1 and 2. (b) Measured amplitude and phase error in S21 vs. phase change in Channel2.

identical measured S-parameters and therefore, the S-parameters of Channel 2 are not shown.

B. On-Chip Isolation Measurements

The on-chip isolation is measured using S-parameter techniques, and the reverse isolation (S12) is better than -50 dB. An accurate measurement of isolation is obtained by measuring Channel 1 (S_{21}) at a fixed phase setting and changing the phase in Channel 2 from 0 to 337.5° [2, 3]. In the 2-beam array, Channel 2 contains the same signal as Channel 1 due to the input power divider, and any leakage from Channel 2 to Channel 1 at the output (after it passes by the Channel 2 phase shifter) can significantly affect the output

978-1-4244-5190-6/09 $25.00 © 2009 IEEE   131

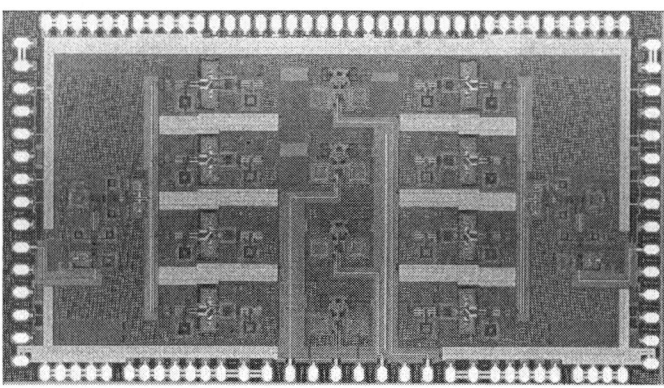

Figure 6. Microphtograph of the 2-antenna/4-beam phased array chip (2.4 mm x 4.3 mm). The antenna inputs are to the left and right, the combiners are in the middle, and the beam outputs are on the bottom)

V. Conclusion

Phased array receivers capable of simultaneous 2- and 4-beams are designed in a 0.18-μm SiGe BiCMOS technology. The 2-beam chip results in excellent channel-to-channel isolation using S-parameters tests.

Acknowledgment

The authors thank Dr. Dave Howard at Jazz Semiconductor, Newport Beach, CA, for his support.

References

[1] J. G. Kim, D. W. Kang, B.W. Min and G. M. Rebeiz, "A 36-38 GHz 4-Element Transmit/Receive Phased-Array with 5-bit Amplitude and Phase Control," *IEEE Int. Microwave Symp.*, Boston, MA, June 2009.

[2] K. Koh and G.M. Rebeiz, "An X- and Ku-Band 8-Element Phased-Array Receiver in 0.18μm SiGe BiCMOS Technology," *IEEE J. Solid-State Circuits*, vol. 43, no. 6, pp. 1360-1371, June 2008.

[3] T. Yu and G.M. Rebeiz, "A 22-24 GHz 4-Element CMOS Phased Array with On-Chip Coupling Characterization," *IEEE J. Solid-State Circuits*, Vol. 43, No. 9, pp. 2134-243, September 2008.

[4] K. Koh and G. M. Rebeiz, "A Millimeter-Wave (40-45 GHz) SiGe BiCMOS 16-Element Phased-Array Transmitter," *IEEE J. Solid State Circuits*, To appear May 2009.

[5] J. P. Comeau, M. A. Morton, W. L. Kuo, T. Thrivikraman, J. M. Andrews, C. M. Grens, J. D. Cressler, J. Papapolymerou, and M. Mitchell,"A Silicon-Germanium Receiver for X-Band Transmit/Receive Radar Modules," *IEEE J. Solid-State Circuits*, vol. 43, no. 9, pp. 1889-1896, September 2008.

[6] B.-W. Min and G. M. Rebeiz, "A Ka-Band BiCMOS T/R Module for Phased Array Applications," *IEEE Compound Semiconductor Integrated Circuits Conf.*, Vol. 1, Monterey, CA, October 2008, pp. 1-4.

[7] X. Guan, H. Hashemi, and A. Hajimiri, "A fully integrated 24-GHz eight-element phased array receiver in silicon," *IEEE J. Solid-State Circuits*, vol. 39, no. 12, pp. 2311–2320, Dec. 2004.

[8] K. Scheir, S. Bronckers, J. Borremans, P. Wambacq, Y. Rolain, "A 52 GHz Phased-Array Receiver Front-End in 90nm Digital CMOS," in *IEEE ISSCC Dig. Tech. Papers*, Feb. 2008, pp. 184-185.

[9] B. Min and G. M. Rebeiz, "Single-Ended and Differential Ka-Band Bi-CMOS Phased Array Front-Ends," *IEEE J. Solid State Circuits*, Vol. 43, No. 10, pp. 2239-2250, October 2008.

amplitude and phase of S_{21}. This can be modeled using the vector representation shown in Fig. 5. It can be seen that the coupling between the channels ((b) in Fig. 5) can be obtained from the amplitude and phase error in S_{21} due to the phase change in Channel 2.

The measured S_{21} vs. phase change in Channel 2 is shown in Fig. 5. The coupling results in a peak gain and phase error of +/- 0.2 dB and < 2° and this can be fitted to a coupling vector magnitude of 0.023 (-32.7 dB). The coupling is quite deterministic, and for a certain phase in Channel 2, a phase change of 180° results in a change of ~1°. This is shown in Fig.5.

IV. 2-Antenna/4-Beam Phased Array chip

The design chip can be extended to 4-beams with two antenna inputs as shown in Fig. 6. The output of LNA is fed to four cascode differential amplifiers to form the power dividing network. Each beam from the left antenna is added to a corresponding beam from the right antenna using an on-chip active combiner. The chip is currently being tested and the results reported at the conference.

978-1-4244-5190-6/09 $25.00 © 2009 IEEE

A 2x22.3Gb/s SFI5.2 SerDes in 65nm CMOS

N. Nedovic, A. Kristensson,
S. Parikh, S. Reddy,
W. Walker, S. McLeod, N. Tzartzanis
Fujitsu Laboratories of America, Inc.
1240 E. Arques Ave. M/S 345,
Sunnyvale, CA 94085 USA

H. Tamura, K. Kanda, T. Yamamoto,
S. Matsubara, M. Kibune, Y. Doi, S. Ide,
Y. Tsunoda, T. Yamabana, T. Shibasaki,
Y. Tomita, T. Hamada, M. Sugawara,
J. Ogawa
Fujitsu Laboratories Ltd.
1-1, Kamikodanaka 4-chome,
Nakahara-ku, 211-8588 Kawaski

T. Ikeuchi, N. Kuwata
Fujitsu Ltd., OITDA
1-1, Kamikodanaka 4-chome,
Nakahara-ku, 211-8588 Kawaski

Abstract—A 2 x 21.5-22.3 Gb/s to 4 x 10.7-11.2 Gb/s SFI5.2
compliant two-chip SerDes for a 40 Gb/s optical transponder
module has been fabricated in 65 nm 12-metal CMOS. The
deserializer receives 2 x 20 Gb/s data from a TIA and outputs
SFI 5.2 4 x 10 Gb/s data and 10 Gb/s deskew channel. The
serializer receives SFI5.2 inputs and outputs 2 x 20 Gb/s for the
optical DQPSK modulator. Although inductor-peaked CML is
needed in the deserializer 20 Gb/s input limiting amplifier (LA)
and the serializer output stages, power reduction to 3 W for both
IC's is effected by deserializing to 16 x 2.5 Gb/s internally and
implementing the core logic using standard CMOS circuits.

Index Terms—SFI 5.2, DQPSK, LA, CDR, Serializer, Deserializer, Mux, Demux.

I. INTRODUCTION

With internet bandwidth demand increasing at a rate of
60% per year, deployment of 40 Gb/s optical trunk line
infrastructure is well underway. To replace legacy 10 Gb/s
transponders in existing racks, strict power budgets must be
met, especially for the SERDES ICs, which represent the
largest component of power consumption. First generation
40 Gb/s SERDES were implemented in SiGe bipolar or
biCMOS [2],[3],[6],[7]. The more recent 90 nm and below
CMOS nodes afford a pure CMOS implementation, as has
been demonstrated in several prototype designs [4],[8],[5].
Adoption of 20 Gb/s x 2 DQPSK modulation, in addition to
providing superior performance in long-haul fiber, makes a
CMOS implemetation of the SERDES even more appealing,
as it halves the required circuit bandwidth compared to 40 Gb/s
NRZ modulation. Finally, transition from the SFI5 to SFI5.2
standard replaces the bulky 16 x 2.5 Gb/s I/O to the framer
with a more compact 4 x 10 Gb/s I/O, affording additional
savings in area, power, and cost.

This paper describes a 2.5 W 2 x 21.5-22.3 Gb/s to 4 x
10.7-11.2 Gb/s SFI5.2 compliant two-chip SerDes solution for
40 Gb/s optical transponders fabricated in 65nm 12-metal bulk
CMOS. It supports a variety of protocols such as SONET OC-
768, ITU G.709 and 4x10G ethernet. The deserializer receives
2 x 20 Gb/s data from a TIA and produces SFI 5.2 4 x 10
Gb/s data and 10 Gb/s deskew channel outputs. The serializer

A part of this work was performed under management of the OITDA supported by New Energy and Industrial Technology Development Organization
(NEDO).

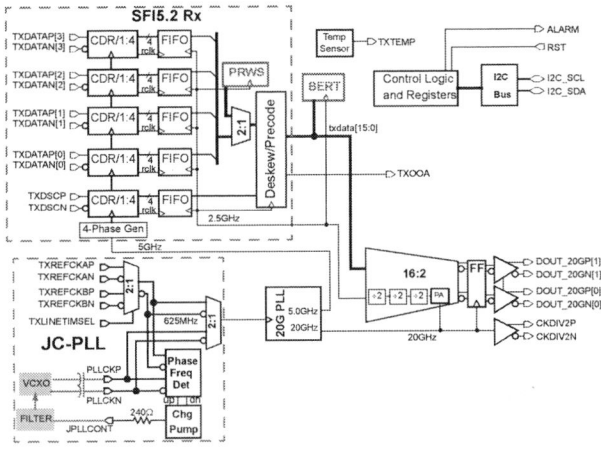

Fig. 1. Serializer Architecture.

receives SFI5.2 inputs and produces 2 x 20 Gb/s output for the
optical DQPSK modulator. Although inductor-peaked CML
is needed in the deserializer 20 Gb/s input limiting amplifier
(LA) and serializer output stages, design simplification and
power reduction is effected by deserializing to 16 x 2.5 Gb/s
internally and implementing the core logic using standard
CMOS circuits. In addition to the required SERDES and
deskew function, internal PRBS, BERT, and an I²C bus are
implemented for configurability and testability.

II. IMPLEMENTATION

A. Serializer

The serializer, shown in Fig. 1, receives five 10Gb/s SFI5.2
inputs (four data channels and deskew), and outputs two 20
Gb/s electrical data channels. It consists of an SFI5.2 Rx block
that deskews input data in compliance with the OIF SFI5.2
standard [1], a reference clock jitter clean-up PLL (JC-PLL),
a 20GHz main PLL, a 16:2 serializer, and a 2x20G retiming
flip-flop and output buffers to generates the transmit data.

A block diagram of the SFI 5.2 Rx is shown in the top left of
Fig. 1. The SFI 5.2 Rx consists of five CDR/1:4 deserializers,
five short FIFO's, and a Deskew/Precoder block that together
deserialize and deskew 4x10Gb/s input data lanes and 1 x 10

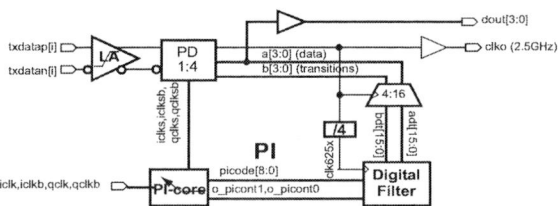

Fig. 2. SFI 5.2 Rx CDR.

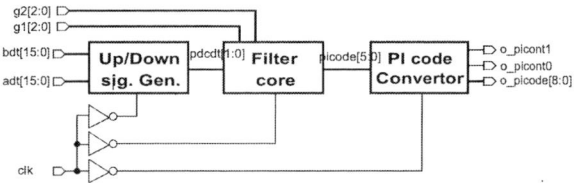

Fig. 3. SFI 5.2 Rx CDR filter.

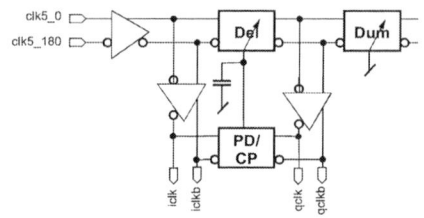

Fig. 4. 4-phase clock generation.

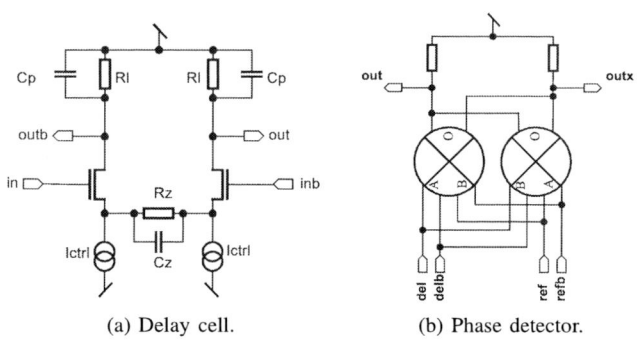

(a) Delay cell.   (b) Phase detector.

Fig. 5. 4-phase clock generation blocks.

TABLE I
FRAME COMBINATION ON THE DESKEW CHANNEL.

Even followed by Odd					Odd followed by Even				
ch3	E	ch0	ch1	ch 2	ch3	O	ch0	ch1	ch 2
ch2	ch3	O	ch0	ch1	ch2	ch3	E	ch0	ch1
ch1	ch2	ch3	E	ch0	ch1	ch2	ch3	O	ch0
ch0	ch1	ch2	ch3	O	ch0	ch1	ch2	ch3	E

Gb/s de-skew lane into a 16 x 2.5 Gb/s CMOS data stream, using a 5GHz differential reference clock frequency-locked to the incoming data. The reference clock is first converted into four phases and distributed to the five half-rate CDR's, which operate independently.

Each CDR (Fig. 2) samples the center and boundary of pairs of adjacent symbols, and these samples are immediately demultiplexed to 4 x 2.5 Gb/s *dout[3:0]* outputs and 4 x 2.5 Gb/s boundaries. The half-rate architecture greatly reduces design effort compared to a full-rate design, as the half-rate samplers operate at 5 GHz. After an additional 4:16 demux, center and boundary samples are processed in the digital filter (Fig. 3), running at 625 MHz. The digital filter processes these samples to create up/down pulses, then filters and converts them to the phase code used by the phase interpolators to match the 5 GHz 4-phase clock to the incomming data. The filter implements the transfer-function $H(s) = (\frac{g1}{s} + g2)/s$, where the coefficients g1,g2 are programable from the I²C interface. The phase interpolators are implemented as differential pairs with tail currents controlled by the the digital filter. Phases are rotated by summing the output currents.

The 4-phase 5 Ghz clock needed by the CDR's is generated from a 2-phase clock using a DLL as shown in Fig. 4. The delay block of Fig. 4 consists of a cascade of four basic delay elements (Fig. 5a). The basic delay block is a differential pair with peaked source degeneration (Rz//Cz) as shown in Fig. 5a. The phase detector (PD) uses two cross coupled gilbert cells, as shown in Fig. 5b, to avoid static phase error on the 4-phase clock signals.

Since the five 2.5 GHz recovered clocks from the CDR/DEMUX blocks are not phase-aligned, FIFO's are needed to transfer recovered data into the global 2.5 GHz clock domain used throughout the rest of the chip. The FIFO's are also able to absorb up to 7UI wander in the input signals. Each FIFO is implemented as a 16 bit deep register written by the recovered clock from a CDR and read by the global 2.5 GHz clock.

The Deskew/Precode block uses synthesized CMOS logic clocked at 2.5 GHz to re-align the data-channels over a 15UI range using the deskew channel per the SFI 5.2 standard. The deskew block operates in following three phases:

Phase 1, Deskew Frame matching: Using the Finite State Machine (FSM) shown in Fig. 6, scan the incoming deskew data every cycle to locate the even and odd parity bits. This is done by looking at five consecutive deskew channel 4-bit words to look for the two possible patterns as show in Tables I. Once locked, the FSM continues to monitor the deskew channel every five cycles.

Phase 2, Delay Determination: Determine the time difference (in UI) between deskew channel and each one of the data channels by comparing the data bits in the deskew channel with the bits being received in the data channels.

Phase 3, Data bit multiplexing: Use the calculated delay values from phase 2 to re-align the 2.5 Ghz 4-bit data from each channel and combine the four channels into a 16 bit word. The optional data inversion functionality as described by the SFI5.2 standard [1] is also performed in this step.

The details of the clock generator and the front-end 2.5

978-1-4244-5190-6/09 $25.00 © 2009 IEEE

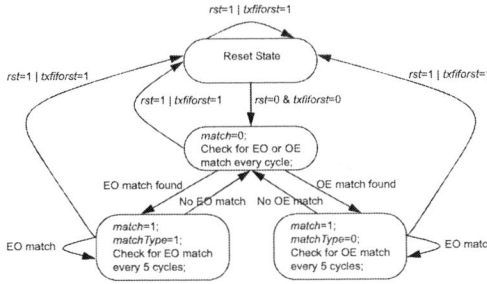

Fig. 6.   Deskew frame match FSM.

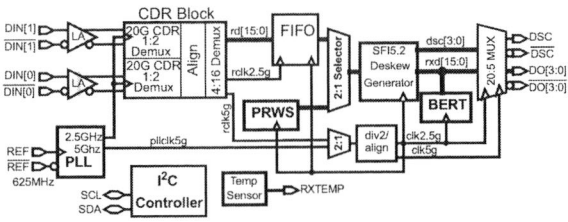

Fig. 7.   Deserializer Architecture.

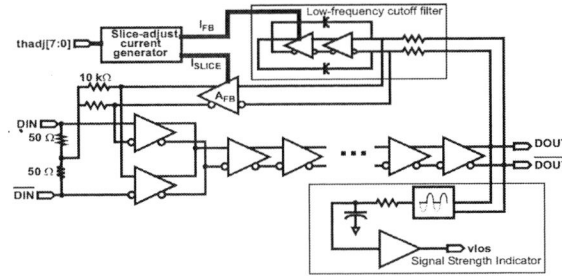

Fig. 8.   20Gb/s limiting amplifier.

Gb/s to 20 Gb/s blocks are described in [10], here we give just a brief synopsis. After a jitter clean-up PLL (JC-PLL, Fig.1) using an external loop filter conditions the 625 MHz reference clock, a secondary PLL uses dual digital/analog control loops operating simultaneously to achieve a tuning range of 19.9-22.3 Gb/s. The fast digital loop is responsible for coarse VCO tuning, while the slower analog loop fine tunes the PLL frequency. The low-gain of the VCO in the analog loop reduces PLL clock jitter. The 16:2 serializer uses a conventional 5-latch structure for serializing data, and a DLL to ensure optimal phase alignment between the 10 GHz and 20 GHz clock.

B. Deserializer

The deserializer is shown in Fig. 7. It consists of a 5 GHz PLL, two 20 Gb/s Limiting Amplifiers (LA), a 2 x 20 Gb/s CDR, a PRWS and selector for test, a SFI5.2 Deskew Generator, a BERT, and a 20 x 2.5 Gb/s to 5 x 10 Gb/s MUX (serializer). Additional details of the deserializer are given in [11].

The LA (Fig. 8) receives an AC-coupled differential signal from a TIA and outputs a minimum 400 mV amplitude differential CML signal to the CDR. It contains eight amplifier forward stages with shunt peaking inductors, a low-frequency offset compensation loop with adjustable threshold levels, a signal strength indicator, and a bias generator. The measured bandwidth of the LA is greater than 20 GHz, and a low-frequency cutoff of 70 KHz is designed in to inhibit DC drift for 72 consecutive identical digits (CID).

The CDR (Fig. 9) uses a half-rate triple loop architecture described in [11]. The half-rate architecture eliminates the need for area-consuming inductors in the design. The triple-loop architecture comprises a digital frequency acquisition loop, an analog frequency acquisition loop, and a phase loop. Splitting frequency acquisition into digital and analog loops allows both a wide data operating range and a low VCO gain while in the analog and phase loops to minimize jitter. The half-rate CDR uses a quadrature LC VCO modified from [12], and configurable phase interpolators (PI) for skewing center clock phases to adjust the position of the data sampling clock in the eyes.

Fig. 10. shows the SFI5.2 deskew and 20:5 serializer block. The core operating blocks are a retiming FIFO, a deskew generator and LSB selector, a 20:5 half-rate serializer

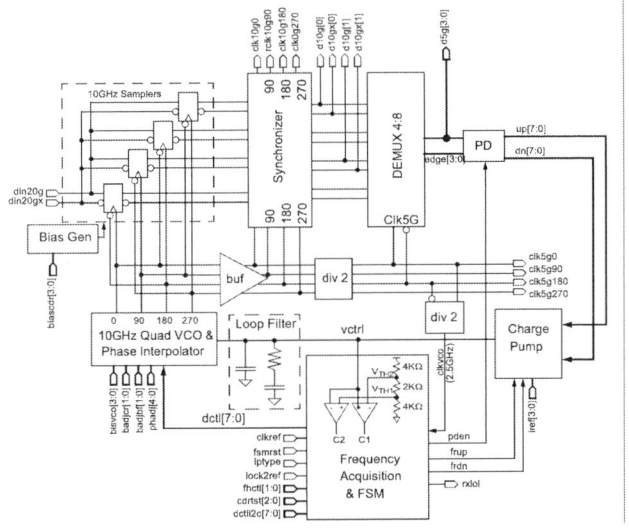

Fig. 9.   CDR for the 20Gb/s data.

incorporating a DLL and clock duty-cycle corrector (DCC). A PRWS generator, test skew generator and BERT are added for testability. The FIFO, DCC, and DLL eliminate the timing uncertainty between 2.5 GHz clock, 2.5 Gb/s data, and 5 GHz clock introduced by the long chain of the repeaters that transport data from CDR to the 20:5 serializer in the presence of process, voltage, and temperature variations.

III. MEASUREMENT RESULTS

Fig. 11 shows a serializer 2 x 20 Gb/s output eye diagram measured at VDD = 1.2V, T = 20 °C for a typical process corner sample. The serializer area is 4 x 4 mm^2 and it consumes 1.6 W at VDD = 1.2V.

Fig. 12 shows two deserializer 10 Gb/s output eyes with and without the DCC circuit enabled, measured at VDD = 1.2V,

978-1-4244-5190-6/09 $25.00 © 2009 IEEE

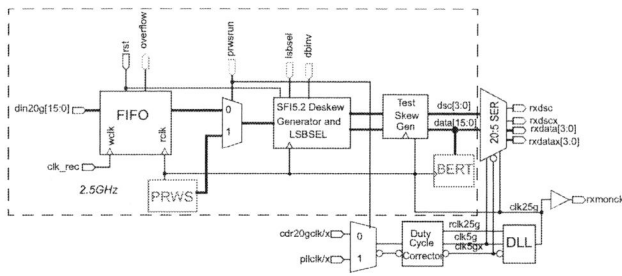

Fig. 10.   SFI 5.2 Deskew generation and 20:5 serializer.

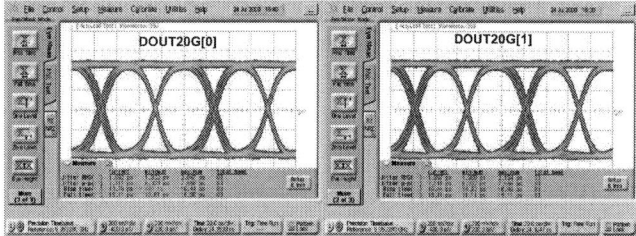

Fig. 11.   Serializer 20Gb/s Output Eyes

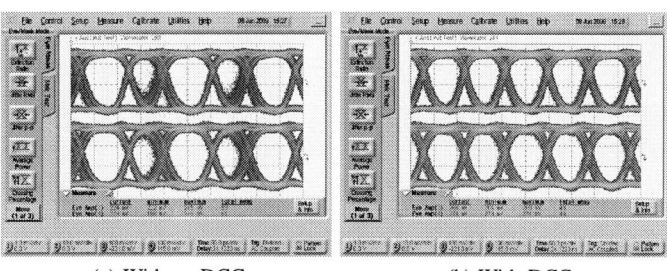

(a) Without DCC.   (b) With DCC

Fig. 12.   Deserializer 10Gb/s recovered signal output eyes.

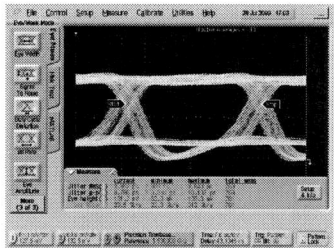

Fig. 13.   Measured 22G LA output eye.

T $= 20\,^{\circ}$C for a typical process corner sample. Fig. 13 shows the output eye from the 20G LA in the deserializer measured with a 50mV$_{pp}$ input, T $= 40\,^{\circ}$C, VDD $= 1.2$V. The LA output eye was measured on a break-out circuit. The deserializer area is 4 x 4 mm^2 and it consumes 1.4W at VDD $= 1.2$V.

Measured bit error rates for both the serializer and deserializer at an aggregate 43-44.6 Gb/s generated by a 2^{31} PRBS are $< 10^{-11}$ for VDD $= 1.14 - 1.26$V, typical process corner and T $= 20 - 50\,^{\circ}$C.

IV. CONCLUSION

A two-chip CMOS SERDES for an aggregated speed of 40Gb/s has been fabricated and measured. This enables the use of DQPSK for communication over longer distances on legacy fiber. The 3W power consumption is less than 25% of a first generation biCMOS SFI5 SERDES chipset [2],[3] and 45% less than a recent CMOS SFI5 SERDES chipset [9]. In addition, minimal use of inductors and adoption of the more recent SFI5.2 standard results in chip area more than 1/3 less than [9].

REFERENCES

[1] *Serdes Framer Interface Level 5 Phase 2 (SFI-5.2): Implementation Agreement for 40Gb/s Interface for Physical Layer Devices*, Optical Internetworking Forum, October 2006.

[2] Adrian Ong, et al, *A 40–43-Gb/s Clock and Data Recovery IC With Integrated SFI-5 1:16 Demultiplexer in SiGe Technology*, IEEE J. Solid-State Circuits, Vol. 38, no. 12, pp. 2155-2168, Dec. 2003.

[3] Hai Tao, et al, *40-43-Gb/s OC-768 16:1 MUX/CMU Chipst With SFI-5 Compliance*, IEEE J. Solid-State Circuits, Vol. 38, no. 12, pp. 2169-2180, Dec. 2003.

[4] J. Lee and B. Razavi, *A 40–Gb/s Clock and Data Recovery Circuit in 0.18-um CMOS Technology*, IEEE J. Solid-State Circuits, Vol. 38, No. 12, pp. 2181-2190, Dec. 2003..

[5] N. Nedovic, et al, *A 40–44 Gb/s 3 Oversampling CMOS CDR/1:16 DEMUX*, IEEE J. Solid-State Circuits, Vol. 42, No. 12, pp. 2726-2735, Dec. 2007.

[6] Mario Reinhold, et al, *A Fully Integrated 40-Gb/s Clock and Data Recovery IC With 1:4 DEMUX in SiGe Technology*, IEEE J. Solid-State Circuits, Vol. 36. No 2, pp. 1937-1945, Dec. 2001.

[7] Martin Wurzer,et al, *A 40–Gb/s Integrated Clock and Data Recovery Circuit in a 50-GHz Silicon Bipolar Technology*, IEEE J. Solid-State Circuits, Vol. 34, No. 9, pp. 1320-1324, Sept. 1999.

[8] Thomas Toifl, et al, *A 72mW 0.03mm^2 Inductorless 40Gb/s CDR in 65nm SOI CMOS*, IEEE ISSCC Dig. Tech. Papers, pp. 226-227, Feb. 2007.

[9] Y. Amamiya, et al, *A 40Gb/s Multi-Data-Rate CMOS Transceiver Chipset with SFI-5 Interface for Optical Transmission Systems*, IEEE ISSCC Dig. Tech. Papers, pp. 358-359, Feb. 2009

[10] K. Kanda et al.,*A Single–40Gb/s Dual-20Gb/s Serializer IC with SFI-5.2 Interface in 65nm CMOS*, IEEE ISSCC Dig. Tech. Papers, pp. 360-361, Feb. 2009.

[11] N. Nedovic et al., *A 2 x 22Gb/s SFI5.2 CDR/Deserializer in 65nm CMOS Technology*, Symp. on VLSI Circuits, Dig. of Tech. Papers, June 2009.

[12] P. Andreani et al., *Analysis and design of a 1.8–GHz CMOS LC quadrature VCO*, IEEE J. Solid-State Circuits, vol.37, no.12, p.1737-1747, Dec. 2002.

A 0.25μm InP DHBT 200GHz+ Static Frequency Divider

M. D'Amore, C. Monier, S. Lin, B. Oyama, D. Scott, E. Kaneshiro, A. Gutierrez-Aitken, and A. Oki

1 Space Park Blvd.
Northrop Grumman Aerospace Systems
Redondo Beach, CA 90278

Abstract—Static frequency dividers are widely used technology performance benchmark circuits. Using a 0.25μm 530GHz f_T / 600GHz+ f_{max} InP DHBT process, a static frequency divider circuit has been designed, fabricated, and measured to operate up to 200.6GHz. The divide-by-2 core flip/flop dissipates 228mW.

Keywords – High-Speed, InP, DHBT, Static Frequency, Divider

I. INTRODUCTION

As high speed communication channel throughput and performance needs continue to evolve so do the technologies used to implement the various system building blocks. Static frequency dividers are widely used as a circuit performance benchmark or figure-of-merit indicator to gauge a particular device technology's ability to implement high speed digital and integrated high performance mixed-signal circuits. Recent developments in highly scaled HBT devices have yielded static frequency divider speeds reported both in SiGe [1,2] and in InP [3,4] over 100GHz. We report measured results for a 200.6GHz static frequency divider demonstration circuit fabricated in a highly scaled 0.25μm InP DHBT process with a f_T = 530 GHz and f_{max} in excess of 600 GHz, which is to the author's knowledge the highest reported frequency of operation for a static frequency divider.

II. TECHNOLOGY

The divider demonstration circuit was fabricated using Northrop Grumman Corporation's 0.25μm InP DHBT process. The advanced InP DHBT process has been developed for device and interconnect scalability with an emphasis on improved manufacturability and yield [5]. Aggressive emitter mesa scaling is achieved down to 0.25μm emitter widths. The emitter size is defined by a combination of photolithography and nitride masking followed by semiconductor dry etches. The emitter formation is terminated by a short selective wet etch to remove plasma damage and produce well controlled and minimum undercutting at the InP emitter ends for reduced base access resistance. The emitter mesa is followed by a self-aligned base metallization process. Base metal and base mesa layouts are designed to produce minimum intrinsic and extrinsic base-collector capacitance (C_{BC}) without significantly impacting base resistance. The fabrication of integrated circuits is completed by a multi-level interconnect process presented in Fig. 1 with four plated Au metal layers and low-k benzocyclobutene (BCB) as the interlayer dielectric between metallization [5].

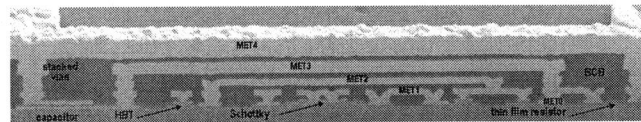

Figure 1. Low-k Dielectric 4-Metal Layer InP Technology

The MBE-grown InGaAs/InP DHBT structure uses a thin n-type doped (5×10^{17} cm^{-3}) InP emitter terminated with an In-rich, heavily doped InGaAs cap layer for low emitter contact resistance. A 220 Å base layer uses a compositionally-graded InGaAs material heavily doped p+ (8×10^{19} cm^{-3}) to maintain low base resistance while reducing base transit time. The total InP collector layer thickness is 1200 Å, and its doping level (8×10^{16} cm^{-3}) is adjusted to support increased current density suitable for high-speed digital applications.

The common-emitter I-V plot for a 0.25×4-μm^2 emitter area HBT is shown in Fig. 2. The breakdown voltage (BV_{CEO}) is over 4 Volts and the peak current gain (β) is 30 (see Gummel characteristics in the inset of Fig. 2). S-parameter measurements have been obtained using an HP8510XF, on-wafer TRL calibration, and no de-embedding techniques. A peak f_T of 530 GHz occurs at V_{CB} = 0.4 V with simultaneous f_{max} in excess of 600 GHz at collector current density (J_C) of 13 mA/μm^2. A MAG/MSG Gain of 15.5 dB is measured at 100 GHz. f_T and f_{max} values versus J_C are plotted from the same 0.25×4-μm^2 device in Fig. 3.

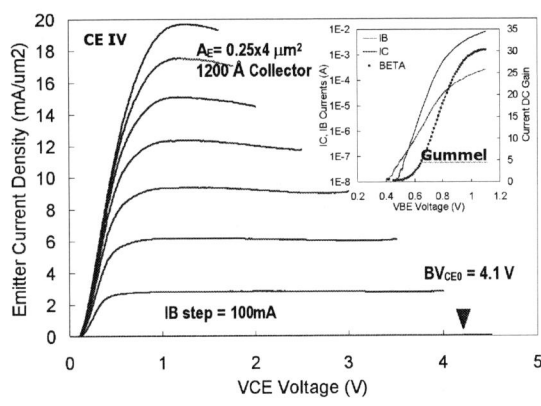

Figure 2. Gummel and Common-emitter I-V Plot for a 0.25×4 μm^2

978-1-4244-5190-6/09 $25.00 © 2009 IEEE

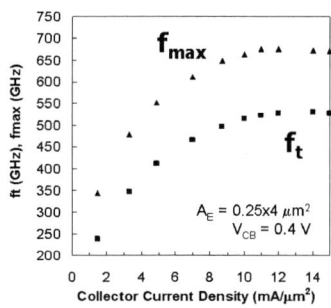

Figure 3. Plot of 0.25×4-μm² Device f_T and f_{max} vs. J_C

III. DESIGN AND OPTIMIZATION

The divider architecture consists of several current-mode logic (CML) circuit stages which were highly optimized to obtain peak performance. Peak divider performance was achieved by maximizing input drive power and minimizing the effective divider output feedback time constant. The circuit stages shown in Fig. 4 consist of front-end bias circuitry, a frequency divider core, and output gain buffers. The demonstration circuit input and output were designed to be single-ended interfaces. The frequency divider core power dissipation was 228mW.

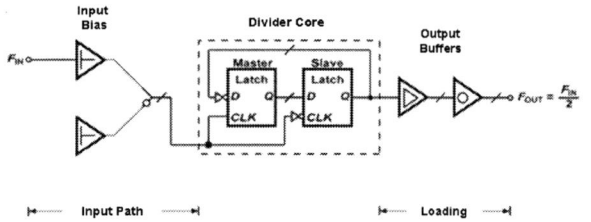

Figure 4. Top Level Block Diagram

A. Front-end Bias Circuitry

The front-end of the divider shown in Fig. 5 simultaneously provides the on-chip 50Ω termination and the bias to the divider circuitry, with no need for intermediate input gain stages or impedance transformations which could limit the input power to the divider. The complementary side bias network was not brought out to bond pads. Series diodes and resistors create the necessary level shifting to provide the voltage bias to the lower differential pair inputs of the latches in the divider core. An ultra-broadband matching network using a small micro-strip inductor, which effectively tunes out the input capacitance of the divider latches as well as on-chip bond pads, was designed to enable maximum input power drive. Better than 10dB return loss was achieved in simulation across the entire divider frequency of operation.

B. Frequency Divider Core

The frequency divider core is implemented using a master-slave flip-flop with an inverted output feedback configuration illustrated in Fig. 6. The master-slave latches use 0.25x3μm

devices biased to ~14mA/μm² with a 6V supply. Emitter followers using 0.25x6μm devices were inserted to provide the impedance transformation for the inductively peaked latch loads and were biased with a 4.5V supply. In order to minimize the effective output time constant of the divider feedback path for the given architecture an optimized latch load and a highly compacted divider layout were necessary.

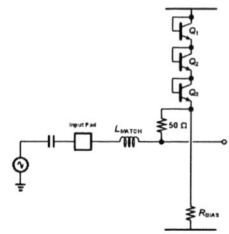

Figure 5. Divider Input Bias Network x2

The inductively peaked latch load optimization was performed using an automated binary search simulation for the maximum frequency of operation for a range of resistances and inductances. The result of the simulation is a divider performance contour plot shown in Fig. 7. As can be seen from the topological plot of one example variant, the peak divider performance occurs at an apex that is near a region of minimized low frequency feedback gain (small resistance) and instability or non static operation. The delicate balance between having just enough low and high frequency feedback gain for broadband operation is the design challenge. It is interesting to note that circular bands exist around the apex where it is possible to achieve a certain maximum frequency of operation with more than one combination of resistance and inductance; however, there is only one optimal value of resistance and inductance at the apex, which represents the absolute maximum frequency for static operation. In order to cover process and modeling uncertainty, variants with different combinations of resistance and inductance around the apex were also fabricated. The measured divider in this work had an optimal resistance of 30Ω and inductance of 22.5pH.

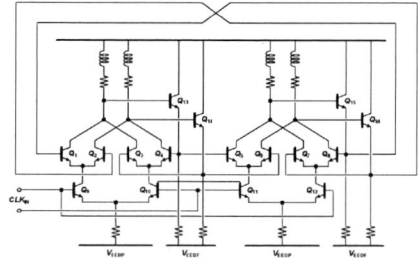

Figure 6. Master-Slave Flip/Flop Circuit Diagram

It was important to construct the divider layout compactly to reduce parasitic capacitance and inductance which ultimately slow down the divider. The frequency divider core layout was symmetrically laid out across the horizontal axis of the die, measuring 50μm x 170μm. The transistors in the latches were oriented to minimize the critical feedback path of the master-slave latches as well as other less critical signal paths. Because of the non-conductive InP substrate, a front-side ground plane was placed over the divider to minimize parasitic inductances which could have caused circuit instabilities.

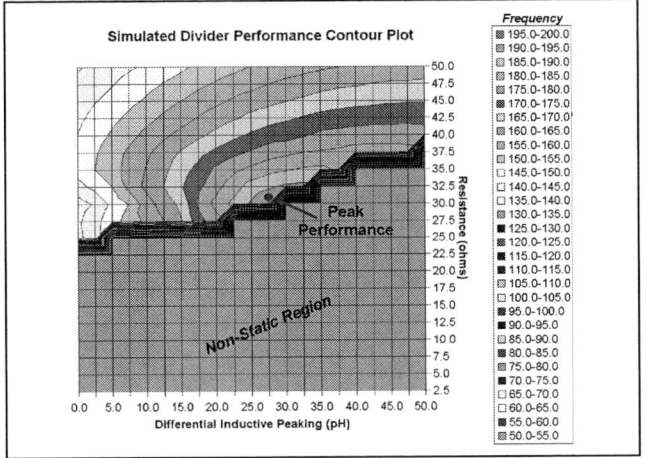

Figure 7. Example Simulated Divider Performance Contour Plot

C. Output Gain Buffers

The output gain buffers provide capacitive load isolation for the frequency divider core. The output gain buffer is a two stage differential pair which runs at approximately half the current density of the master-slave latches. The pre-amplifier stage isolates the divider from the final output buffer which provides the impedance transformation to 50Ω. A broadband match similar to the front-end is applied to the output buffer but only needs to be optimized to one half the operating range of the front-end divider input. Only one of the outputs is brought out to the bond pad. The complementary side is terminated on-chip. Better than 10dB return loss was achieved in simulation for this interface.

IV. TEST AND MEASUREMENT

After fabrication testing was performed at room temperature on a wafer probe station using a series of sources to span the full bandwidth of the divider. Nominal dual supply voltages of 4.5V and 6V were used for all divider tests and no bias optimization tuning was performed. Many test setups were needed to cover the full bandwidth of the divider which resulted in many hours of setup changes and careful data collection. Low frequency testing down to 5GHz was done with a conventional Agilent 50GHz CW source. Testing beyond 50GHz in the V, W, D, and G-bands was performed with backward wave oscillators (BWO) from ELVA-1 or a combination of lower frequency sources, in-house amplifiers, and mixers from Virginia Diodes, Inc depending on the band needed. Table 1 provides a more detailed summary of the RF source test setup. An Agilent E4448A 50GHz spectrum analyzer was used to view the divider output. V-band or W-band harmonic mixer modules were needed on the divider output when testing above 100GHz and 150GHz RF inputs, respectively. Fig. 8 and Fig. 9 illustrate 5GHz and 200.6GHz measured operation. It should be noted that Fig. 9 includes the

conversion loss (42dB) of the W-band harmonic mixer since it could not be calibrated out. In addition we verified that over G-band operation the divider did not have any divide-by-4 tones which can be a common failure mechanism.

The divider measured vs. simulated input sensitivity plot is presented in Fig. 10. A variable mechanical attenuator is used to find the minimum input power. Input power calibration was performed up to the RF probe inputs and do not include the RF probe losses. The simulated input sensitivity consisted of using an extrapolated ~600GHz f_T transistor model, lumped-element transmission line parasitics, and the corresponding latch load values of 30Ω and 22.5pH. Our simulated input sensitivity is in good agreement with the observed measured data.

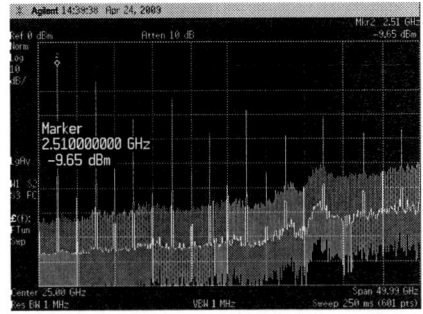

Figure 8. Measured 5GHz Divider Output Spectrum

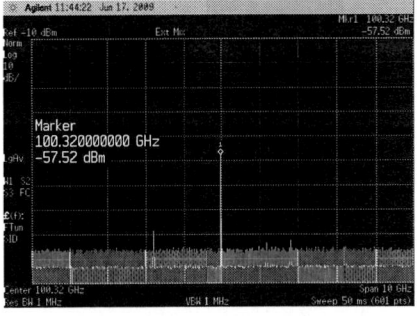

Figure 9. Measured 200.6GHz Divider Output Spectrum. Output W-band Harmonic Mixer Conversion Loss is 42dB at 100GHz.

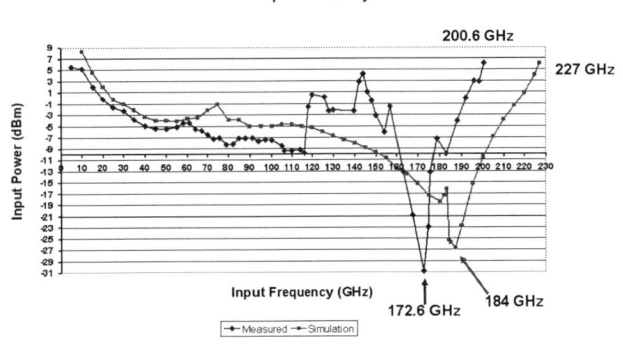

Figure 10. Measured vs. Simulated Divider Input Sensitivty Plot

TABLE I.   RF Source Test Setup Summary

Frequency Range (GHz)	RF Source Test Setup Description
<50	Agilent 50GHz frequency synthesizer connected to RF probe with coaxial V-cable. An Agilent power meter was used for calibration.
50-75	Agilent 50GHz frequency synthesizer with Agilent V band MMIC module which outputs WR-15. WR-15 waveguides are used up to a WR-15 interface wafer probe. An Agilent V band power meter was used for calibration.
75G-110	ELVA-1 W-band BWO outputs WR-10 and connects to a variable mechanical attenuator. The attenuator outputs WR-10. WR-10 is used up to the probe which is also WR-10. An Agilent W band power meter was used for calibration.
110-130	Agilent 50GHz frequency synthesizer with a V band MMIC module outputs WR-15 to a RF amplifier. Two different amplifiers had to be used; they each cover 105-116 GHz and 116-130 GHz. The amplifier output drives a Virginia Diodes X2 multiplier which outputs WR-8. Due to lack of WR-8 waveguides we used WR-10 instead. There is a waveguide mismatch at the output of the frequency doubler. A WR-10 variable attenuator is used to vary the input power and the probe interface is also WR-10. An Agilent W band power meter was used for calibration.
140-150	Agilent 50GHz frequency synthesizer with a W band MMIC module outputs WR-12 to a RF amplifier that is WR-12 at both input and output (no waveguide transition was used). The amplifier output drives a Virginia Diodes X2 multiplier with a WR-12 input and WR-6 output. Due to the high loss of WR-5 and WR-6 waveguids we used WR-10 instead and operate it out of band. There is a waveguide mismatch between the WR-6 at the output of the doubler with a WR-5 to WR-10 waveguide transition. After a length of WR-10 there is another WR-10 to WR-5 transition to a WR-5 interface wafer probe. An ELVA-1 D band power meter was used for calibration.
154-220	The ELVA-1 G-band BWO goes as low as 154 GHz. The BWO has been custom configured to output on WR-10 size waveguide operating in the TE$_{20}$ mode. WR-10 to WR-5 transitions join the BWO with a WR-5 variable mechanical waveguide attenuator and then a segment of WR-10. One more WR-10 to WR-5 transition connects the WR-10 to a WR-5 interface wafer probe. An ELVA-1 G band power meter was used for calibration.

V.   CONCLUSION

In conclusion we report a 200.6GHz static frequency divider demonstration circuit fabricated in Northrop Grumman Corporation's advanced 0.25µm InP DHBT process with four levels of metal interconnects. The divide-by-two core was implemented using a master-slave flip-flop with inverted feedback configuration with 0.25x3µm devices biased to ~14mA/µm^2. The divider core flip/flop power dissipation was 228mW. The overall die size was 700µm x 850µm (Fig. 11).

The authors would like to acknowledge and thank the DARPA TFAST program under Contract HR0011-08-C-0065 for support of this work (Approved for Public Release, Distribution Unlimited).

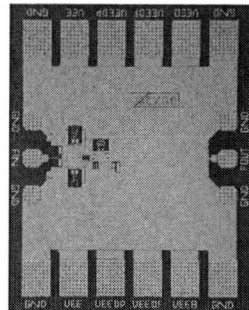

Figure 11. Divider Die Photo (700µm x 850µm)

REFERENCES

[1]   S. Trotta et al., "110-GHz Static Frequency Divider in SiGe Bipolar", Compound Semiconductor IC Digest, pp. 291-294 (2005)

[2]   E. Laskin et al., "Low-Power, Low-Phase Noise SiGe HBT Static Frequency Divider Topologies up to 100 GHz", IEEE BCTM Digest, pp. 235-238 (2006)

[3]   N. Phan et al., "A 154-GHz Static Divider in 0.25µm InP DHBT", Proc. Device Research Conference, pp. 205-206 (2006)

[4]   D. A. Hitko et al., "A low power (45mW/latch) static 150GHz CML divider", Compound Semiconductor IC Digest, pp. 167-170 (2004)

[5]   C. Monier et al., "High-Speed InP HBT technology for advanced mixed-signal and digital applications", IEEE IEDM Digest, pp. 671-674 (2007)

A 32-GS/s 6-bit Double-Sampling DAC in InP HBT Technology

Munehiko Nagatani, Hideyuki Nosaka, Shogo Yamanaka*, Kimikazu Sano, and Koichi Murata

NTT Photonics Laboratories, NTT Corporation
3-1 Morinosato Wakamiya, Atsugi-shi, Kanagawa, 243-0198, Japan
*Now with NTT Network Innovation Laboratories, NTT Corporation

Abstract — This paper describes the circuit design and measured performance of a high-speed digital-to-analog converter (DAC) for a coherent optical transceiver based on digital signal processing (DSP). To achieve a high sampling rate, a novel double-sampling technique was devised. A 6-bit DAC test chip was fabricated with InP HBT technology, which yields a peak f_t of 175 GHz and a peak f_{max} of 260 GHz. Measured DNL and INL were within +0.49/-0.17 LSB and +0.97/-0.06 LSB, respectively. Measured SFDR was 45.0 dB for a sinusoidal output of 72.5 MHz at a sampling rate of 13.5 GS/s, which is the limit of our measurement setup. Ramp-wave output at a sampling rate of 32 GS/s was obtained. Total power consumption was 1.4 W with a supply voltage of -4.0 V. To our knowledge, this is the first 6-bit DAC that can operate at a sampling rate of over 30 GS/s.

Index Terms — **Digital-to-Analog Converter (DAC), R-2R, Current-Steering, InP, HBT**

I. INTRODUCTION

Coherent optical data transmission schemes based on digital signal processing (DSP) are now attracting a great deal of attention. Coherent schemes using multi-level or multi-carrier modulation formats (e.g., QAM or OFDM) are candidates for use in the next generation of long-haul optical communications networks because of their high spectrum efficiency and robustness with regard to fiber chromatic dispersion (CD) and polarization mode dispersion (PMD) [1]. In transmitters for such systems, a digital-to-analog converter (DAC) is a key component. The DAC performance requirements for 112-Gb/s coherent optical transmission systems are now being examined, and they might be a sampling rate of up to 28-56 GS/s and a resolution of up to 6 bits [2]. Though several high-speed DACs have been reported [3]-[5], there is no 6-bit DAC that can operate at a sampling rate of over 28 GS/s.

This paper describes the design of a high-speed DAC with a novel double-sampling technique. The fabricated DAC can operate at a sampling rate of over 28 GS/s. Section II presents design considerations for a high-speed 6-bit DAC. Section III describes measurement results for a test chip. Section IV summarizes the paper.

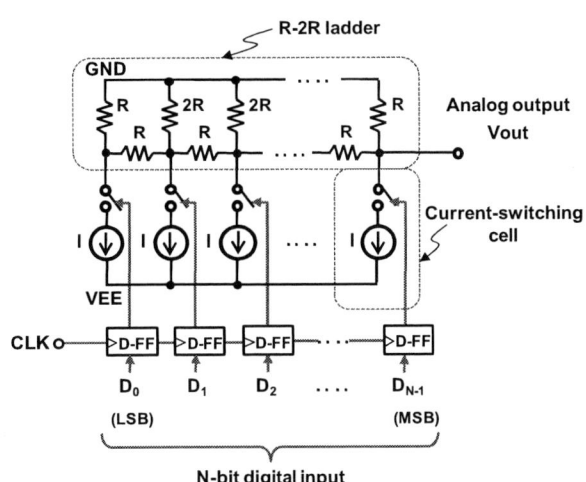

Fig. 1.   R-2R current-steering DAC.

II. CIRCUIT DESIGN

A. DAC Architecture

Most high-speed DACs are based on a current-steering architecture. In such a DAC, an R-2R ladder can be used to implement binary weighting [6]. Figure 1 illustrates a conventional current-steering architecture that employs an R-2R ladder. This circuit consists of identical current-switching cells, an R-2R ladder, and D-type flip-flops (D-FFs). The D-FFs are placed just before the current switches to reduce the timing skew among the input digital data and to suppress output glitches. This DAC provides high-speed operation and has lower power consumption and a smaller die area than other types of DACs.

B. Double-Sampling Technique

An R-2R current-steering DAC provides fast operation. Still, a sampling rate of over 28 GS/s is a tough target. The main factor that limits the sampling rate of this DAC is the timing skew among the digital data that drive the current-switching cells. As the sampling rate increases, the amplitude and jitter of the clock signal, which drives the D-FFs, decreases and gets worse. In other words, when the sampling rate is very high, the D-FFs do not suppress the timing skew, which results in distortion of the analog output.

978-1-4244-5190-6/09 $25.00 © 2009 IEEE

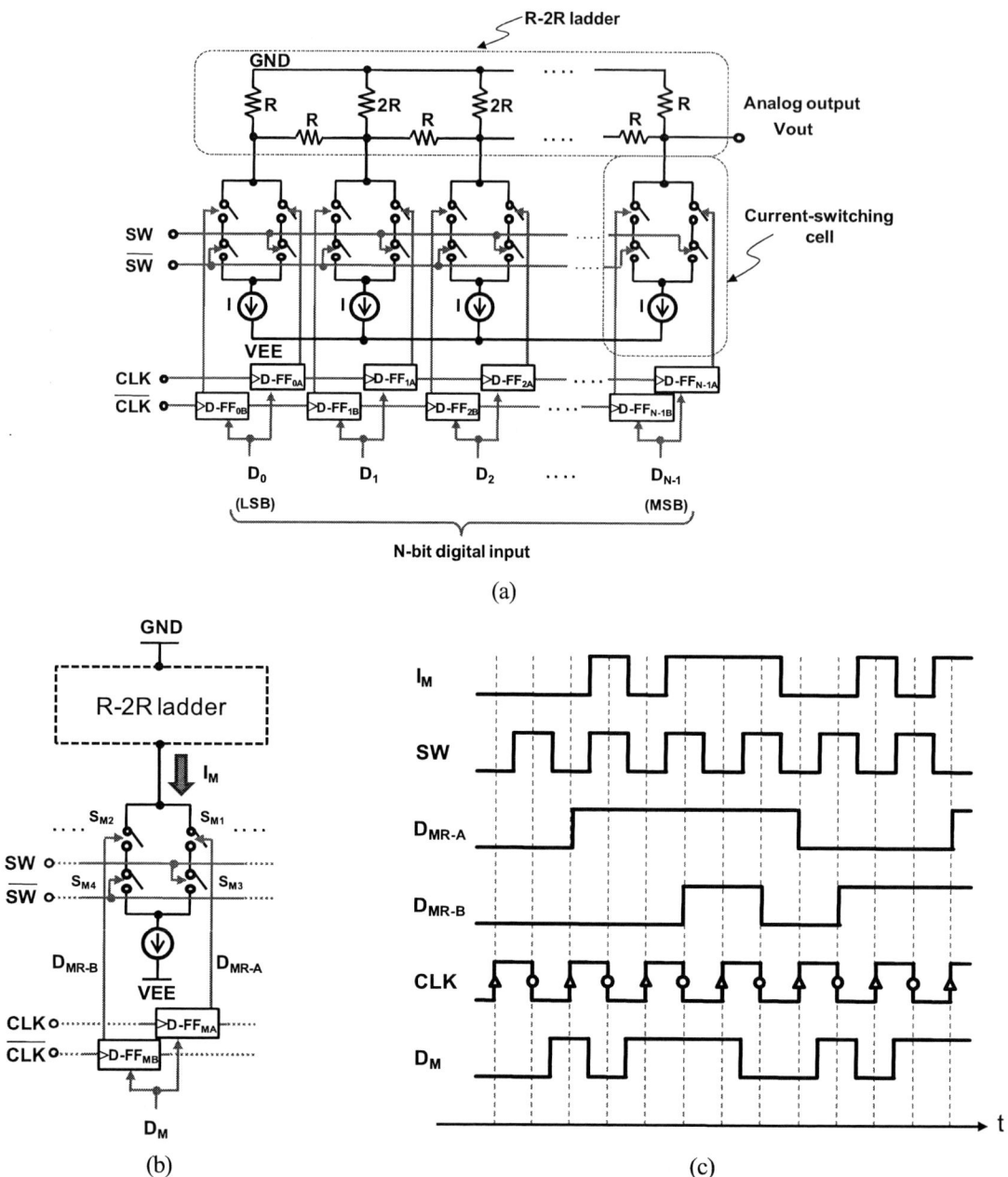

Fig. 2. (a) R-2R current-steering DAC with double-sampling technique. (b) Current-switching cell. (c) Timing chart.

To achieve a sampling rate of over 28 GS/s, we devised a new sampling technique, which we call double-sampling. Figure 2(a) shows a new architecture for a DAC that employs this technique. This DAC has a pair of D-FFs just before each current-switching cell and four switching functions in each current-switching cell [Fig. 2(b)]. One D-FF latches input signal at every rising edge of a half-rate clock signal, and the other one latches input signal at every falling edge. As a result, the input signal (D_M) is demultiplexed into two half-rate signals (D_{MR-A} and D_{MR-B}), as shown in Fig. 2(c). These half-rate signals (D_{MR-A} and D_{MR-B}) drive the two switches (S_{M1} and S_{M2}) in the current-switching cell, respectively. The other switches (S_{M3} and S_{M4}) in the cell are driven by the differential switching signals (SW

and $\overline{SW}$). Consequently, two half-rate signals are multiplexed in the cell, and an output current (I_M) that corresponds to the input signal (D_M) is generated. Thus, this DAC operates with a half-rate clock. For example, to achieve a sampling rate of 28 GS/s, 14-GHz clock signal is needed. This technique relaxes the speed restraints on the timing alignment of the digital data and on the routing of the clock and data signals. As a result, this DAC provides a higher sampling rate than the conventional one.

C. Layout Considerations

DAC performance also depends on the layout. To reduce the timing skew for the current switches and eliminate the noise from the analog output, we designed the layout with the focus on low wiring capacitance and high symmetry. In addition, the power supplies were separated into analog power supplies for the DAC core and digital power supplies for the other building blocks. Notably, the analog power supplies for the DAC core were routed in wide lines to reduce the parasitic resistance of the lines and consequently the voltage loss.

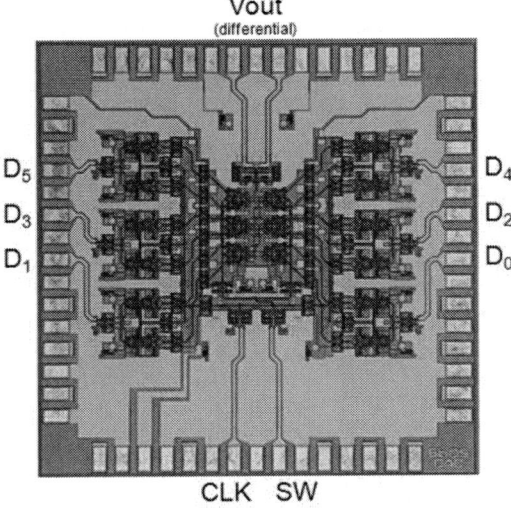

Fig. 3. Microphotograph of the 6-bit DAC test chip (3 mm x 3 mm).

III. MEASUREMENT RESULTS

A. Process Technology

A 6-bit DAC test chip was fabricated with our InP HBT technology [7]. The HBT has a 70-nm-thick undoped InP emitter, a 50-nm-thick carbon-doped InGaAs base, and a 300-nm-thick InGaAs collector. The lateral emitter dimension of the standard HBT is 1 μm x 4 μm. The fabricated HBTs have a peak cutoff frequency (f_t) of 175 GHz and a peak maximum oscillation frequency (f_{max}) of 260 GHz. The technology also features two metal interconnect layers. A microphotograph of the chip is shown in Fig. 3. The chip size is 3 mm x 3 mm. The number of elements is about 2,000.

B. Measurement Setup

The DAC was tested on wafer with a probe station and high-frequency probes. Figure 4 shows the test setup. The 8-ch pulse-pattern generator (PPG) generated programmed 6-bit pulse-patterns for the digital input (D_5-D_0), a clock signal (CLK), and a switching signal (SW). Then, the differential analog outputs of the DAC were analyzed with an oscilloscope and a spectrum analyzer. Unfortunately, due to the limitations of the PPG, we could not measure the dynamic characteristics for sinusoidal output of the DAC at speeds of over 13.5 GS/s. To measure the DAC operations at speeds of over 13.5 GS/s, we used another test setup (Fig. 5). The PPG and one synthesizer generated over-13.5-Gbps digital ramp sequence for the upper 4 bits (D_5-D_2). The lower 2 bits (D_1, D_0) were kept high state. The other synthesizer generated a CLK and a SW.

C. Measured Performance

The DAC operates with a supply voltage of -4.0 V. It consumes a total power of 1.4 W. The data and clock input buffers, the clock distributers, and the retiming D-FFs account for almost 80% of the total power dissipation. Figure 6 shows measured differential and integral non-linearity (DNL/INL). DNL and INL were within +0.49/-0.17 LSB and +0.97/-0.06 LSB, respectively. These results ensure the monotonicity of the DAC.

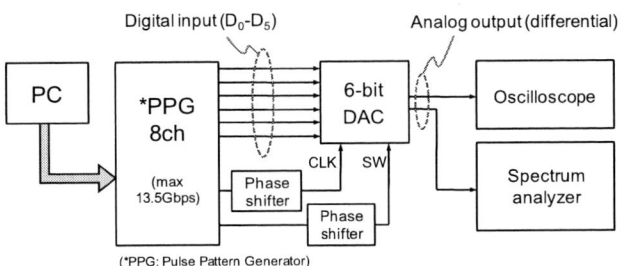

Fig. 4. Measurement setup for testing the static and dynamic characteristics of the DAC (~13.5 GS/s).

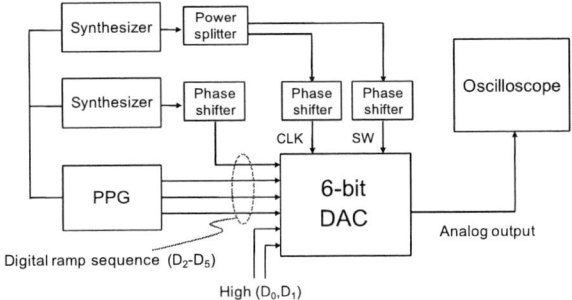

Fig. 5. Measurement setup for testing the 16-step ramp-wave (> 13.5 GS/s).

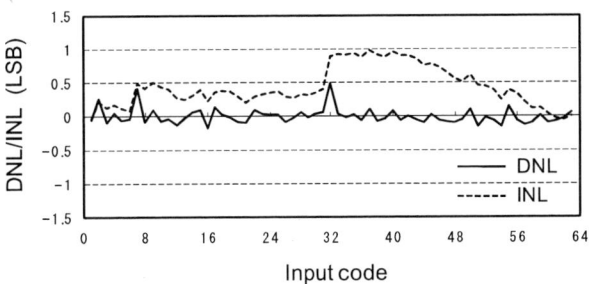

Fig. 6. Measured DNL and INL.

Figure 7 shows the measured sinusoidal output waveform and spectrum of 72.5 MHz at a sampling rate of 13.5 GS/s. The largest spurious signal was third-harmonic distortion (3rd HD), resulting in a spurious-free dynamic range (SFDR) of 45.0 dB. Figure 8 summarizes measured SFDR versus output frequency at a sampling rate of 13.5 GS/s. The SFDR remains about 30 dB up to an output frequency of 6.66 GHz (near-Nyquist frequency). These results show the DAC achieves good dynamic linearity over the entire Nyquist bandwidth. Degradation of the SFDR was caused by an output buffer inserted after the DAC core. It could be suppressed by using an improved linear buffer.

978-1-4244-5190-6/09 $25.00 © 2009 IEEE   143

Figure 9 shows a measured 16-step ramp-wave output at a sampling rate of 32 GS/s. Though a small glitch is seen in the 8th step, where the digital input change from "0111(11)" to "1000(11)", the output of the DAC changes every sampling period (31.25 ps). Thus, the expected waveform at a sampling rate of 32 GS/s was obtained. This result demonstrates the DAC can operate at a sampling rate of over 28 GS/s. Table I summarizes the performance of the DAC.

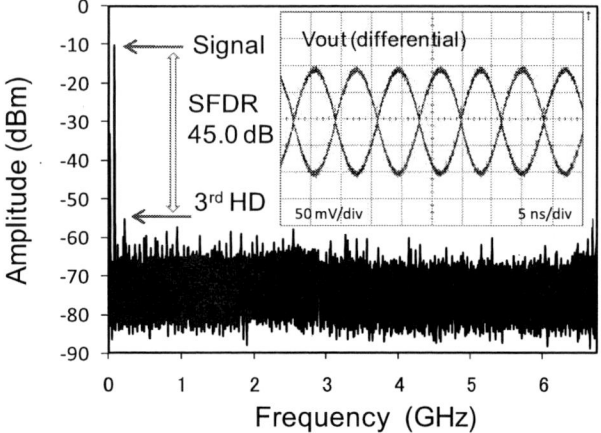

Fig. 7.   Measured sinusoidal output waveform and spectrum (fout=72.5 MHz @ 13.5 GS/s).

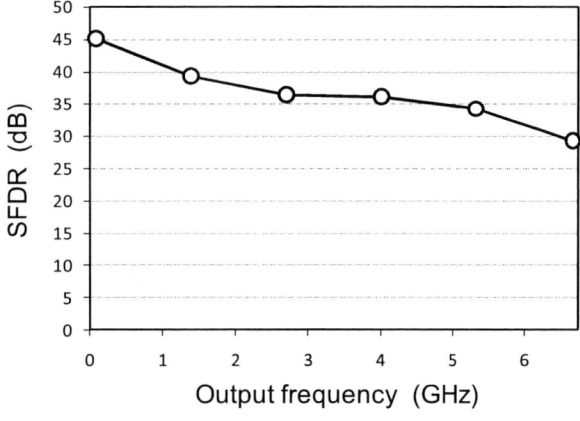

Fig. 8.   Measured SFDR versus output frequency at 13.5 GS/s.

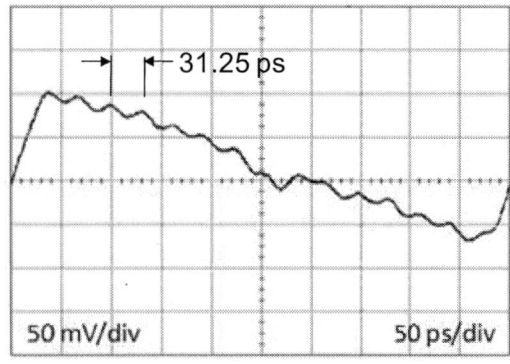

Fig. 9.   Measured 16-step ramp-wave output at 32 GS/s.

TABLE I.
DAC PERFORMANCE SUMMARY

Resolution	6 bits
Sampling rate	~ 32 GS/s
DNL	+0.49/-0.17
INL	+0.97/-0.06
SFDR (@ 13.5 GS/s)	45.0 dB (fout=72.5 MHz)
	29.2 dB (fout=6.66 GHz)
Power consumption	1.4 W
Supply voltage	-4.0 V
Chip size	3 mm x 3 mm
Technology	InP HBT (ft=175 GHz, fmax=260 GHz)

IV. CONCLUSION

The design of a high-speed DAC fabricated with InP HBT technology has been described. To achieve a resolution of 6 bits and a sampling rate of over 28 GS/s, we introduced R-2R current-steering architecture and a novel double-sampling technique. A 6-bit DAC test chip was fabricated with our InP HBT technology. The chip size is 3 mm x 3 mm. The DAC consumes 1.4 W with a supply voltage of -4.0 V. Measured DNL and INL were within +0.49/-0.17 LSB and +0.97/-0.06 LSB, respectively. The measured SFDR was 45.0 dB for a sinusoidal output of 72.5 MHz at a sampling rate of 13.5 GS/s. The ramp-wave output at a sampling rate of 32 GS/s was obtained. The measurement results obtained so far demonstrate that our DAC can provide a resolution of 6 bits and a sampling rate of over 28 GS/s. To our knowledge, this is the first 6-bit DAC that can operate at a sampling rate of over 28 GS/s.

ACKNOWLEDGMENT

The authors thank T. Enoki for his encouragement throughout this work, and A. Sano and N. Kikuchi for their continuous support and helpful advice.

REFERENCES

[1] Akihide Sano et al., "No-Guard-Interval Coherent Optical OFDM for 100-Gb/s/ch Long-Haul Transmission Systems," *Optical Fiber Communication Conf. (OFC)*, OTuO3, Mar. 2009.

[2] Peter J. Winzer et al., "100G Ethernet – A Review of Serial Transport Options," *IEEE LEOS Summer Topical Meetings*, pp. 7-8, July 2007.

[3] Tobias Ellermeyer et al., "DA and AD Converters for 25GS/s and Above," *IEEE LEOS Summer Topical Meetings*, pp. 117-118, July 2008.

[4] Peter Schvan et al., "A 22GS/s 6b DAC with Integrated Digital Ramp Generator," *IEEE Int. Solid-State Circuits Conf. (ISSCC) Deg. Tech. Papers*, pp. 122-123, Feb. 2005.

[5] Dalius Baranauskas et al., "A 0.36W 6b up to 20GS/s DAC for UWB Wave Formation," *IEEE Int. Solid-State Circuits Conf. (ISSCC) Deg. Tech. Papers*, pp. 2380-2389, Feb. 2006.

[6] Behzad Razavi, *PRINCIPLES OF DATA CONVERSION SYSTEM DESIGN*, IEEE Press, 1995.

[7] M. Ida et al., "Undoped-Emitter InP/InGaAs HBTs for High-Speed and Low-Power Applications," *IEEE Int. Electron Device Meeting (IEDM)*, pp. 854-856, Dec. 2000.

Ultra-Low-Power 500-MSPS 12-bit A/D Converter Using Interleaving and CMOS Charge-Domain Technology

Michael P. Anthony[1], G. Sollner[2]

[1] Intersil Corporation, Woburn, MA, 01801, USA

[2] Formerly with Intersil Corporation, Woburn, MA, 01801, USA

Abstract — **A 12-bit analog-to-digital converter (ADC) has been developed using a unique charge-domain method of handling the analog signals. By interleaving two 250-MS/s unit ADCs on a single chip, an aggregate sample rate of 500 MS/s is achieved. Performance is comparable or superior to all existing ADCs at this sample rate, with power consumption less than 1/5th of that needed by other available designs. Signal-to-noise ratio (SNR) of 65.6 dBFS and spurious-free dynamic range (SFDR) of 78 dBc are obtained at an input frequency of 250 MHz. Total power consumption is 432 mW from a single 1.8-V supply. Added sampling jitter is 60 fs.**

Index Terms — **Analog-digital conversion, CMOS analog integrated circuits.**

I. INTRODUCTION

As the need for higher signal bandwidths increases, the requirement for ADCs with high sample rates grows concurrently. Applications such as software radios, power amplifier linearization, data acquisition and test equipment all need higher bandwidths and high accuracy. The usual architecture for wideband ADCs with the best accuracy is a pipelined successive-approximation design. These designs until recently have been implemented with op-amps for inter-stage signal transfer and gain. These op-amps must settle accurately at the sample rate, so they tend to consume substantial power.

The authors of this paper recently reported a new charge-domain technique for processing the analog signals in a pipelined successive-approximation ADC [1]. This boosted bucket-brigade approach transfers signal charge between stages without op-amps and without amplification. The resulting ADC consumes much lower power than the op-amp approach at comparable performance. This low power consumption, together with compact size and high substrate-noise immunity, support implementation of multiple ADCs on a single chip.

Time-interleaving of N individual ADCs provides an equivalent sampling rate N times the sampling rate of the individual ADC, while retaining conversion accuracy which can approach that of the individual ADCs. Such accuracy requires that the input offset, overall gain, and sampling phase of the individual converters be well-matched, or that the artifacts resulting from mismatch be corrected. There has been substantial analytical work [2] on correcting interleave artifacts using digital techniques. These have been employed to varying degrees in previously-reported interleaved ADCs [3]. However, the best of these reported designs have signal-

to-noise-and-distortion ratios (SINAD) in the 55-57 dBc range. The ADC reported here achieves typical SINAD of 65.7 dBc for inputs up to its Nyquist frequency. This performance is obtained by using digital estimation of unit-ADC mismatch combined with analog correction.

II. OVERVIEW OF THE ADC

Figure 1 shows a block diagram of a dual-interleaved 12-bit 250-MS/s ADC pair providing aggregate sampling at 500 MS/s.

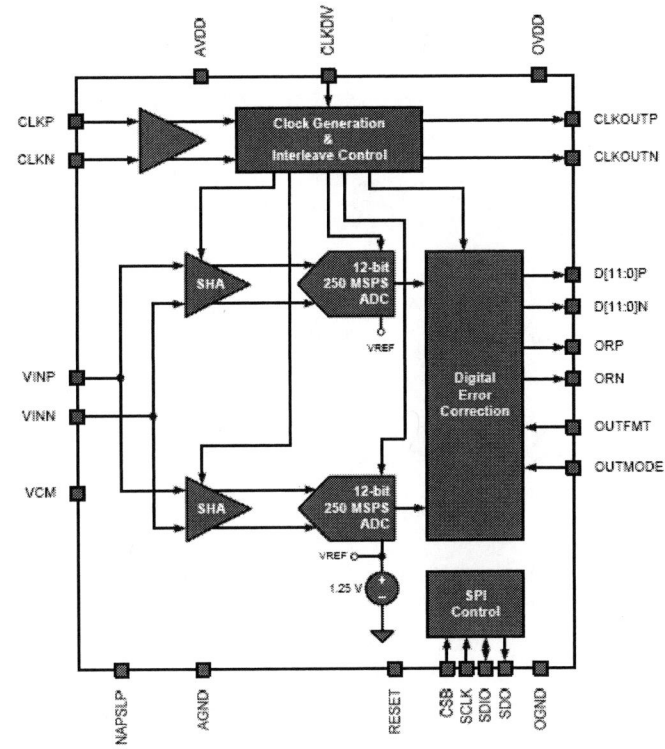

Fig. 1. Block diagram of dual interleaved ADC.

At power-up, an internal calibration process adjusts the two unit ADCs to have zero input offsets and equal gains. The clock-generation block divides the input clock by two, providing two sampling clocks at half the input clock rate and nominally uniform spacing. The interleave-control functionality of this block allows adjustment of the sampling clocks of the two unit ADCs to exact uniform spacing (zero sampling skew). The skew, offset and gain match can be adjusted during operation via the SPI port. In the reported design there is no tracking and adjustment of mismatch on the chip, although the calibration routine can be rerun. In many applications, the ADC output data is received by an external FPGA. In this case, an algorithm can be programmed into the FPGA that estimates the offset, gain, and relative sampling phase of the unit ADCs and periodically re-adjusts them via the SPI port. In future versions of this ADC, this tracking algorithm could be incorporated on the chip.

Figure 2 shows a photograph of the dual-ADC chip. It was fabricated in TSMC's 0.18-μm mixed-signal process. The two ADC pipelines are clearly visible above and below the horizontal centerline, with inputs on the left edge. The chip is approximately 5×5 mm, although the active area is much smaller.

Fig. 2. Die photograph showing the two pipelined ADCs.

A block diagram of one of the unit ADCs is shown in Fig. 3. The pipeline structure is typical of such designs, resolving 2.5 bits in the first stage, 1.5 bits in each of the next 6 stages, 1 bit per stage in the next three, and 3 bits in a final flash block. Redundant comparators alleviate comparator threshold accuracy requirements; the redundancy is resolved in the "digital error correction" block, providing a 13 bit result. 12 bits are presented as the final off-chip output, with the two unit-ADC outputs alternating. The block labeled QCM Sensor monitors the common-mode charge at the final flash and provides a feedback signal to an upstream stage to maintain an appropriate common-mode charge level through the pipeline.

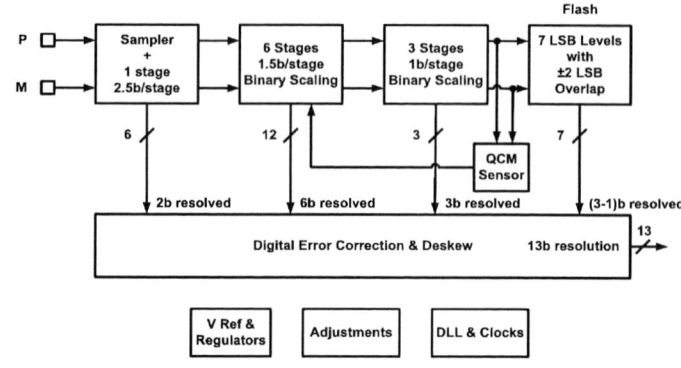

Fig. 3. Block diagram of unit ADC.

III. INTERLEAVING ADJUSTMENTS

As mentioned above, gain and offset of the two unit ADCs are adjusted to match during an internal startup calibration. Any sampling skew can be removed by adjustment via the SPI interface. One straightforward approach to matching the unit ADCs is to input a sine-wave test tone and fit a sine wave to alternate output samples. Any differences in offset, gain, and phase are easily estimated from the fitting parameters, and can be corrected through the SPI interface. This is the procedure we use in production testing. Background algorithms, such as those described by Seo et al. [2], can also be used to estimate ADC mismatches including changes due to temperature and supply voltage variation over time, and to drive appropriate adjustments via the SPI interface.

Figure 4 shows a spectrum resulting from conversion of a sine wave input when the offset, gain, and phase adjustments have been set to mid-scale. The spurious tones due to unit-ADC mismatch are clearly visible.

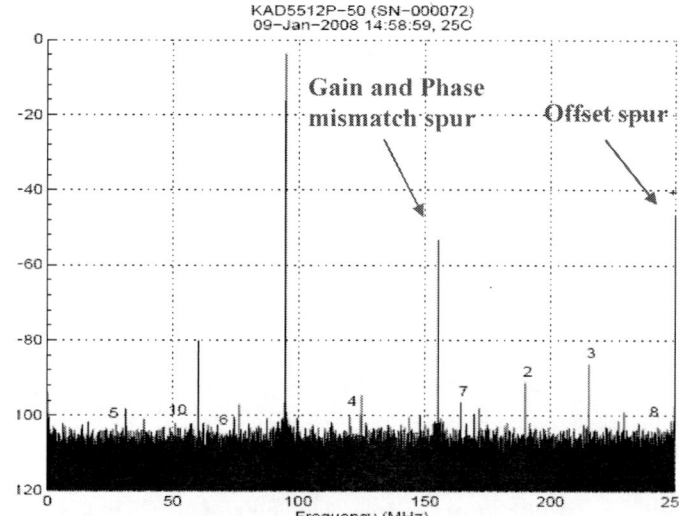

Fig. 4. Spectrum showing the interleaving spurs due to mismatch of the unit ADCs.

After running an Intersil-proprietary blind background calibration algorithm, these tones are reduced to levels below those produced by ADC nonlinearities, as can be seen in Fig. 5. This background calibration algorithm converges quickly for both sine-wave and broadband signals.

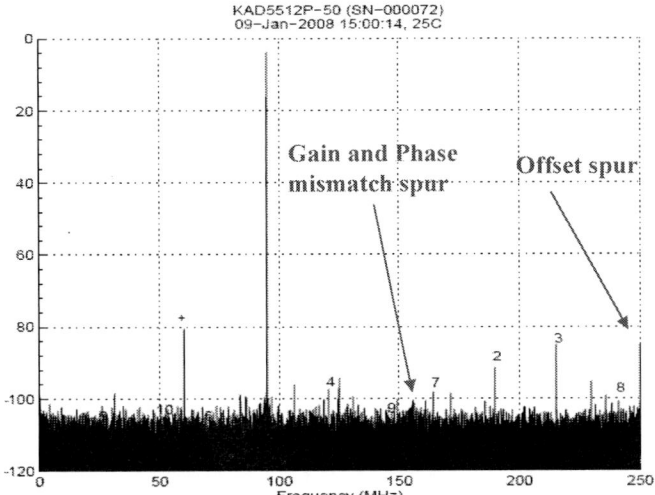

Fig. 5. Spectrum after running the algorithm for matching the unit ADCs.

IV. MEASURED PERFORMANCE AND COMPARISON

Louwsma shows a survey of several high-speed interleaved ADCs [3]. This work has been added to Louwsma's summary in Table I. The figure of merit (FoM) is defined as ADC power divided by the product of 2^{ENOB} and the minimum of the sample rate or 2 times the effective resolution bandwidth ERBW. The ERBW is defined as the bandwidth over which the SINAD decreases by 3 dB (or the ENOB decreases by 0.5). Among interleaved ADCs, the charge-domain approach presented here is clearly superior in ENOB, jitter, and FoM.

TABLE I

Design Parameter	Poulton 2003	Gupta 2006	Hsu 2007	Louws-ma 2008	This Work
Sample rate, GS/s	20	1.0	0.8	1.35	0.5
ENOB low frequency	6.5	8.85	9.0	7.7	10.6
ERBW, GHz	2.0	0.4	0.4	1.0	0.5
Input Bandwidth, GHz	6.6	-	-	6	1.3
Power consumption, W	10	0.25	0.35	0.18	0.43
FoM, pJ	28	0.7	0.9	0.6	0.39
Jitter, ps RMS	0.6	-	0.43	0.2	0.06

Fig. 6 shows typical SNR, SFDR, and SINAD for our 500-MS/s ADC. The best-in-class jitter allows for sustained high SNR at high input frequencies.

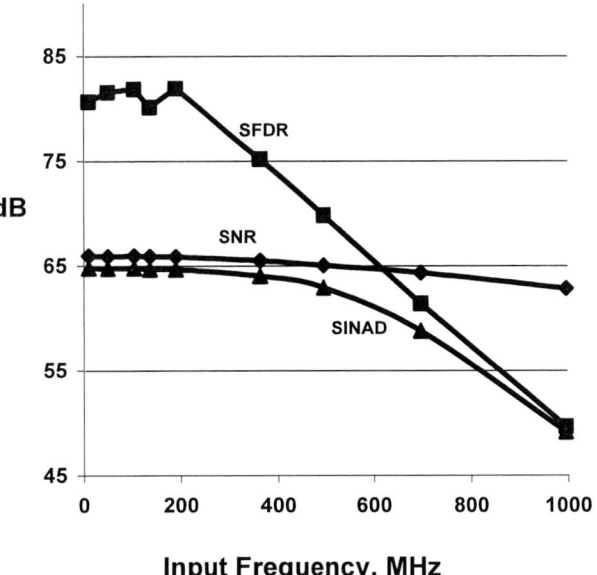

Fig. 6. Typical SNR, SFDR, and SINAD at 500 MS/s as a function of input frequency.

Fig. 7 shows typical SINAD in the first two Nyquist zones for this and three other commercially-available 12-bit ADCs at 500 MS/s, showing superior performance for the presented interleaved ADC to 500 MHz.

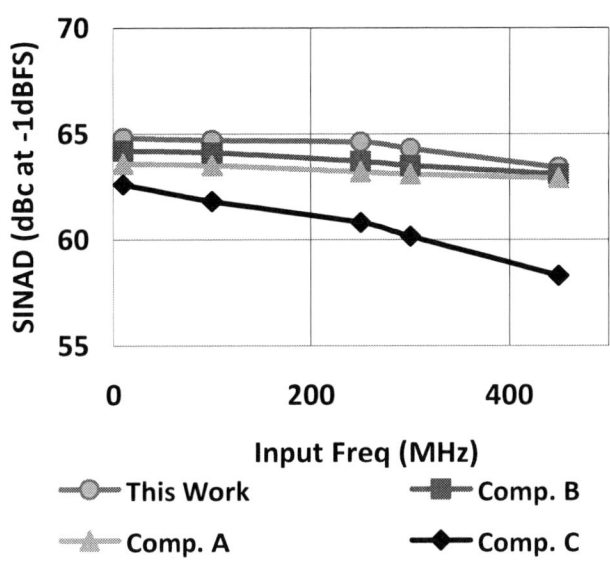

Fig. 7. Signal-to-Noise-and-Distortion (SINAD) vs. input frequency for several 500-MS/s ADCs.

Fig. 8 shows the operating power of the same four chips. The power consumption of the interleaved ADC is about 1/5 that of the traditional approach. This reduced power is primarily due to the charge-domain architecture. Interleaving does not, in principle, reduce power, although it is enabled in this case by the low power of each unit ADC.

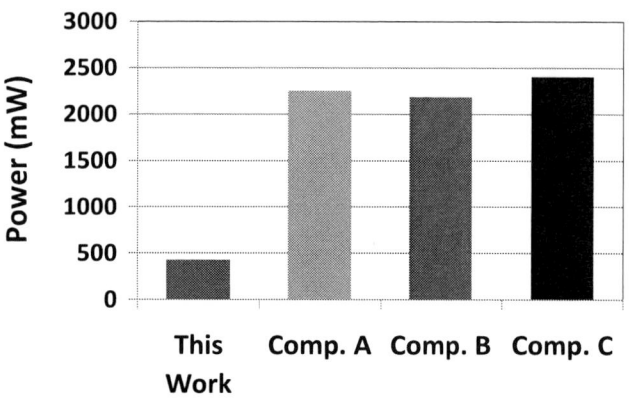

Fig. 8. Power consumption for the ADCs shown in Fig. 7

V. CONCLUSIONS

This paper has reported results for a 500-MS/s 12-bit ADC achieved by interleaving samples from two 250-MS/s unit ADCs on the same chip. The performance is comparable or superior to all ADCs designed with traditional approaches, but the power consumption is substantially lower because of the charge-domain approach. This approach and the associated on-chip calibration will be discussed in more detail in the presentation.

ACKNOWLEDGEMENTS

The authors would like to thank all the current and former employes of Kenet Inc. (now part of Intersil Corp). The following made important contributions to this work: E. Kohler, J. Kurtze, L. Kushner, R. Addiss, D. Barlas, T. Dodd, J. Fisher, M. Gilbride, B. Harris, R. Harwood, G. Huang, N. Kayathi, S. Kidambi, M. Lamenza, A. Lawrence, J. Messier, M. Murphy, D. Terranova, J. Toker, W. Washkurak, P. Rivenburgh, M. Schelling, J. Boucher, Thai Tran, Loi Tan Nguyen, and E. Rubin.

REFERENCES

[1] M. Anthony, E. Kohler, J. Kurtze, L. Kushner, and G. Sollner, "*A process-scalable low-power charge-domain 13-bit pipeline ADC,*" IEEE Symposium on VLSI Circuits, pp. 222-223, 18-20 June 2008.

[2] See for example: S.M. Jamal et al., *IEEE Trans. Circuits Sys.*, vol. 51, pp 130-139, Jan. 2004. M. Seo et al., *IEEE Trans. Microwave Theory Techn.*, vol. 53, pp 1072-1082, Mar. 2005. J. Elbornsson et al., *IEEE Trans. Sig. Proc.*, vol. 53, pp 1413-1424, Apr. 2005.

[3] Poulton et al., "A 20GS/s 8b ADC with a 1MB Memory in 0.18µm CMOS," *IEEE Int. Solid State Circuits Conf.,* paper 18.1, pp. 318-319, Feb. 2003. Gupta, et al., "A 1GS/s 11b Time-Interleaved ADC in 0.13µm CMOS," *ISSCC Dig. Tech. Papers*, pp. 576-577, Feb., 2006. Hsu, et al., "An 11b 800MS/s time-interleaved ADC with Digital Background Calibration," *IEEE Int. Solid State Circuits Conf.,* paper 25.7, pp. 464-465, Feb. 2007. Louwsma et al., "*A 1.35 GS/s, 10 b, 175 mW time interleaved A/D converter in 0.13 µm CMOS.*

Ultra Low-Loss 50-70 GHz SPDT Switch in 90 nm CMOS

Mehmet Uzunkol and Gabriel M. Rebeiz

Electrical and Computer Engineering

The University of California, San Diego

muzunkol@ucsd.edu, rebeiz@ece.ucsd.edu

Abstract— **This paper presents an ultra low-loss 50-70 GHz single-pole double-throw (SPDT) switch built using standard 90 nm CMOS process. The switch is based on λ/4 transmission lines with shunt inductors at the output matching network. High substrate resistance contacts are used to achieve low insertion loss. The SPDT switch results in a measured insertion loss of 1.5 dB at 55 GHz and < 2 dB at 50-70 GHz. The measured isolation is > 25 dB, and the output port-to-port isolation is > 27 dB at 50-70 GHz. The measured P_{1dB} is 13.5 dBm with a corresponding IIP3 of 22.5 dBm at 60 GHz. The return loss is better than -8 dB at 50-70 GHz. The active chip area is 0.5x0.55 mm^2 and can be reduced in future designs by folding the on λ/4 transmission lines. To our knowledge, this paper presents the lowest insertion loss 60 GHz SPDT in any CMOS technology to-date.**

Index Terms—**SPDT switch, millimeter-wave, 90-nm CMOS**

I. INTRODUCTION

MM-WAVE CMOS switches are used for 60 GHz high data rate communication systems for switched beam antennas, multi-band receivers on chip, built-in-self-test (BIST) circuits, and instrumentation systems [1]-[2]. Previously, mm-wave single-pole multiple-throw switches have been demonstrated using 0.13 μm, 90 nm, and 65 nm CMOS circuits, and have shown wideband performance [3]-[6]. Recently, a 60 GHz SPDT switch has been demonstrated with excellent performance using metamorphic HEMTs [7]. However, the InGaAs process is expensive and is not suitable of integration with the analog and digital circuits. This paper presents a high-performance 60 GHz SPDT realized in a standard 90 nm CMOS technology. The resulting SPDT shows state-of-the-art insertion loss and isolation from 50 to 70 GHz.

II. DESIGN

Schematic and cross section view of SPDT is shown in Fig. 1. The SPDT schematic is based on a tuned λ/4 transmission-line approach with a single shunt transistor. In the off-state, the transistor gate and junction capacitances together with the substrate resistance transform into an equivalent resistance and capacitance network. The output matching network consists of a shorted stub that acts as an inductor and cancels the equivalent capacitance of the transistor in the off-state.

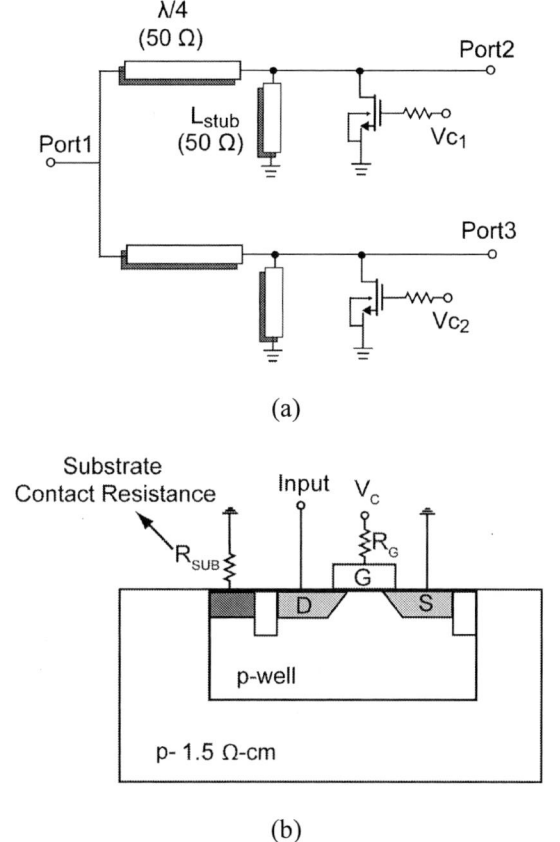

Fig. 1. (a) Schematic and (b) cross sectional view of SPDT

978-1-4244-5190-6/09 $25.00 © 2009 IEEE

High substrate resistance contacts are used to reduce the RF loss and substrate coupling through the junction capacitances. For a high-substrate resistance, the junction capacitances can be resonated out by adjusting the value of the shunt inductor [8]. The high substrate resistance is realized by using very small substrate contacts close to the CMOS transistors.

The isolation could be improved using a wide transistor. However, the required shunt inductor to resonate out the equivalent capacitance of the transistor in the off-state will be quite large and result in a reduced bandwidth. A FET width of 250 μm is therefore chosen for an isolation of 25 dB, and results in ~ 20 GHz bandwidth. The corresponding shunt inductor is 80 pH and is synthesized using a shorted CPW stub with impedance of 50 Ω, a length of 200 μm, and a simulated Q of ~17 at 60 GHz. The λ/4 transmission lines are 610 μm long, and are implemented using a 12/10/12 μm grounded CPW line with a simulated loss of 0.65 dB/mm at 60 GHz. All transmission lines and matching stubs are simulated using full-wave EM modeling (Sonnet [9]). The insertion loss is mostly determined by the Q of inductor and the λ/4 transmission lines. The simulated insertion loss of SPDT is 1.5 dB with an impedance match (see Fig. 2) better than -10 dB from 50-70 GHz.

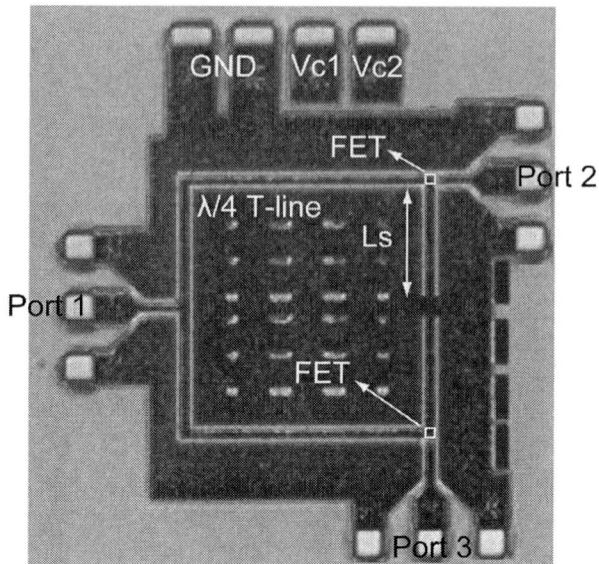

Fig. 3. Microphotograph of SPDT switch. The size is 0.5x0.55 mm², not including the pads.

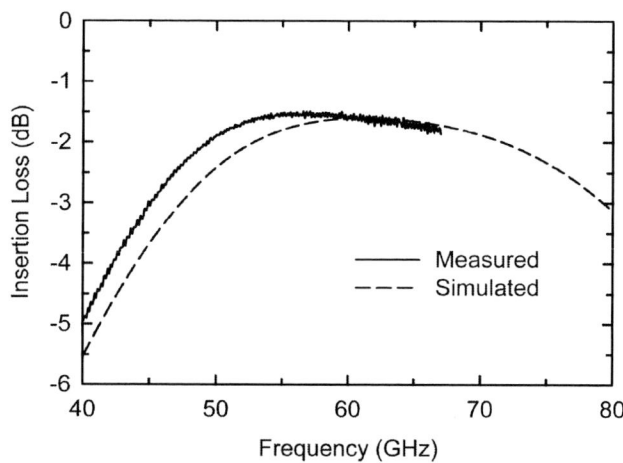

Fig. 4. Measured and simulated insertion loss of SPDT.

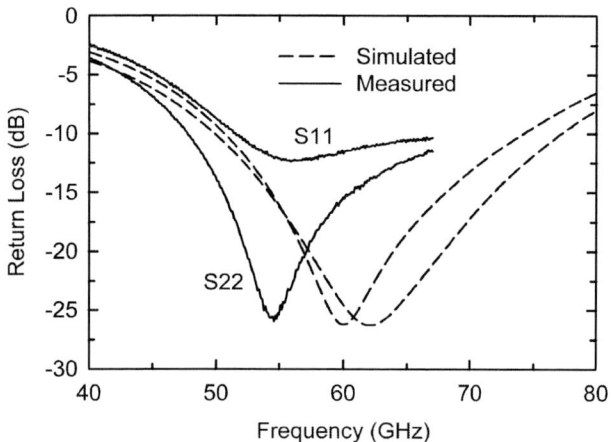

Fig. 2. Measured and simulated return loss of SPDT.

III. MEASUREMENTS

The SPDT switch is fabricated in a 90 nm CMOS process (IBM 9RF) and is shown in Fig. 3. The chip area not including the pads is 0.55x0.5 mm² and could be reduced by folding the λ/4 transmission lines. S-parameter measurements are done using the 67 GHz Agilent E8361A PNA. The pad loss is de-embedded using on-chip TRL calibration.

The measured insertion loss of SPDT switch is 1.5 dB at 55 GHz (Fig. 4). The insertion loss remains less than 2 dB from 50 to 70 GHz, showing very wideband performance, and is identical for Ports 2 and 3. The measured isolation is > 25 dB and the measured isolation between the output ports is > 27 dB from 50 to 70 GHz and again both output ports result in almost identical responses (Fig. 5). In the isolation measurement set-up, one of the output ports is terminated with a fixed termination waveguide connected to the 1.85 mm CPW probe with a waveguide-to-coaxial adaptor (50-75 GHz). This results in the ripples especially below 50 GHz. The measured return losses shifted slightly, but both S_{11} and S_{22} are better than -8 dB from 50 to 70 GHz.

978-1-4244-5190-6/09 $25.00 © 2009 IEEE

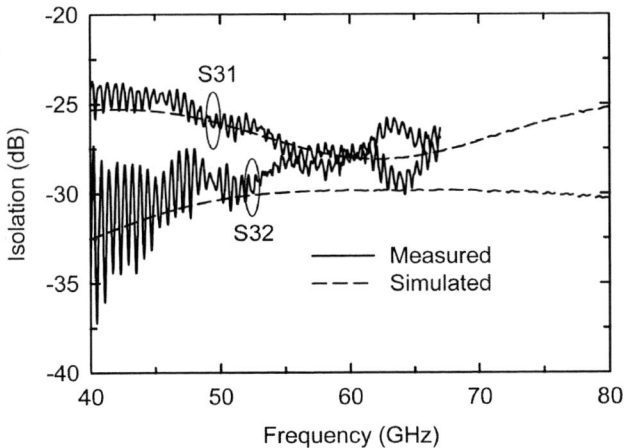

Fig. 5. Measured and simulated isolation and output to output isolation of SPDT.

The SPDT switch has a simulated P_{1dB} of 13.5 dBm, but this could not be measured due to power limitations at 60 GHz (compression is not observed up to 10 dBm). Fig. 6 shows the measured IIP3 which is 22.5 dBm at 60 GHz and agrees well with simulation. Power handling is mainly limited by the voltage swing (coupled from drain to gate through junction capacitances) at the gate of the transistors [10].

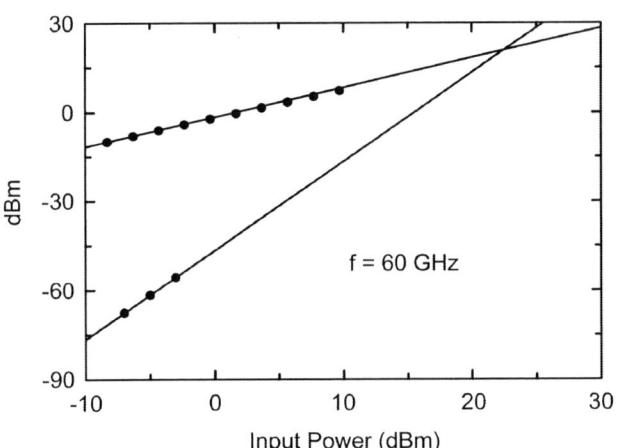

Fig. 6. Measured IIP3 of SPDT at 60 GHz.

The switching time of the SPDT is mainly determined by the gate biasing resistor (20 kΩ) and the junction capacitance at the gate. The simulated switching time at 60 GHz is ~3 ns and is shown in Fig. 7. The switching time could be improved to ~1 ns by using a smaller value gate biasing resistor at the expense of a slight increase in loss (0.-0.2 dB).

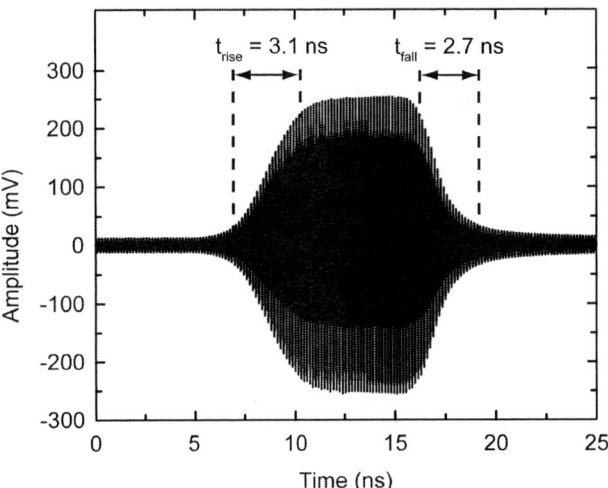

Fig. 7. Simulated switching time performance of SPDT at 60 GHz.

IV. CONCLUSION

The paper presented wideband SPDT switch from 50 to 70 GHz with excellent insertion loss and isolation performance. The power handling is limited to 13-14 dBm which is sufficient for most short-range 60 GHz communication systems.

ACKNOWLEDGMENT

This work was funded by Intel Corp. and the UC-Discovery Program, Drs. Ian Young and Jad Rizk, Program Monitors. We also thank Yusuf A. Atesal and Berke Cetinoneri for technical discussions.

REFERENCES

[1] S. Reynolds et al., "A Silicon 60 GHz Receiver and Transmitter Chipset for Broadband Communications," *IEEE J. Solid-State Circuits*, vol. 41, no. 12, pp. 2820-2831, Dec. 2006.

[2] T. Mitomo et al., "A 60 GHz CMOS Receiver Front-End With Frequency Synthesizer," *IEEE J. Solid-State Circuits*, vol. 43, no. 4, Apr. 2008.

[3] S.-F. Chao et al., "A 50-94 GHz CMOS SPDT switch using traveling-wave concept," *IEEE Microwave Wireless Compon. Lett.*, vol. 17, no. 2, pp. 130-132, Feb. 2007.

[4] A. Tomkins et al., "A 94 GHz SPST switch in 65 nm bulk CMOS," *IEEE CSICS Dig.*, pp. 139-132, Oct. 2008

[5] Y.-A. Atesal et al., "Low-Loss 0.13-μm CMOS 50-70 GHz SPDT and SP4T switches" *IEEE RFIC Symp. Dig.*, pp. 43-46, June 2009.

[6] B. Cetinoneri et al., "A Miniature DC-70 GHz SP4T Switch in 0.13-μm CMOS" *IEEE IMS Symp. Dig.*, pp. 1093-1096, June 2009.

[7] I. Kallfass et al., "Multiple-throw millimeter-wave FET switches for frequencies from 60 up to 120 GHz," in Proc. 38th European Microwave Conference, Amsterdam., pp. 1453-1456, Oct. 2008.

[8] B.-W. Min et al., "Ka-band low-loss and high-isolation 0.13 μm CMOS SPST/SPDT switches using high substrate resistance," *IEEE RFIC Symp. Dig.*, pp. 569-572, June 2007.

[9] Sonnet, ver. 12.02, Sonnet Software Inc., Syracuse, NY, 1986-2008.

[10] B.-W. Min et al., "Ka-band low-loss and high-isolation switch design in 0.13-μm CMOS," *IEEE Trans. Microwave. Theory &Tech.*, vol. 56, no. 6, pp. 1364-1371, June 2008

A 4-bit Passive Phase Shifter for Automotive Radar Applications in 0.13 μm CMOS

Sang Young Kim and Gabriel M. Rebeiz

ECE Department,
University of California, San Diego,
La Jolla, CA 92093-0407
sangykim@ucsd.edu, rebeiz@ece.ucsd.edu

Abstract— **This paper presents a 67 – 78 GHz 4-bit passive phase shifter using CMOS switches available in the 0.12 μm SiGe BiCMOS process (IBM 8HP). The phase shifter is based on a low-pass π-network and CMOS passive transistors. The phase shifter achieves -19.2+/-3.7 dB of gain including pad loss at 77 GHz. The RMS phase error is less than 11.25° and the RMS gain error is less than 2.5 dB over the 67 – 78 GHz range. The total chip size is 450 x 300 μm² (0.135 mm²), excluding pads, and the chip consumes virtually no power. The measured P_{1dB} is > +8 dBm at 77 GHz and the simulated IIP3 is > 22 dBm, making it possible to precede this design with a high gain LNA without loss of system linearity or input power handling.**

Index Terms – phase shifter, passive, phased array, millimeter-wave, CMOS switch

I. INTRODUCTION

Phased array systems have been widely used in defense applications such as radars and communication systems to achieve electronic beam forming and fast beam scanning [1]. At millimeter-wave frequency, these have been implemented with GaAs and/or InP based discrete modules, resulting in high cost and low integration density. However, recent development in silicon technologies have led to Si-based phased array on a single chip. This does not only reduce the cost but also multiple elements can be integrated on a single chip with excellent uniformity. A major issue in building phased array is the phase shifter design, especially at millimeter-wave frequencies. The phase shifter can be built using a switched delay [2], loaded reflection [3], loaded line [4] and vector modulation [5]. In this paper, low-pass passive networks with MOSFET switches are used [6].

II. DESIGN AND IMPLEMENTATION

A. Process, Transmission Line, Capacitor and RF pads

The 4-bit digital phase shifter is designed using CMOS switches available in the IBM 8HP, 0.12 μm SiGe BiCMOS process. All the inductors used in this design are implemented with transmission lines as shown in Fig. 1. The small capacitors are built using two interconnect metal layers in the process stack-up (Fig. 1). Each bit of phase shifter is designed with the input and output impedance of 50 Ω and connected using short

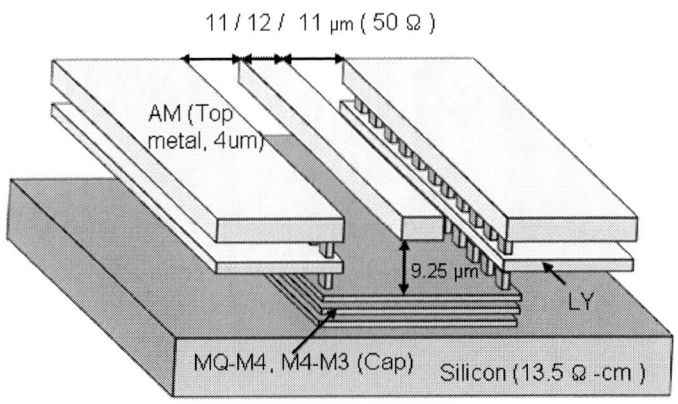

Fig. 1. The grounded-CPW transmission line and Metal-Oxide-Metal capacitor.

50 Ω transmission lines. A tapered transition between G-S-G pad transition and the transmission line is designed to provide 50 Ω input and output impedance. All the design including transmission lines, capacitors, pads and interconnections are simulated using Sonnet[1], a full-wave EM solver.

B. Phase Shifter

A 4-bit digital phase shifter with 22.5° phase resolution is designed using CMOS passive switches. The digital phase shifter consumes virtually no power and requires simple digital control circuits. Fig. 2 presents the schematic of the 4-bit digital phase shifter. Two 90° phase shifters are tied together to build the 180° phase shifting element. Then, 22.5°, 45° and 90° phase shifter are placed afterwards. Each phase shifter is based on a low-pass π-network which consists of one series inductor and two shunt capacitors [6]. The network can switch between a phase-delay state and a bypass state using CMOS passive switches. When T1 is off and T2 is on, Ls and Cp form a low-pass π-network which results in a phase delay given by $\phi = \sin^{-1}(w_o L_s / Z_o)$. However, when T1 is on and T2 is off, Ls and Cp / 2 has minimal effect on the phase which gives a bypass state [6]. Because the junction capacitors of the shunt

---

[1] Sonnet, ver.11.52, Sonnet Software Inc., Syracuse, NY, 1986-2007

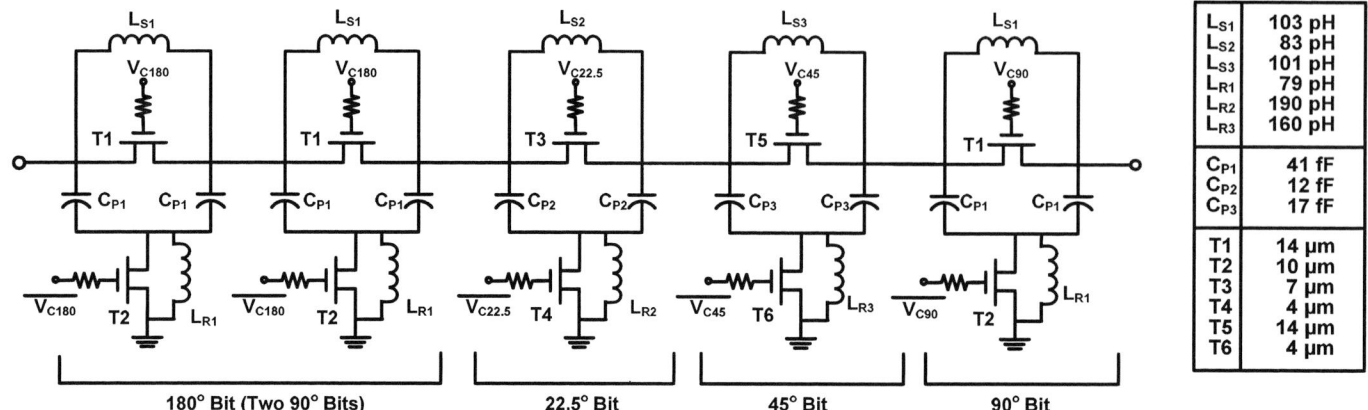

L_{S1}	103 pH
L_{S2}	83 pH
L_{S3}	101 pH
L_{R1}	79 pH
L_{R2}	190 pH
L_{R3}	160 pH
C_{P1}	41 fF
C_{P2}	12 fF
C_{P3}	17 fF
T1	14 μm
T2	10 μm
T3	7 μm
T4	4 μm
T5	14 μm
T6	4 μm

Fig. 2. Schematic of the 4-bit digital phase shifter.

CMOS switch (T2) degrades the isolation to ground at the bypass state, a shunt inductor Lr is added in parallel with the shunt switch to resonate out the parasitic capacitor at the desired frequency. This is a standard design and has been implemented before at Ka-band [6]. The challenge in this work is to scale it to W-band using 0.13 μm CMOS transistors.

A large CMOS transistor results in a small series resistance, which is desirable for a small insertion loss. However, as the transistor size increases, the capacitive coupling to the substrate increases due to the increased shunt junction capacitances of the source and drain, resulting in an increase of the signal loss. Therefore, the transistor size must be optimally chosen to minimize the insertion loss at the desired frequency [7]. Simulations show that the insertion loss is minimized at 77 GHz when the gate width is 14 μm. Also, the substrate resistance, (Rsub), between source/drain junction and substrate

depends on the size and distance of substrate contact [8]. Therefore, large substrate contacts (81 x 54 μm²) are placed closely around each transistor to minimize the uncertainty and to ensure that Rsub is close 50 Ω, assumed by the IBM model [9]. Also, the CMOS gate node is biased using a large resistor, Rc = 22 kΩ, which prevents signal leakage and oxide breakdown [10].

Fig. 3 presents the chip photograph of the designed phase shifter. The values of the inductors are shown in Fig. 2 and each is translated to an equivalent short transmission-line stub. The chip size is 450 x 300 μm² without pads. The inductors and capacitors are designed using a full-wave electromagnetic simulator and therefore the interconnect and mutual coupling effects are taken into account.

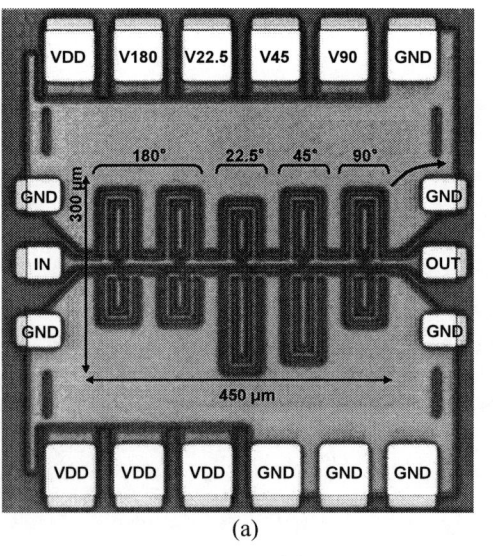

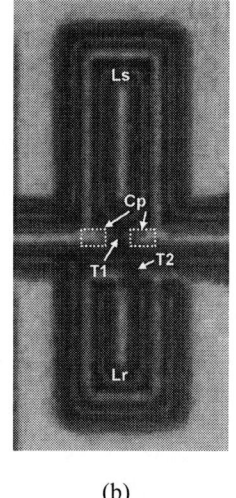

(a)　　　　　　　　(b)

Fig. 3. Chip photograph of (a) 4-bit digital phase shifter and (b) 90° phase shifter cell.

Fig. 4. Measured phase response of 16 different phase states.

978-1-4244-5190-6/09 $25.00 © 2009 IEEE

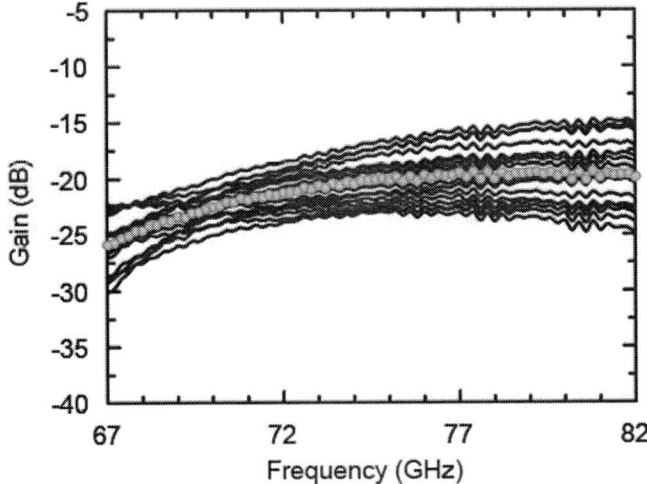

Fig. 5. Measured gain response of 16 different phase states and average gain.

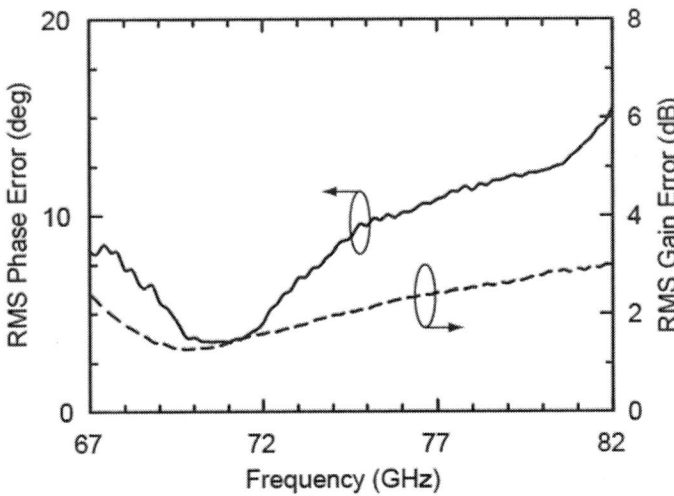

Fig. 6. RMS phase error and RMS gain error of 4-bit phase shifter.

III. MEASUREMENTS

The 4-bit digital phase shifter was measured using an Agilent E8361A 67 GHz PNA with extenders up to 110 GHz. All the measurements were done with SOLT calibration to the probe tips. The probe loss and transition to the CPW line are included in the measurements. Fig. 4 presents the measured phase response of the phase shifter for all 16 different phase states over 67-82 GHz. It is seen that 180° and 90° phase shifters result in excellent phase performance. The measured gain of all 16 different states and the average gain are shown in Fig. 5. The phase shifter results in -19.2+/-3.7 dB of gain at 77 GHz, which includes 1 dB pad loss. The RMS phase error and RMS gain error are 10.8° and 2.4 dB at 77 GHz as shown in Fig. 6, respectively. The input return loss is < -10 dB over 67 – 82 GHz for all 16 states (Fig. 7). The output return loss is < -10 dB over 67 – 81 GHz as shown in Fig. 8 for all 16 states.

The measured and simulated gain are shown in Fig. 9 vs. phase state. One can clearly notice that the transistor insertion loss (or Ron) is not accurately modeled since the measured insertion loss at the 0° state is 5 dB higher than the simulated gain. In this case, all the series transistors are biased ON, and therefore each stage has approximately 1 dB higher insertion loss than simulated. On the other hand, when all the shunt transistors are turned ON, 337° (phase delay state), then the measured and simulated insertion loss are nearly the same. Also notice the measured large insertion loss variation when the 90° state is toggled. This is currently being investigated and is not available in the simulations.

The measured 1 dB gain compression point is > 8 dBm as shown in Fig. 10 and is limited mainly by the available test power. The simulated IIP3 is > 22 dBm, and is not measured.

IV. CONCLUSION

A 67 – 78 GHz 4-bit passive phase shifter using CMOS

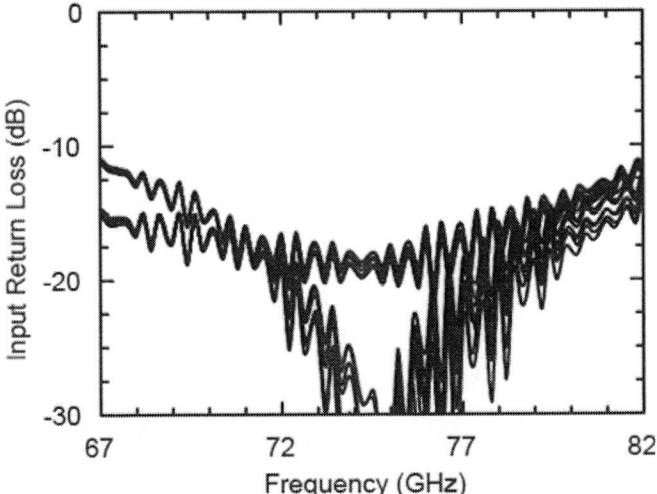

Fig. 7. Measured input return loss of 16 different phase states.

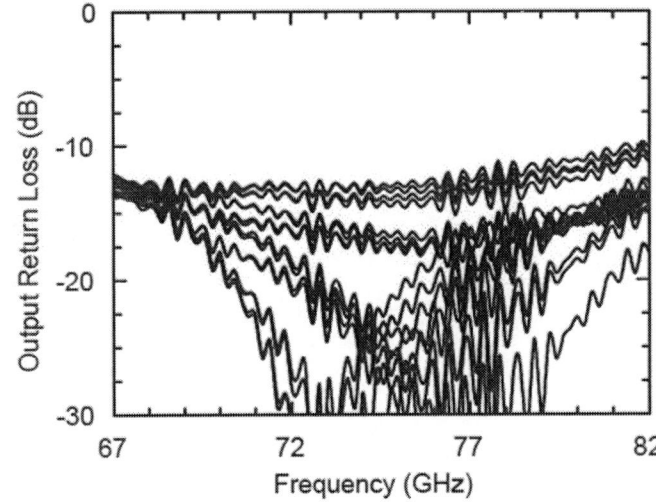

Fig. 8. Measured output return loss of 16 different phase states.

978-1-4244-5190-6/09 $25.00 © 2009 IEEE

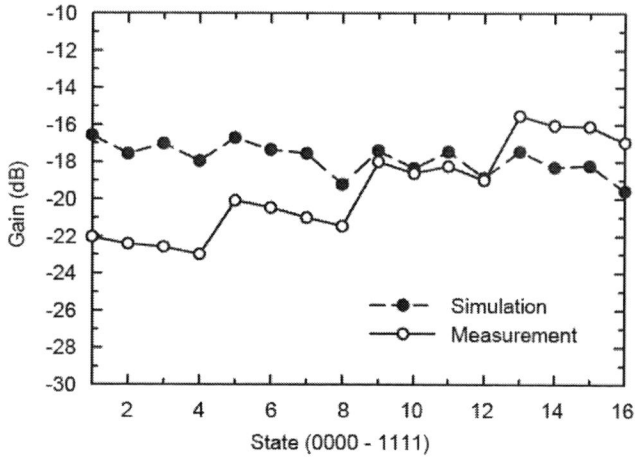

Fig. 9. Measured and simulated gain vs. phase states.

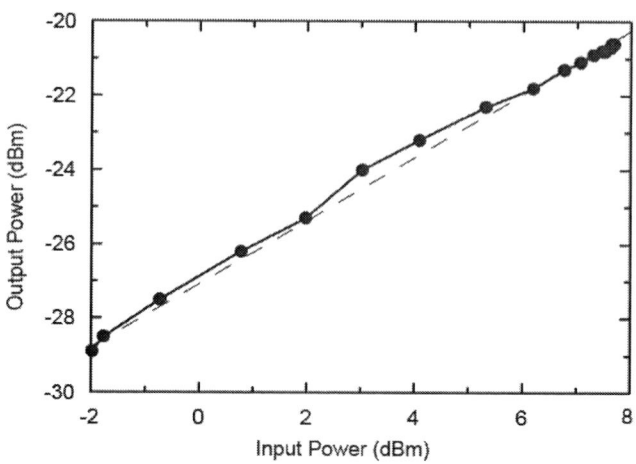

Fig. 10. Measured output power vs. input power.

switches available in 0.12 μm SiGe BiCMOS process is presented. The RMS phase error is < 11.25° over 67 – 78 GHz, respectively. The phase shifter must be combined with a 20 dB gain LNA/VGA to compensate for its loss and gain variation for millimeter-wave applications. The passive phase shifter is very small and is therefore excellent for array applications without taking a lot of space on the RFIC wafer.

ACKNOWLEDGMENT

This work was supported by Toyota Research Laboratories, Ann Arbor, Michigan.

REFERENCES

[1] D. Parker and D. Z. Zimmermann, "Phased Arrays – Part I: Theory and architectures," IEEE Trans. Microwave Theory & Tech., vol. 50, no. 3, pp. 678 – 687, Mar. 2002.

[2] C. Campbell, S. Brown, T. Inc, and O. Beaverton, "A compact 5-bit phase shifter MMIC for K-band satellitecommuncaion systems," IEEE Trans. Microwave & Theory Tech., Vol.48, no. 12, pp.2652 – 2656, Dec. 2000.

[3] A. E. Ashtiani, S. Nam, S.Lueyszyn, and I. D. Robertson, "Monolithic Ka-band 180-degree analog phase shifter employing HEMT based varactor diodes," IEE Colloq. Microwave and Millimetre-Wave Oscillators and Mixers, Vol. 48, no. 12, pp. 7/1 – 7/6, Dec. 1998.

[4] F. Ellinger, H. Jackel, and W. Bachtold, "Varactor-loaded transmission-line phase shifter at C-band using lumped elements," IEEE Trans. Microwave & Theory Tech., vol. 51, no. 4, pp. 1135 – 1140, Apr. 2003.

[5] K. Koh and G. M. Rebeiz, "0.13-μm CMOS phase shifters for X-, Ku-, and K-band phased array," IEEE J. Solid-State Circuits, vol. 42, no. 5, pp.1414 – 1424, May 2004.

[6] B. Min and G. M. Rebeiz, "Single-Ended and differential Ka-Band BiCMOS Phased Array Front-Ends," IEEE J. Solid State Circuits, Vol. 43, No. 10, pp.2239-2250, October 2008.

[7] B. Min and G. M. Rebeiz, "Ka-Band Low-Loss and High-Isolation Switch Design in 0.13 μm CMOS," IEEE Trans. Microwave Theory & Tech., Vol. 56, No. 6, pp. 1364-1371, June 2008.

[8] J. Han and H. Shin, "A scalable model for the substrate resistance in multi-finger RF MOSFETs," IEEE MTT-S Int. Microwave Symp. Dig., Philadelphia, PA, Jun. 2003, pp.2105-2108.

[9] "BiCMOS-8HP Model Reference Guide," V1.1.0.0HP ed Compact Model Develop. LZVV Dept., Microelectron. Div., IBM, Essex Junction, VT, Jul. 2007.

[10] B. Min and G. M. Rebeiz, "Ka-band BiCMOS 4-bit phase shifter with integrated LNA for phased array T/R modules," in IEEE MTT-S Int. Microwave Symp. Dig., Honolulu, HI, Jun. 2007, pp.479-482.

Broadband, Thin-Film, Liquid Crystal Polymer Air-Cavity Quad Flat No-Lead (QFN) Package

Morgan J. Chen and Seyed A. Tabatabaei
Endwave Technology Center
Endwave Corporation
San Jose, CA, USA
morgan.chen@endwave.com and seyed.tabatabaei@endwave.com

Abstract—We have developed a proprietary package that is fully compatible with variously sized chips. In this paper, we present design and development of a Quad Flat No-Lead (QFN) package. We will show how we have built and characterized low-loss packages using standard Printed Circuit Board (PCB) laminate materials. In particular, this package has been developed using Liquid Crystal Polymer (LCP). These packages are unique in that they fully account for and incorporate solder joint and ball bond wire parasitic effects into design. The package has a large cavity section that allow for a variety of chips and decoupling capacitors to be quickly and easily packaged. Insertion loss through a single package transition is measured to be less than 0.4 dB across DC to 40 GHz. Return losses are measured to be better than 15 dB up through 40 GHz. Further, a bare die low noise amplifier (LNA) is packaged using this technology and measured after being surface mounted onto PCB. The packaged LNA is measured to show 19 dB gain over 32 GHz to 44 GHz. Return loss for both bare die and packaged version show no difference, and both measure 15 dB. The LCP package LNA exhibits 4.5 dB noise figure over 37 GHz to 40 GHz.

Keywords-Hybrid integrated circuit packaging, liquid crystal polymer, and microwave devices

I. INTRODUCTION

We have developed an approach where raw die from any manufacturer may be dropped into our high performance, Liquid Crystal Polymer (LCP) Quad Flat No-Lead (QFN) packages. LCP packages have been emerging as a convenient way to achieve low insertion loss performance in an air-cavity package format through high frequencies [1-7]. Air-cavity packaging ideally preserves MMIC functionality in regards to providing identical phase, loss, and transistor gain that are otherwise affected by dielectric packaging materials. LCP offers near-hermetic capabilities, which in turn mean low leak rates and opportunity for low cost moisture sensitivity level 1 (MSL1) rated parts.

Broadband interconnect design is limited by parasitic elements encountered along the signal path between the PCB motherboard and the active portion on chip. The most substantial parasitic elements include the solder joint between QFN and PCB, via pad through QFN, and bond wire between package and chip. In this paper, we present the design and development of an LCP package that operates from DC to past 40 GHz. Section II will describe the process steps required to form the package and the surface mount onto a PCB for characterization. Section III will describe the package design. We will demonstrate how our packages employ capacitive compensation and varied feature dimensions to successfully achieve a design for manufacturability. Electrical measurement will show that at 40 GHz, insertion loss is measured to be better than 0.4 dB, and return loss is better than 15 dB. Section IV will demonstrate the package with a bare die LNA example. Electrical measurements will show that the package gain is reduced by less 1.1 dB during nominal 40 GHz operation. Lastly, section V will provide conclusions.

II. FABRICATION AND ASSEMBLY PROCESS

Thin-film LCP packages are manufactured as shown in Fig. 1. Processing begins with a 4 mil thick LCP laminate that is copper clad on both sides. Secondly, copper is etched to define circuit features. Thirdly, blind vias and a chip cavity are formed by laser ablation. Vias are copper plated in order to form an electrical interconnect from bottom-side pads onto top-side metal. Lastly, the bottom metal is plated up to the desired copper thickness to ensure mechanical rigidity. An LCP QFN base part is shown from the top and bottom view in Figs. 1a and 1b, respectively. For assembly, bare die components are epoxy mounted into QFN packages and wire bonded. Parts are sealed with LCP lids constructed using non-conductive B-staged epoxy.

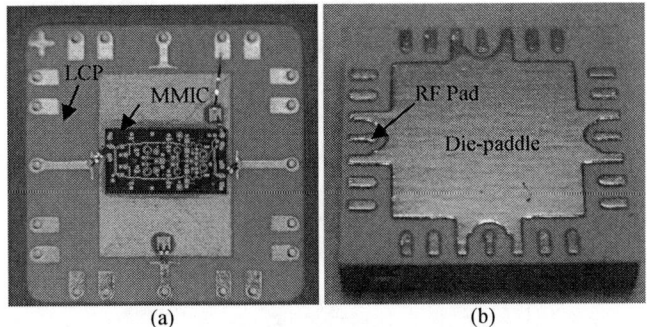

Figure 1. LCP package showing (a) exposed view of an example QFN assembly, and (b) bottom-side QFN footprint

After final inspection, sealed packages are surface mounted onto PCB with reflowed solder. LCP softens at temperatures above 220 °C, but this does not present problems as long as

978-1-4244-5190-6/09 $25.00 © 2009 IEEE

mechanical forces are not applied. We have demonstrated that using a soldering iron will cause irreparable damage to the package. Hence, solder paste must be applied through a reflow process. We have successfully performed non-optimized QFN surface mounting using a 5 minute profile at 200 °C resulting in solder joints less than 50 μm thick. Thinner standoffs provide better electrical performance but are challenging to manufacture in large QFN formats. Our QFN is 7 mm x 7 mm x 1 mm thick. A potential problem of QFN lifting during solder reflow can cause open connection and is known to occur when significantly different solder paste areas are applied to the same component. For this reason, solder paste areas are sectioned on PCB, which limit solder paste to 50%-80% of the die-paddle area.

III. PACKAGE DESIGN & MEASUREMENT

A cross-sectional diagram of a thin-film LCP package on printed circuit board (PCB) is shown in Fig. 2. Locations of significant EM features include 1) the reflowed solder joint between PCB and QFN, 2) the RF via through LCP, and 3) the RF bond wire connection. Surface mounting signal pads along the bottom QFN surface are made to have 50 ohms of characteristic impedance in a grounded coplanar waveguide (CPWG) configuration. The CPWG transition into the chip is further enhanced by designing in a continuous connection from neighboring ground leads directly to the die paddle. Care has been taken to provide a smooth transition into the LCP package. Both the microvia and the corresponding annular via pad diameters are made as small as possible to minimize parasitic effects. The top-side RF metal path on 4 mil thick LCP is made to approximate a microstripline with 50 ohms characteristic impedance. Minimizing the distance between the bottom signal-pad and die-paddle to minimum design rules allows an efficient CPWG-to-microstrip transition. Further, a compensation capacitance has been placed in the location of the package bond stitch to directly offset the inductive bond wire parasitic.

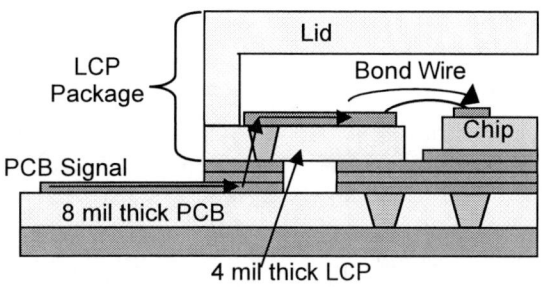

Figure 2. Diagram cross-section view of interconnect from PCB to chip-packaged in an LCP QFN

Package performance has been simulated with full-wave Finite Element Modeling (FEM) using Ansoft HFSS. The package effect is simulated with PCB, solder, and typical chip bond pad effects, as shown in Fig. 3. Simulation results are plotted in Fig. 4 to show better than 24 dB return loss and less than 0.4 dB to 40 GHz.

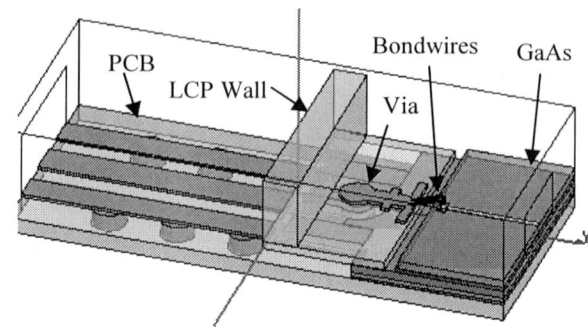

Figure 3. Simulation of an LCP QFN transition

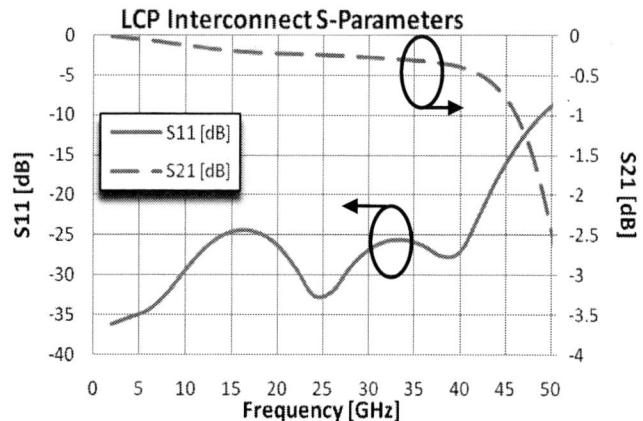

Figure 4. Simulated return losses in a double wire bond configuration

S-parameter measurements are performed on a Cascade probe station, GGB Picoprobes, and an Agilent E8361A Performance Network Analyzer (PNA). A Load-Reflect-Match (LRM) calibration establishes a reference plane to be at the probe tips. Photograph of the measurement is provided in Fig. 5. An unlidded LCP QFN base is surface mounted mounted onto a PCB. A probe is applied at port 1 directly onto an 8 mil thick Rogers 4003c gold plated PCB. An alumina adapter is used in place of a chip for CPW probing at port 2. Measurements are provided in Fig. 6. Insertion losses are measured to be less than 0.4 dB through 40 GHz. Return loss is better than 15 dB across the band. Insertion loss measurements vary from simulation by less than 0.05 dB. Hence, this package shows excellent broadband performance over DC to 40 GHz frequencies and adds minimal loss to transceiver modules.

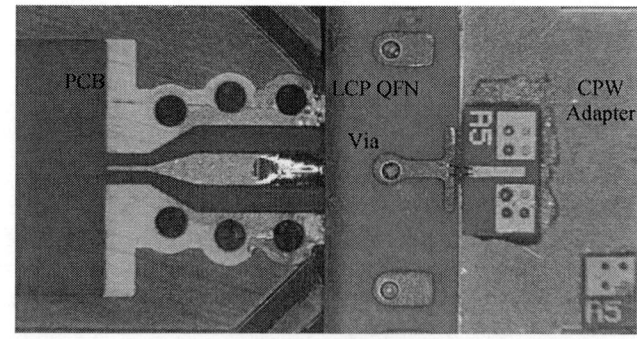

Figure 5. Measured LCP package on PCB motherboard to include wirebond and solder joint effects

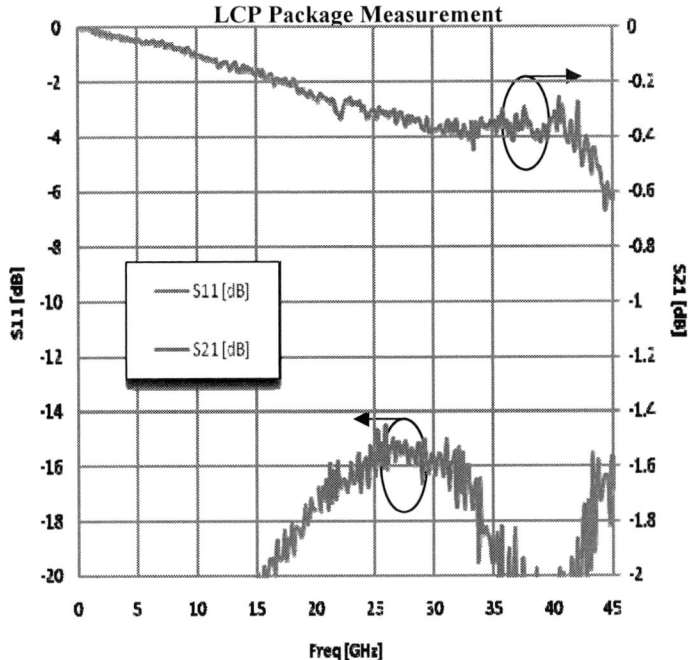

Figure 6. Measured LCP package s-parameters as mounted onto PCB

IV. LNA IN PACKAGE

In order to demonstrate excellent performance offered by the LCP package, a 40 GHz LNA is packaged into our thin-film LCP package. This die is chosen in order to characterize the upper frequency package limitation. The bare LNA die is attached using Diemat 5030 conductive epoxy. A short alumina transmission line is inserted to close the gap between the chip and internal package contact. This scheme allows for many different chip form factors to be implemented into the same package. Decoupling chip capacitors are also implemented along DC connections. Each RF connection is double ball bonded. The packaged LNA assembly is shown below in Fig. 7.

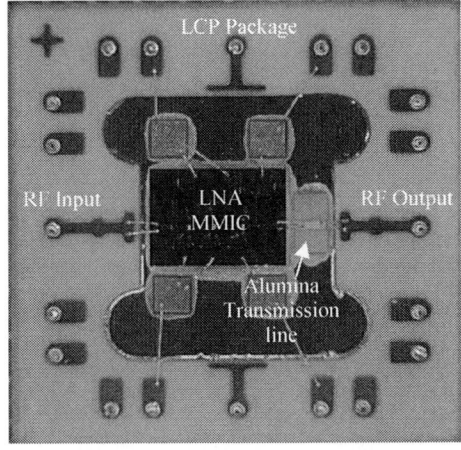

Figure 7. Exposed Top-side view of an LCP packaged LNA

The packaged LNA is surface mounted on PCB and measured by directly probing the PCB at each port. Measurement results are plotted in Fig. 8 to provide comparison between raw die and the packaged version. The packaged part provides 19.3 dB ± 0.6 dB gain over 37 GHz to 43 GHz. Return loss in package is measured to be better than 14 dB over the same frequency range. Although, the package is intended to be used up through 40 GHz, we design and simulate excellent performance up through 45 GHz. This allows the package to accommodate any shift that may occur in cut-off frequency while still allowing excellent performance through 40 GHz. This packaged LNA example shows that excellent measurements may be obtained on this package through 44 GHz. The package is shown to add in the bare die by ~1.1 dB. This is attributable to ~0.4 dB loss at input and output of the package and ~0.3 dB due to an extra set of double bond wires. The alumina transmission line itself is expected to contribute less than 0.1 dB.

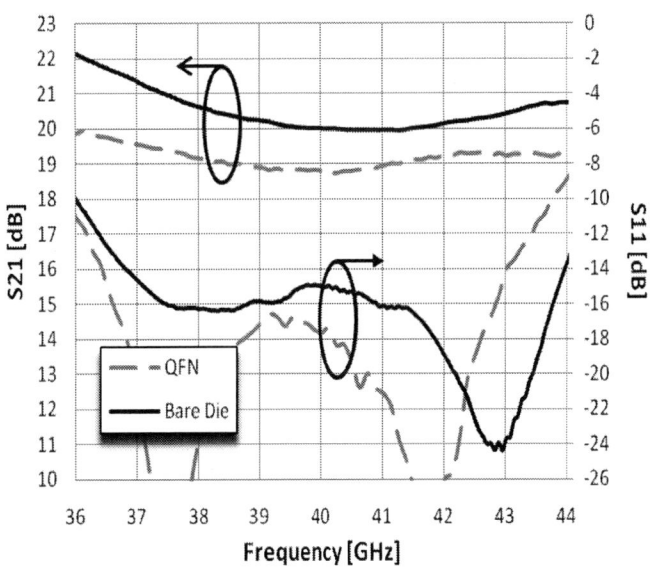

Figure 8. Measured LCP package s-parameters

In addition, noise figure has been measured on the packaged LNA and compared against bare die, as shown in Fig. 9. A Friis cascaded NF calculation shows that packaging effects on the output side contributes less than 0.1 dB noise figure. This is due to 20 dB chip gain that minimizes noise figure in stages that follow. Hence, noise figure is expected to increase by ~0.4 dB, which is contributed by the package interconnect at RF In. Measurement on the full package shows less than 4.5 dB in package over 37 GHz to 40 GHz with typical values at ~4.5 dB NF in package. We are currently working to minimize ripple in the measurement, i.e. 38 GHz, which may indicate that measurement accuracy is currently limited to ±0.5 dB. However, even with this level of ripple, we can see that the LCP QFN package appears to introduce only ~0.4 dB noise figure, which is contributed by the interconnect loss at the input as expected.

978-1-4244-5190-6/09 $25.00 © 2009 IEEE   159

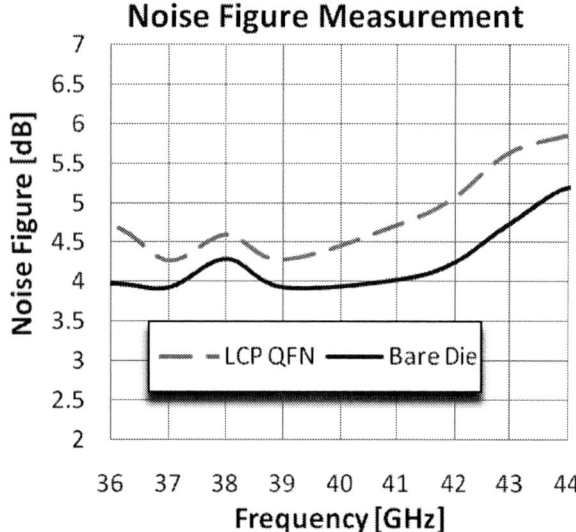

Figure 9. Measured noise figure on an LCP packaged LNA

V. CONCLUSIONS

An air-cavity QFN package has been successfully developed and demonstrated using thin-film LCP. The package has been characterized to show a low insertion loss of less than 0.4 dB and return loss greater than 15 dB measured over DC to 40 GHz. A bare LNA die is LCP packaged to show 4.5 dB NF with greater than 19 dB gain and a 14 dB return loss over 37 GHz to 43 GHz. Compared against bare die performance, packaging adds 1.1 dB gain attenuation and ~0.4 dB NF attenuation. This package example includes an alumina transmission line that allows us to quickly package different chip form factors using the same LCP package.

ACKNOWLEDGMENTS

The authors wish to acknowledge Steve Avery, Bob Jabor, David Zeeb, Ed Brown, Dan Teuthorn, Anthony Sweeney, and Douglas Okamoto of Endwave Corporation for providing contributions, feedback, and/or advice on this work.

REFERENCES

[1] K. Aihara, M. J. Chen, A-V. Pham, "Development of Thin-Film Liquid Crystal Polymer Surface Mount Packages for Ka-band Applications" IEEE Transactions on Microwave Theory and Techniques, Vol. 56, No. 9, pp. 2111-2117, September 2008.

[2] M. Chen, A-V. Pham, C. Kapusta, J. Iannotti, W. Kornrumpf, N. Evers, and J. Maciel, "Design and development of a hermetic package using LCP for RF/Microwave MEMS Switches," IEEE Trans. Microwave Theory and Tech., vol. 54, no. 11, pp. 4009-4015, Nov. 2006.\

[3] M. J. Chen, A. Pham, N. A. Evers, C. Kapusta, J. Iannotti, W. Kornrumpf, J. Maciel and N. Karabudak, "Development of multilayer organic modules for hermetic packaging of RF MEMS circuits," in IEEE MTT-S Int. Microwave Symp. Dig., San Francisco, CA, June 2006, pp. 271-274.

[4] D. Thompson, N. Kingsley, G. Wang, J. Papapolymerou, M.M. Tentzeris, "RF characteristics of thin film liquid crystal polymer (LCP) packages for RF MEMS and MMIC integration," in IEEE MTT-S Int. Microwave Symp. Dig., Long Beach, CA, June 2005, pp. 857-860.

[5] Z. Aboush, J. Benedikt, J. Priday, and R. J. Tasker, "DC-50 GHz Low Loss Thermally Enhanced Low Cost LCP Package Process Utilizing Micro Via Technology," in IEEE MTT-S Int. Microwave Symp. Dig., Philadelphia, PA, June 2003, pp. 1159-1162.

[6] K. Aihara and A. Pham, "Development of thin-film liquid crystal polymer surface mount packages for Ka-band applications," IEEE International Microwave Symposium, San Francisco, pp 956 – 959, June 2006.

[7] K. Aihara, A. Pham, D. Zeeb, T. Flack and E. Stoneham, "Development of Multi-Layer Liquid Crystal Polymer Ka-band Receiver Modules," Microwave and Optical Technology Letters, vol. 51, no. 2, pp. 364-367, Feb. 2009.

MMIC LNAs for Radioastronomy Applications Using Advanced Industrial 70 nm Metamorphic Technology

W. Ciccognani[1], E. Limiti[1], P.E. Longhi[1&2], M. Renvoisè[3]

1 Electronic Engineering Department, University of Roma Tor Vergata, Via del Politecnico 1
00133, Roma, ITALY - limiti@ing.uniroma2.it

2 HW Development Department, Elettronica S.p.A., Via Tiburtina Valeria Km 13,700
00131, Roma, ITALY

3 OMMIC, 2 Chemin Du Moulin, B.P. 11
94453, Limeil-Brévannes Cedex, FRANCE

Abstract — **A set of monolithic low-noise amplifiers have been designed and realised making use of an advanced industrial GaAs 70 nm metamorphic technology with high indium content. Operating frequencies of the realised amplifiers span from C- to W-band, with resulting performance in line with the requirements of many advanced Radioastronomy applications. As an example, 2.7 dB NF from 75 to 90 GHz associated to 25 dB gain and 1.3 typical NF from 30 to 50 GHz with 30 dB gain have been achieved.**

AIndex Terms — **Amplifier Noise, HEMT, Millimetre wave FET amplifiers, HF FET amplifiers.**

I. INTRODUCTION

Radioastronomy observation at microwave and millimetre wave frequencies is indeed demanding from the point of view of absolute performance of the electronic front-end. In particular, state-of-the-art requirements are demanded, in terms of noise temperature (noise figure), gain and gain stability. Such requirements have been translated into technological advances regarding mainly the semiconductor compound to be adopted and aggressive gate length reduction.

Being the latter a key feature for maximum operating frequency extension and low noise operation, InP-based HEMT devices have been addressed as the natural extension of GaAs-based ones, thanks to the superior intrinsic frequency and noise performance.

On the other hand, advanced radioastronomy applications require the development of versatile and complex front-end systems: focal plane arrays or millimetre-wave cameras typically adopt an extremely large number of channels, together with an high integration level. Such further requirements, joined with the companion request for an high yield and low cost of the technological process, lead to investigate the viability of industrial-grade robust technologies based on GaAs. The performance requirements clearly dictate the adoption of small gate features, together with high indium content in the active channel for gain and noise optimisation. In this contribution an advanced 70 nm

process has been adopted for the design and realisation of four low-noise amplifiers, spanning in frequency from C- to W-Band, in order to address the needs of several radioastronomy applications. The amplifiers have the common technological process, being however featured by different specifications and goals. In the following the resulting performance is reported, demonstrating the usability and effectiveness of the GaAs-based approach here proposed.

II. TECHNOLOGY

The four MMIC LNAs (W, Q/Ka, K and C-Band) reported in the following have been realised making use of OMMIC's foundry facilities. The D007IH process is based on a high Indium content epitaxial active layer, grown on a metamorphic buffer layer creating a smooth transition with the GaAs substrate. The active components are based on a GaInAs-InAlAs-GaInAs-InAlAs heterostructure containing 52% of Indium and 70% of Indium in the conductive channel. A special type of graded buffer (Metamorphic) is used to ensure a good transition between the GaAs substrate and the active layer, which are not lattice matched due to the high Indium content.

The main features of the process are: gate length 0.07 μm, leading to a f_t of 300GHz and a f_{max} in the region of 350GHz. The threshold voltage is -0.5V while the gate-to-drain breakdown condition is 3.0V. Finally, the I_{DSS} is approximately 240mA/mm. This process shows very high cut off frequency and extremely low noise.

III. DEMONSTRATOR CIRCUITS

A. W-Band LNA

The millimetre and sub-millimetre bands offer a unique window through which Radioastronomers can "see" and study features of the universe which are otherwise invisible. The reasons for this are that such components contain over 3000 radio spectral lines of interstellar and circumstellar molecules. Millimetre and sub-millimetre bands can detect the emission of cool dust in space and are the only bands in which Radioastronomers can detect the emission from dust and molecules in young galaxies at high redshift in the early universe.

The W-Band LNA here presented was designed and realised to serve as a technology and methodology demonstrator in this context. The adopted design flow was oriented towards obtaining high gain and limited ripple over a bandwidth as broad as possible, associated to low noise behaviour and prescribed port impedance. A 4-stage topology was adopted to obtain an average gain around 25dB in the W-Band range. The four stages are very similar to each other since a minimum noise measure technique was adopted to design the single stage: the first stage's gain is not very high, approximately 6dB, and therefore unable to conceal the subsequent stages' noise contribution. More information regarding this component can be found in [1]. Resulting die size is 3.0x2.0mm2 and its layout and microphotograph is depicted in Fig. 1.

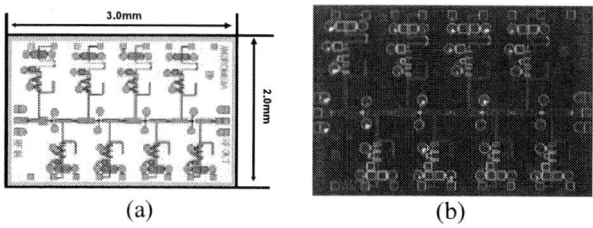

(a)   (b)

Fig. 1:   W-Band LNA layout (a) and microphotograph (b).

Fig. 2 depicts the LNA measured linear parameters. The average gain is 25dB, with 2dB ripple from 75 to 95GHz and the |S21| is greater than 21dB on the entire 75-95GHz measured frequency band. The measured |S22| is better than -10dB from 72 to 90GHz while the |S11| is better than -5dB from 70 to 88GHz.

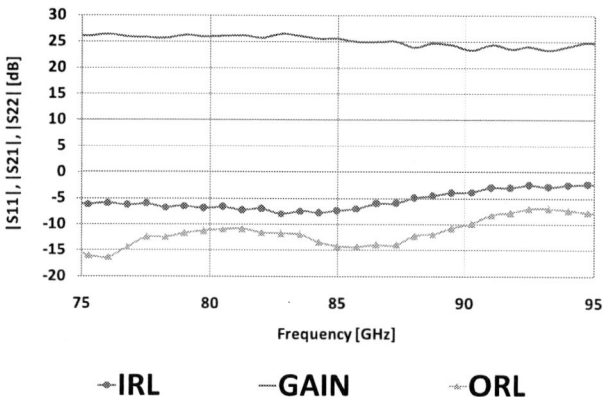

Fig. 2.   W-Band LNA measured linear parameters

Finally, the preliminary noise characterisation on the W-Band LNA, has shown that NF is 2.7dB between 80 and 95GHz.

B. Q/Ka-Band LNA

The frequency band inside the 30-50GHz region is practically free from significant atmospheric absorption and is therefore important for Radioastronomy observations. Some noteworthy observed particles are briefly recalled: the frequency band 36.4-36.5 GHz is exploited for the search of HC_3N and OH lines. 42.5GHz is subject of extensive radio astronomy single dish and VLBI measurements. At this frequency the lines of SiO often indicate maser emission, the mechanism of which is not understood but which extends over a wide range of excitation in the SiO molecule.

The region between 42.5 and 49 GHz contains important spectral lines of some diatomic and other molecules. The lines of carbon monosulphide (CS) and its less common isotopes have been shown to be constituents of both giant molecular clouds and cool dark clouds.

The key requirements for a wideband LNA operating inside the Q/Ka-band frequency range, for the targeted application, are 30dB small signal gain, NF around 1.5dB and adequate I/O return loss.

The chosen design approach is based on the combined use of both lumped components and distributed elements. This choice has led to the fulfilment of the required operating bandwidth performance in terms of noise, gain and adaptation and has produced a reduced amplifier area in order to fall within the prescribed limits of $3x2mm^2$.

The chip design consists of a four-stage amplifier, properly sized to minimize the total noise figure and to enable to obtain the required gain of 30dB from throughout the Q/Ka- band. During the synthesis of networks intensive use of electromagnetic simulation software adapt has made, to take into account possible couplings between electromagnetic discontinuity, which in these frequencies can always be present, and to confirm the foundry model of passive elements.

Fig. 3 shows the amplifier layout and microphotograph.

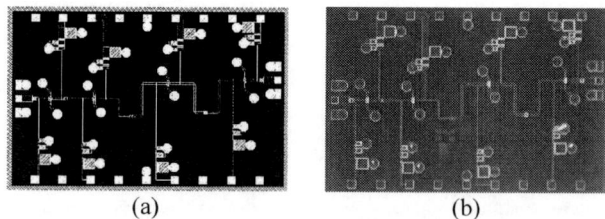

(a)   (b)

Fig. 3:   Q/Ka-Band LNA layout (a) and microphotograph (b).

Fig. 4 shows the MMIC Q/Ka-LNA linear and noise behaviour measured at room temperature.

978-1-4244-5190-6/09 $25.00 © 2009 IEEE   162

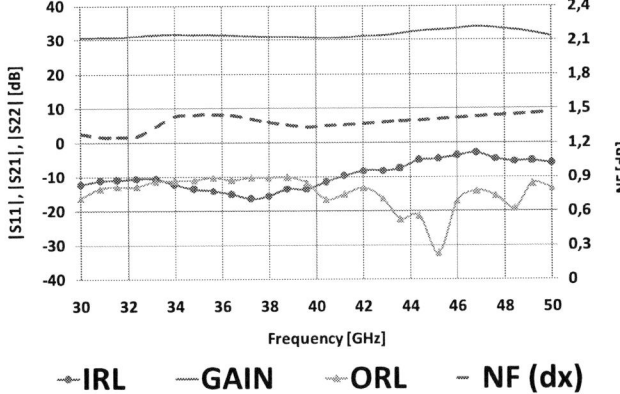

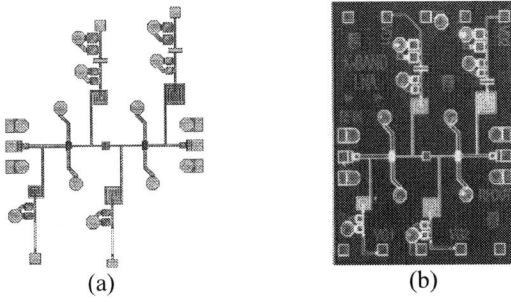

(a)   (b)

Fig. 5:   K-Band LNA layout (a) and microphotograph (b).

The measured NF is slightly above 1dB as expected by applied the design procedure.

Fig. 4.   Q/Ka-Band LNA measured linear and noise parameters

C.   K-Band LNA

The band from 18 to 30 GHz is densely packed with observed lines. Lovas (1986) recorded a list of 173 transitions within this spectral range, only 37 of which are covered by the four spectral lines that entered the ITU-R Recommendation RA.314. But also many other lines are of continuous interest for the determination of astrophysical parameters of celestial sources. The LNA here presented has a narrow band operating frequency around 22GHz, which corresponds to Radioastronomy observations of the redshifted H_2O particularly important for radio astronomy spectroscopy. Moreover, observation of maser at 22GHz is attractive, towards possible radio astronomy SKADS (Square Kilometer Array Design Studies) applications [2].

A two-stage design topology was adopted to fulfil the design specifications: 15dB gain and 1dB Noise Figure. Another major constraint was the total occupied area: $1.5 \times 2.0 mm^2$.

The theory proposed in [3] and [4] was extended to the double-stage case. The first step consists in selecting the impedance seen by the input of the first ($\Gamma_{S,1}$) and second ($\Gamma_{S,2}$) stage for optimum noise behaviour. The criterion for selecting the source impedance ($\Gamma_{S,1}$) and the impedance seen by the second FET ($\Gamma_{S,2}$) is briefly described: $\Gamma_{S,1} = \Gamma_{S,2} = \Gamma_{opt,M}$ where $\Gamma_{opt,M}$ is the input impedance which minimizes the active device's noise measure. This criterion is formally the most accurate since it implies the accomplishment of the minimum noise figure of the cascade. The reason for the choice of minimizing the measurement noise is described in [5] and [6].

The next Fig. 5 shows the layout and the photograph of the microscope stage LNA 2-band K. The MMIC is 2mm in height to 1.5mm in length.

Fig. 6 reports the linear and noise characterisation carried out on the 2-stage K-band LNA. At 22GHz, the targeted frequency for the Radioastronomical observations reported before, the gain is about 18dB, while both the input and output match consistent with the results expected from the employed design methodology: 14dB and 21dB.

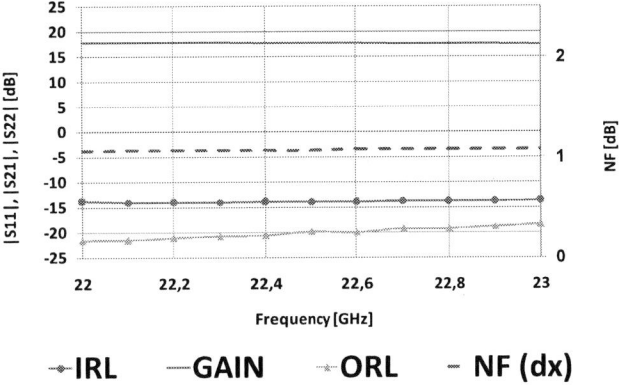

Fig. 6:   K-Band LNA measured linear and noise parameters.

D.   C-Band LNA

The spectral region around 5 GHz has been one of the most widely used frequency ranges in Radio Astronomy during the last decade. Astronomers have made use of this frequency range in order to study the detailed brightness distributions of both galactic and extragalactic objects. One of the most important uses of the band around 5 GHz is the study of the formaldehyde (H_2CO) interstellar clouds. The H_2CO line at this frequency is considered to be one of the most important radio lines in the entire spectrum, primarily because it can be detected in absorption in almost any direction where there is a continuum radio source. Finally, the 6.7 GHz band is important for observations of methanol (CH_3OH). This transition of methanol is a very powerful cosmic maser found exclusively in regions where massive stars form. It is widely observed in Europe using single dishes, MERLIN interferometry and VLBI [7]-[8].

The C-Band LNA was designed using the theory proposed in [3] and reprised in [4]. The first contribution defines the input/output matching limitations that constrain LNA design. The second contribution provides guidelines on how to synthesise the output impedance that simultaneously fulfils the prescribed I/O matching conditions. The layout

978-1-4244-5190-6/09 $25.00 © 2009 IEEE   163

(left) and a microphotograph (right) of the single stage LNA is depicted in Fig. 7.

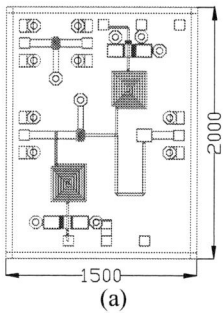 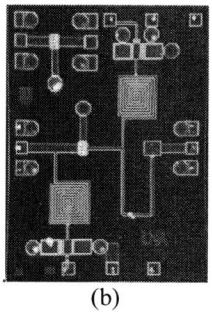

(a)   (b)

Fig. 7:   C-Band LNA Layout (a) and microphotograph (b)

The measured performances in terms of input/output match (S11, S22), gain (S21) and noise (NF) are reported in Fig. 8. The LNA's behaviour is satisfactory in the targeted 5.0-6.6GHz bandwidth, proving that the employed technology and design methodology is gainfully applicable for the targeted application.

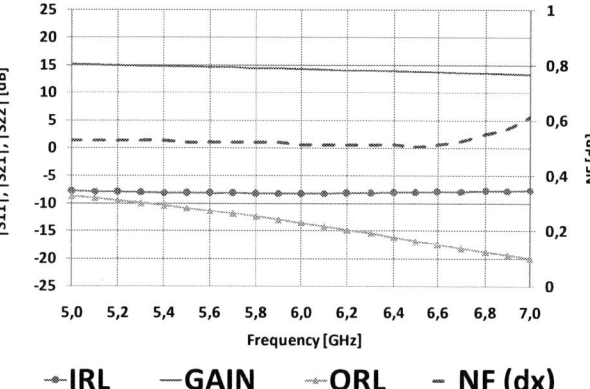

Fig. 8:   C-Band LNA measured linear and noise parameters.

IV. FREQUENCY TREND

The four MMIC LNAs (W, Q/Ka, K and C-Band) here proposed equivalent noise temperature measured at 290K vs frequency is reported in Fig. 9. A linear behaviour is quite evident. The gradient of such dependence is approximately 2.4K/GHz.

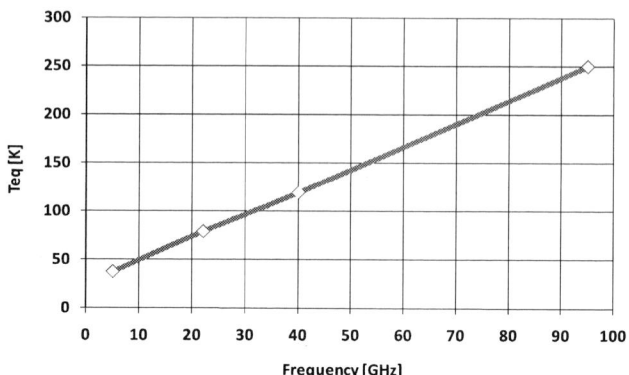

Fig. 9:   Measured equivalent noise temperature (at 290K) *vs.* frequency for the four proposed LNAs.

VII. CONCLUSION

In this contribution the design and performance of four LNAs operating from C- to W-Band has been presented. Measured results demonstrate the usability of 70 nm GaAs-based technology for advanced Radioastronomy applications, even not being fully released for commercial use. The industrial grade of the process, being featured by a good uniformity and yield, may actually allow to afford the realization of multi-channel and high performance receivers, well-suited for focal-plane arrays and microwave/millimeter-wave cameras featured by an high integration level.

ACKNOWLEDGEMENT

This work was funded by the PHAROS-RadioNet Project (FP6 − R113CT 2003 5058187) and Radionet FP7 (grant agreement no.227290) whose support is gratefully acknowledged. Visit the website www.radionet-eu.org.

REFERENCES

[1]   W. Ciccognani, F. Giannini, E. Limiti, P.E. Longhi, "Full W-Band High-Gain LNA in mHEMT MMIC Technology," *Proc. 38th Euro Micr Conf*, Amsterdam (NED), 27-31 Oct. 2008.

[2]   www.skads.-eu.org

[3]   M. Sierra, "Matching, Gain and Noise Limits on Linear Amplifier Four Poles," *Microwave Opt. Tech. Lett.*, vol. 2, pp. 29-34, Jan. 1989

[4]   W. Ciccognani, F. Giannini, E. Limiti, P.E. Longhi, "Determining Optimum Load Impedance for Noisy Active 2-Port Networks," *Proc. 37th Euro Micr Conf*, October 2007, Munich (GER).

[5]   H.A. Haus, R.B. Adler, Circuit Theory of Linear Noisy Networks, New York, Technology Press of MIT and John Wiley Ed

[6]   W. Ciccognani, F. Di Paolo, F. Giannini, E. Limiti, P.E. Longhi, A. Serino, "GaAs cryo-cooled LNA for C-band radioastronomy applications," *Electronics Letters* Volume 42, Issue 8, 13 April 2006. Pages:471–472.

[7]   www.merlin.ac.uk

[8]   www.evlbi.org

A Low Power Ka-Band Receiver Front-End in 0.13μm SiGe BiCMOS for Space Transponders

Firooz Aflatouni and Hossein Hashemi

Department of Electrical Engineering – Electrophysics

University of Southern California

Los Angeles, CA 90089 USA, Email: aflatoun@usc.edu, hosseinh@usc.edu

Abstract— **In direct-to-earth or deep space communication links, providing energy is very expensive and therefore it is essential for the link receiver to consume very low power while providing high sensitivity and gain. This paper presents a low power 0.13μm SiGe Ka-band receiver where the current sharing technique is used to reduce the power consumption of the front-end to 5.8mW while providing 17dB of RF-IF conversion gain, double side band noise figure less than 8.9dB across the bandwidth, and -1dB input referred compression point of -25dBm.**

Index Terms— SiGe BiCMOS receivers, mm-wave integrated circuits, receiver architecture.

I. INTRODUCTION

Smaller and lighter antenna as well as higher data rate make the Ka-band a good candidate for the direct-to-earth and deep space communication system design. In such a system, blocks must consume low power since providing energy in the space can be challenging and expensive. Multi-chip module based approaches have been used to design space transponders where the large size and mass as well as high power consumption (due to standard interstage matching) increases the cost. Integrated circuit approach is more cost effective since it reduces the size and power consumption and results in a lighter design.

mm-Wave receivers with high gain and sensitivity and relatively low power consumption have been reported [1][2][3] in which innovative architectures and variety of circuit techniques have been used. A careful low power design strategy combined with use of vertical multi-layer inductors, while maintaining the desired noise figure and conversion gain, can help further improve the performance of a space transponder.

II. RECEIVER ARCHITECTURE AND BUILDING BLOCKS

Figure 1 shows the fabricated receiver architecture. The conventional quadrature mixing is replaced with two-step quadrature down conversion to improve the image rejection. Note that only one local oscillator signal is used in this architecture. Assuming the same in-phase (I) and quadrature (Q) amplitude and phase mismatches for both down conversion steps, the image rejection ratio (IRR) for this

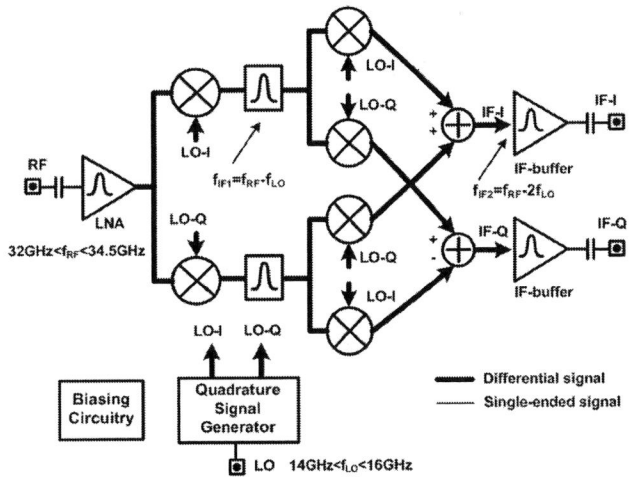

Fig. 1. Receiver block diagram.

architecture can be approximated as

$$IRR \approx (\frac{4}{\varepsilon^2 + \theta^2})^2, \qquad (1)$$

where ε and θ are the I and Q signal amplitude and phase mismatches. This indicates that the image rejection of a conventional quadrature down converter [4] is improved by a factor of two in logarithmic scale. Another advantage is that unlike the direct down conversion scheme, the LO feed-through is not problematic.

The single-ended LO signal is provided from an existing carrier tracking PLL loop and/or a PLL chip. In future, the system power consumption and area can be further improved by integrating the LO generation circuitry with the receiver front-end.

The low noise amplifier (LNA) amplifies the single-ended input RF signal (with frequency range of 32-34.5 GHz) and converts it to a differential signal. The single-ended 14-16 GHz off-chip signal is converted to differential in-phase and quadrature components in the quadrature generator circuit and is used for two-step down-conversion of the LNA output to generate quadrature IF signals.

978-1-4244-5190-6/09 $25.00 © 2009 IEEE

A. Common Gate LNA with Transformer Feedback

Figure 2(a) shows the fabricated two-stage single-ended to differential LNA. The first stage is a common gate LNA (CGLNA) in which its transconductance, g_m, is boosted using a transformer [5]. The advantage of using a transformer over active amplifier for g_m boosting is to eliminate the feedback amplifier power consumption and noise contribution. For a CGLNA with transformer feedback, considering only the device noise, the noise figure can be written as

$$NF = 1 + \frac{1}{2g_m R_s (1 + k\sqrt{\frac{L_2}{L_1}})}, \qquad (2)$$

where g_m and R_s are the device transconductance and the source impedance, respectively. It can be shown that for the CGLNA, the device DC current is primarily set by the linearity requirement. Therefore given the LNA -1dB input referred compression point and considering a desired input matching and NF, the required loop gain, $k\sqrt{L_2/L_1}$, can be estimated. Note that the loop gain sets the real part of the LNA input admittance while L_1 is used to tune out the imaginary part of the input admittance. Although high coupling and large L_2/L_1 is desired, these values are limited due to the effect of the transformer parasitic components at the mm-wave frequencies.

The second stage of the LNA performs the single- ended to differential conversion while sharing the same DC current with the first stage. A shunt inductor and a series capacitor are used to perform the impedance matching between two LNA stages. The feedback transformer which is depicted in 2(b), is simulated using Zeland IE3D electromagnetic simulator. In order to reduce the chip area, all inductors are designed as multi-turn vertical inductors in the top four metal layers.

B. Current Sharing Quadrature Mixers

One potential issue of the two-step quadrature down conversion receiver can be the increase in the power consumption (in case of active mixing), loss (in case of passive mixing), and chip area due to addition of the second set of the down converting mixers. Mixing in the current domain allows stacking mixers and therefore DC current can be re-used. The schematic of the quadrature two-step down conversion circuitry is shown in Fig. 3. The input differential pair converts the input RF signal to a current and sends it to cascaded switching mixers. The output current of the mixers at IF is added and converted to a voltage. A band pass filter with center frequency at $f_{IF1} = f_{RF} - f_{LO}$ is placed after the first mixer to eliminate the unwanted signals. The minimum DC current required by the circuit in Fig. 3 depends on the compression point and the noise figure. There are two mechanisms that cause the two-step quadrature down converting mixers to compress.

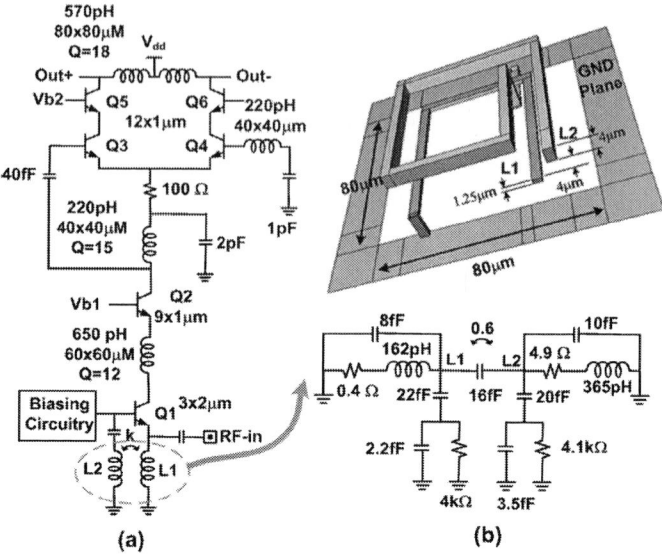

Fig. 2. (a) The two stage current sharing single-ended to differential LNA and (b) the feedback transformer and its approximated lumped model.

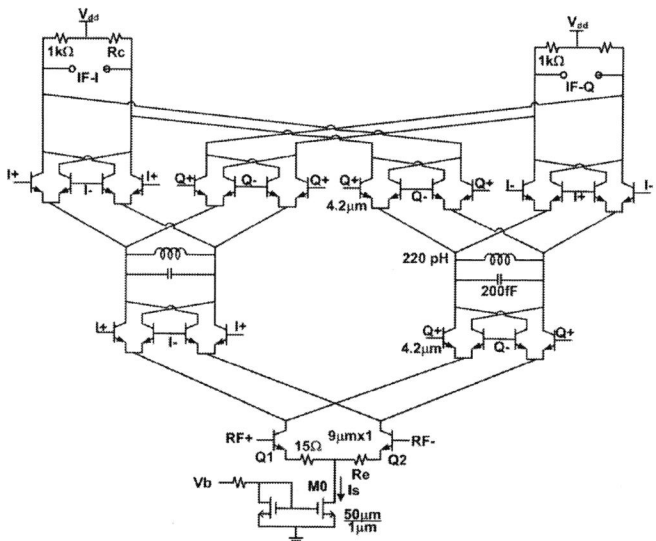

Fig. 3. The current sharing two-step quadrature down converting mixers.

First, a large voltage swing at the input of the RF differential pair (Q_1 and Q_2), may compress theses devices. The -1dB voltage compression point at the input node can be approximated as

$$A_{-1dB, in} \approx 2.5 V_T (1 + \frac{I_s}{2V_T} R_e), \qquad (3)$$

where V_T, I_s, and R_e are the device thermal voltage, the differential pair tail current, and the degeneration resistor, respectively. As expected, large bias current and degeneration resistance improves the linearity. Another mechanism that causes compression is a large voltage swing at the output of the two-step quadrature down converting mixers.

978-1-4244-5190-6/09 $25.00 © 2009 IEEE

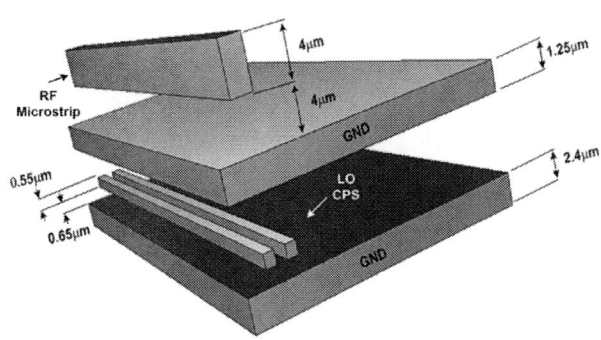

Fig. 4. The vertically stacked shielded transmission lines.

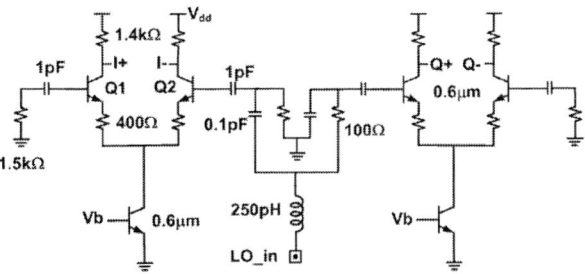

Fig. 5. The quadrature generator circuit.

In this case, the output voltage swing that causes compression can be approximated as

$$A_{out} \approx \frac{V_{dd} - 4V_{od} - V_{R_e} - V_M - V_{R_c}}{\frac{1}{2}g_{m1,2,eff}R_c}, \qquad (4)$$

where V_{od}, V_{R_e}, V_M, and V_{R_c}, are the voltage drops across HBTs, R_e, M_0, and R_c, respectively. Also $g_{m1,2,eff} = g_{m1,2}/(1 + g_{m1,2}R_c)$ is the effective transconductance of Q_1 and Q_2. Assuming proper switching of the mixers, the noise figure of the circuit in Fig. 3 can be approximated as

$$F \approx 1 + \frac{r_b + R_e}{R_s} + \frac{(1 + g_{m1,2}R_e)^2}{g_{m1,2}R_s}\left(\frac{1}{2} + \frac{4}{g_{m1,2}R_c}\right), \qquad (5)$$

where r_b and R_s are the base and source impedances, respectively. Knowing the required voltage compression point, from equations 3 and 4, $g_{m1,2}R_e$ and $g_{m1,2}R_c$ can be calculated. For a given noise figure, from equation 5 and using calculated values for $g_{m1,2}R_e$ and $g_{m1,2}R_c$, the $g_{m1,2}$ and the minimum required tail current, satisfying desired compression point and NF, can be calculated. Although these calculations were done under low frequency conditions, the calculated values were used as initial values for circuit high frequency simulations and mm-wave design.

At mm-wave frequencies, the parasitics of the interconnects between devices in the mixer core may change the performance dramatically. In order to reduce the chip area while minimizing the parasitic effect of these interconnects, vertically stacked shielded transmission lines were used. As an example, Fig. 4 shows a section of a mixer layout where a microstrip line is running on top of a CPS. Two transmission lines are separated by ground plane. The electromagnetic simulation shows that the crosstalk between the shielded transmission lines is negligible at the desired operating frequencies.

The single-ended LO input is converted to differential in-phase and quadrature signals using an RC poly-phase

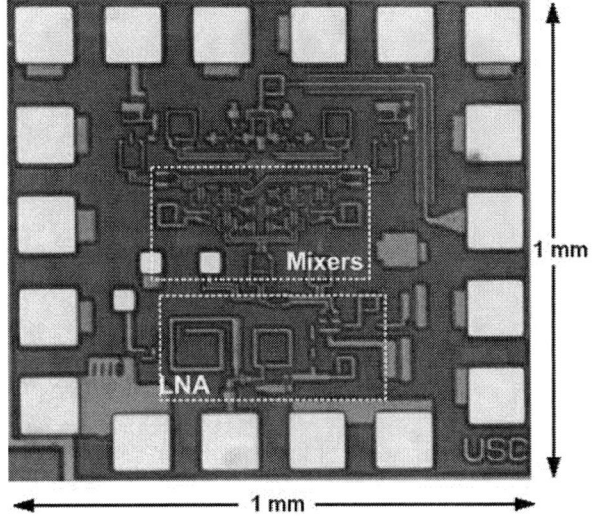

Fig. 6. The chip microphotograph.

filter followed by a pair of single-ended to differential converters (Fig. 5). The output of the mixer drives the single-ended measurement instrument load through differential to single-ended converting output buffers.

III. MEASUREMENT RESULTS

The receiver is fabricated in the IBM8HP 0.13μm BiCMOS SiGe process. The chip microphotograph is shown in Fig. 6. Single-ended probed measurements are reported. The chip is mounted on a PCB and the single-ended LO pad is wirebonded to an on-board microstrip line. The RF and IF pads are probed. The measured return loss of RF and IF ports are shown in Fig. 7.

The scalar measurements were performed to measure the receiver conversion gain (Fig. 8). Note that in order to achieve high gain while consuming low power, the inductors in the receiver front-end were designed to have a relatively large quality factor. As it is shown in Fig. 1, there are few tuned circuits in series in the RF to IF path. Thus, small frequency misalignment in these tuned circuits (due to either inaccuracy in EM simulations or un- accounted layout parasitics) may cause the overall gain to drop noticeably.

978-1-4244-5190-6/09 $25.00 © 2009 IEEE   167

TABLE I
PERFORMANCE SUMMARY AND COMPARISON WITH OTHER WORKS

Ref.	Gain [dB]	NF [dB]	$CP_{-1,in}$[dBm]	BW_{-3dB}[GHz]	S_{11}[dB]	LO to IF Leakage [dB]	LO to RF Leakage [dB]	Front-End Power [mW]	Chip area [mm²]	Technology
This Work	17	8.9 (DSB)	-25	31-33.4	-12	-42	-48	5.8 @ 3V	0.26 (core)	0.13μm SiGe
[1]	19.8-22	5.7-7.1	-27.5	57-61	-	-	-65	32 @ 1.2V	0.19 (core)	90nm CMOS
[6]	13	7	-12	39-41	-	-	-45	15.2 @ 1.6V	1.04	0.13μm SiGe
[2]	-8-55	6.1-6.3	-26	57-63	-15	-	-77	24 @ 1V	1.55	90nm CMOS

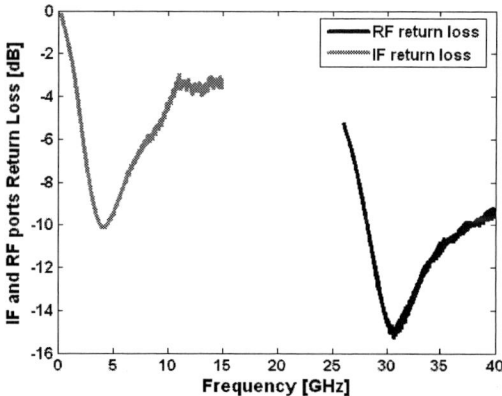

Fig. 7. The measured return loss of the receiver at the RF and IF ports.

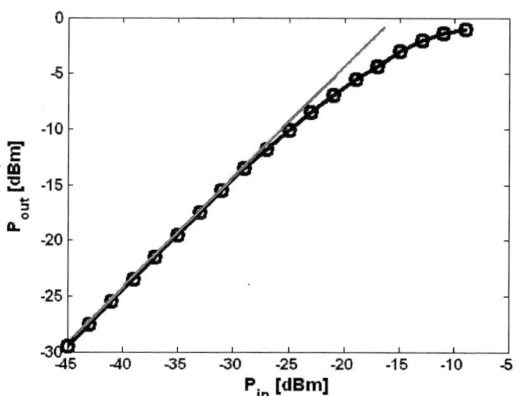

Fig. 9. The measured receiver linearity for $f_{LO} = 14.5\text{GHz}$.

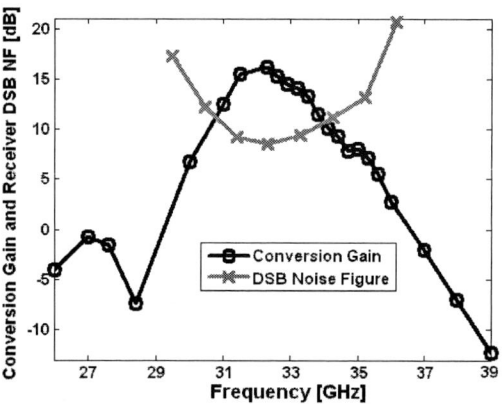

Fig. 8. The measured receiver conversion gain and DSB NF for $f_{LO} = 14.5\text{GHz}$.

In future designs, automatic calibration loops may be used to line up the frequencies.

The receiver noise figure was measured using a spectrum analyzer with noise figure measurement personality and a noise source. The measurement result is reported in Fig. 8. Figure 9 shows the measured -1dB input compression point of the receiver. SpectrRF simulations show that the linearity is limited by the two-step quadrature down converting mixers and not the output buffers. At V_{dd} of 3 volts, the LNA consumes 2.8 mW, the mixers consume 3 mW, the quadrature generator circuit consumes 4.5mW, and the output buffers consume 35.5 mW each.

The performance of a few recently reported receivers is compared with this work in Table I.

IV. CONCLUSION

This paper demonstrates a low-power Ka-band receiver front-end. The current re-use technique and low power design strategy, in both common gate LNA with transformer feedback and two-step down converting mixers, further reduce the chip power consumption. Vertical multi-turn inductors were used to decrease the chip area.

ACKNOWLEDGMENT

This work was supported by the JPL under SURP program. The authors would like to acknowledge the support of Drs. Venkatesan and Mysoor from JPL.

REFERENCES

[1] A. Parsa, B. Razavi, "A 60GHz CMOS receiver using a 30GHz LO," IEEE ISSCC, Feb 2008.

[2] B. Afshar, Yanjie Wang, Ali M. Niknejad , "A robust 24mW 60GHz receiver in 90nm standard CMOS," IEEE ISSCC, Feb 2008.

[3] B. W. Min and G.M. Rebeiz, "Ka-band SiGe HBT low noise ampli- fier design for simultaneous noise and input power matching,"IEEE microwave and wireless components letters, Vol. 17, No. 12, Dec. 2007.

[4] B. Razavi, RF microelectronics. Englewood Cliffs, NJ: Prentice Hall, 1998.

[5] Xiaoyong Li, et al., "Gm-boosted Common-Gate LNA and differen- tial colpitts VCO/QVCO in 0.18μm CMOS," IEEE JSSC, Vol. 40, no. 12, Dec. 2005.

[6] S. Pruvost et. al., "A 40GHz superheterodyne receiver integrated in 0.13 um BiCMOS SiGe:C HBT technology," IEEE BCTM, Oct. 2005

AUTHOR INDEX

Abbasi, Morteza	76, 117
Aflatouni, Firooz	165
Ambacher, O.	56
Anthony, Michael P.	145
Barratt, C. A.	9
Benz, W.	56
Berdaguer, Philippe	105
Blanck, H.	12
Borremans, Jonathan	68
Bourdoux, André	68
Bowers, John E.	44
Brebels, Steven	68
Bronckers, Stephane	68
Bronner, W.	56
Brückner, P.	121
Carroll, M.	97
Cavus, A.	101, 109
Chan, Beckie	101
Chang, Pablo	101
Chen, Cheng-Duan	113
Chen, Morgan J.	157
Cheng, Chia-Shih	40
Chevalier, P.	28
Chiu, Hsien-Chin	40, 84
Chou, Y. C.	109
Chung, Jin Wook	36
Ciccognani, W.	161
Clausen, William	88
Costa, J.	97
D'Amore, M.	101, 137
De Graauw, Anton	76
Doi, Y.	133
Enayati, Amin	68
Ferndahl, Mattias	76, 117
Floriot, D.	12
Fu, Jeffrey S.	40, 84
Gaquière, Christophe	105
Garcia, P.	72
Gavell, Marcus	117
Godin, Jean	105
Green, D.	52
Gunnarsson, Sten E.	117
Guthoerl, M.	16
Gutierrez-Aitken, A.	101, 137
Hamada, T.	133
Hashemi, Hossein	165
Haupt, C.	56
Heijden, Edwin V. D.	76
Hirata, Akihiko	64
Hoogland, J. A.	121
Hoversten, John	48
Huang, Xiaojun T.	44
Hung, Shih-Han	125
Ide, S.	133
Ikeda, Nariaki	32
Ikeuchi, T.	133
Irino, Akihiko	64
Iversen, C.	97
Jardel, Olivier	105
Kado, Yuichi	64
Kallfass, I.	16, 24
Kambayashi, Hiroshi	32
Kanda, K.	133
Kaneshiro, E.	101, 137
Kang, Dong-Woo	129
Kato, Sadahiro	32
Ke, Po-Yu	84
Kerr, D.	97
Kibune, M.	133
Kiefer, R.	56
Kim, Sang Young	153
Kjellberg, Torgil	76
Koch, S.	16, 24
Koh, Kwang-Jin	129
Kosugi, Toshihiko	64
Krishnamurthy, K.	52
Kristensson, A.	133
Kukutsu, Naoya	64
Lai, R.	109
Lai, Yu-Ann	125
Laskin, E.	28
Lee, L. S.	109
Leuther, A.	16, 24
Liao, Shuhsien	60
Libois, Michael	68
Limiti, E.	161
Lin, C. H.	109
Lin, S.	137
Lin, Shao-Wei	40
Lin, Steven	101
Liu, P. H.	109
Longhi, P. E.	161
López, Néstor D.	48
Lu, Bin	36
Macdonald, Noel C.	44
Maroldt, S.	56
Martin, J.	52
Mason, P.	97
Matsubara, S.	133
Matsuzuka, Takayuki	80
McLeod, S.	133
Mei, X. B.	109
Miyahara, K. Jay	1
Monier, C.	101, 137
Moser, Brian	88
Mueller, S.	56
Murata, Koichi	141
Nagatani, Munehiko	141
Nagatsuma, Tadao	64
Nakayama, Masatoshi	80

AUTHOR INDEX

Nakayama, Toshihiro 64
Nallatamby, Jean-Christophe 105
Nedovic, N. 133
Nicolson, S. T. 28
Niiyama, Yuki.......................32
Nishikawa, Hiroshi.......................64
Nishimoto, M. Y. 109
Nodjiadjim, Virginie 105
Nomura, Takehiko 32
Nosaka, Hideyuki 141
Ogawa, J. 133
Ogawa, Nobuyuki.......................80
Okamura, Atsushi 80
Oki, A. 101, 137
Ootomo, Shinya 32
Oyama, B. 101, 137
Palacios, Tomás 36
Parikh, S. 133
Parvais, Bertrand 68
Pavageau, Christophe 68
Pierres, J.-B. 97
Piotrowicz, Stéphane.......................105
Popovic, Zoya 48
Poulton, M. 52
Quay, R. 56, 121
Raczkowski, Kuba 68
Ramon, Valéry 68
Rebeiz, Gabriel M. 20, 92, 129, 149, 153
Reddy, S. 133
Renvoisè, M. 161
Riet, Muriel 105
Rodenburg, M. 121
Saito, S. 16, 24
Sano, Kimikazu 141
Sarkas, I. 28
Sato, Ken.......................101
Sautreuil, B.......................28
Scavennec, André 105
Scheir, Karen.......................68
Schlechtweg, M.......................24
Scott, D.......................101, 137
Seelmann-Eggebert, M. 121
Shibasaki, T. 133
Shimura, Teruyuki.......................80
Shin, Donghyup.......................92
Shin, Woorim 20
Sollner, G.......................145
Splettstößer, J.......................12
Sudo, Naohiro.......................64
Sugawara, M. 133
Suijker, E. M. 121
Tabatabaei, Seyed A. 157
Takahashi, Hiroyuki.......................64
Takatani, Shinichiro 113
Tamura, H. 133

Tasker, Paul J. 5
To, R. 109
Todd, Shane T. 44
Tombak, A. 97
Tomita, Y. 133
Tomkins, A. 28, 72
Tsai, R. 109
Tsunoda, Y. 133
Tzartzanis, N. 133
Uzunkol, Mehmet 20, 149
Van Heijningen, M. 121
Van Thillo, Wim 68
Van Vliet, F. E. 121
Vasylchenko, Alexander 68
Verbruggen, Bob 68
Vetury, R. 52
Voinigescu, S. P. 28, 72
Walker, W. 133
Wambacq, Piet.......................68
Wang, Yeong-Her 125
Wang, Yuanxun Ethan 60
Weber, R. 24
Werquin, Matthieu.......................105
Wojtowicz, M.......................109
Yamabana, T. 133
Yamamoto, Kazuya.......................80
Yamamoto, T. 133
Yamanaka, Shogo.......................141
Yang, J. M. 109
Yoshida, Naohito 80
Yoshii, Yutaka 80
Zirath, Herbert.......................76, 117